T0400077

Toxic Metals Contamination
Generation, Disposal, Treatment and Valuation

Editors

Jeferson Steffanello Piccin
University of Passo Fundo (UPF)
Faculty of Engineering and Architecture (FEAR)
Passo Fundo/RS, Brazil

Aline Dettmer
University of Passo Fundo (UPF)
Faculty of Engineering and Architecture (FEAR)
Passo Fundo/RS, Brazil

Natarajan Chandrasekaran
Center for Nanobiotechnology
Vellore Institute of Technology [VIT]
Vellore, India

CRC Press
Taylor & Francis Group
Boca Raton   London   New York

CRC Press is an imprint of the
Taylor & Francis Group, an **informa** business

A SCIENCE PUBLISHERS BOOK

First edition published 2023
by CRC Press
6000 Broken Sound Parkway NW, Suite 300, Boca Raton, FL 33487-2742

and by CRC Press
4 Park Square, Milton Park, Abingdon, Oxon, OX14 4RN

Library of Congress Cataloging-in-Publication Data (applied for)

ISBN: 978-0-367-68745-8 (hbk)
ISBN: 978-0-367-68748-9 (pbk)
ISBN: 978-1-003-13890-7 (ebk)

DOI: 10.1201/9781003138907

Typeset in Times New Roman
by Radiant Productions

Preface

Toxic metals are a class of chemical elements that, in certain concentrations, can cause damage to the health of humans and animals, in addition to impacts on ecosystems. This book is intended for anyone with an interest and/or concern about toxic metals. The main toxic metals found in ecosystems are antimony, arsenic, cadmium, chromium, copper, lead, mercury, nickel, selenium, tellurium, thallium, vanadium, and tin. These metals, when present in the human body, can cause damage to health, such as respiratory, cardiovascular, central nervous system, gastrointestinal problems, skin, reproductive system, among others. Thus, the present work will be showing the state of the art regarding technologies for the management of systems contaminated with toxic metals. There are ten chapters covering different topics related to toxic metals, from the contamination of surface and deep water intended for human consumption to the genotoxic effects of nanoparticles of these metals. The stress caused by arsenic in plants and the possibilities of mitigating the problem are also addressed, as well as the reuse of waste containing metals such as mercury and chromium. In the sequence, the main methods for removing toxic metals from liquid effluents are discussed, including membrane separation processes, adsorption and bioadsorption and treatment using ionic liquids. In this context, the main sources of emissions will be presented, including their forms of speciation and their impacts on the health of the environment. In addition, some techniques for disposal and valuation of solid waste containing toxic metals are also discussed. We hope that the book can contribute to science in order to show alternatives for the correct disposal and treatment of this hazardous waste.

Contents

Preface　　iii

1. **Toxic Metals: An Overview of Main Sources, Exposure Routes, Adverse Effects and Treatment Approaches**　　1
   Cesar Vinicius Toniciolli Rigueto, Mateus Torres Nazari, Ionara Regina Pizzutti, Natarajan Chandrasekaran, Aline Dettmer and Jeferson Steffanello Piccin

2. **Groundwater Quality Indexing for Drinking Purpose from Arsenic Prone Areas, West Bengal: A Health Risk Assessment Study**　　10
   Madhurima Joardar, Nilanjana Roy Chowdhury, Antara Das, Deepanjan Mridha and Tarit Roychowdhury

3. **Genotoxicity and Mutagenicity of Metal-based Nanomaterials, with an Emphasis on using *Drosophila***　　33
   Mohamed Alaraby, Doaa Abass and Ricard Marcos

4. **Alleviation of Arsenic Stress in Plants using Nanofertilizers and its Extent of Commercialization: A Systemic Review**　　47
   Iravati Ray, Deepanjan Mridha, Madhurima Joardar, Antara Das, Nilanjana Roy Chowdhury, Ayan De and Tarit Roychowdhury

5. **Foam Glasses from Glasses of Fluorescent Lamps Waste**　　72
   Isaac dos S. Nunes, Venina dos Santos and Rosmary N. Brandalise

6. **Recovery and Disposal of Tannery Waste Containing Toxic Metals**　　94
   Caroline Agustini, Taysnara Simioni, Nadini Pinheiro, Éverton Hansen, Victória Kopp and Mariliz Gutterres

7. **Treatment of Water Contaminated by Heavy Metal using Membrane Separation Processes**　　117
   Wendel Paulo Silvestre and Camila Baldasso

8. **Adsorption as an Efficient Alternative for the Removal of Toxic Metals from Water and Wastewater**　　146
   Yasmin Vieira, Juliana Machado Nascimento dos Santos, Jeferson S. Piccin, Ádrian Bonilla-Petriciolet and Guilherme Luiz Dotto

9. Biosorption of Toxic Metals from Multicomponent Systems and Wastewaters   172

Heloisa Pereira de Sá Costa, Giani de Vargas Brião, Talles Barcelos da Costa, Cléophée Gourmand, Caroline Bertagnolli, Meuris Gurgel Carlos da Silva and *Melissa Gurgel Adeodato Vieira*

10. Ionic Liquids Applied in Removal of Toxic Metals from Water and Wastewater   203

Carolina Elisa Demaman Oro, Victor de Aguiar Pedott, Rogério Marcos Dallago and *Marcelo Luis Mignoni*

Index   223

1

Toxic Metals
An Overview of Main Sources, Exposure Routes, Adverse Effects and Treatment Approaches

Cesar Vinicius Toniciolli Rigueto,[1] *Mateus Torres Nazari,*[2]
Ionara Regina Pizzutti,[1] *Natarajan Chandrasekaran,*[3]
Aline Dettmer[4] and *Jeferson Steffanello Piccin*[2,4,*]

1. Introduction

In the last decades, several terms have been used around the world to refer to metals of environmental interest, such as metals, metalloids, semimetals, heavy metals, essential metals, toxic metals, trace metals, micronutrients (Ali et al. 2019, Duffus 2002). This lack of terminology standardization may be associated with the fact that, for example, heavy metals can be toxic, essential, or non-essential, and that not all heavy metals are toxic under all conditions (Ali and Khan 2018). In the present book, the use of "toxic metals" was standardized, which can be essential (performing important biochemical and physical functions for living organisms) or non-essential (having no known biological role), depending on their concentration (Ali et al. 2019, Edelstein and Ben-Hu 2018, Al Osman et al. 2019, Rai et al. 2019). The toxicity of these elements varies according to the dose and duration of exposure (Ali et al. 2019).

---

[1] Federal University of Santa Maria (UFSM), Rural Science Center, Postgraduate Program in Food Science and Technology (PPGCTA), Roraima Avenue, 97105-900, Santa Maria/RS, Brazil.
[2] University of Passo Fundo (UPF), Faculty of Engineering and Architecture (FEAR), Postgraduate Program in Civil and Environmental Engineering (PPGEng), Passo Fundo/RS, Brazil.
[3] Center for Nanobiotechnology, Vellore Institute of Technology [VIT], Vellore, India.
[4] University of Passo Fundo (UPF), Faculty of Engineering and Architecture (FEAR), Postgraduate Program in Food Science and Technology (PPGCTA), Passo Fundo/RS, Brazil.
* Corresponding author: jefersonpiccin@upf.br

Elements commonly classified as toxic metals are Mercury (Hg), Lead (Pb), Chromium (Cr), Cadmium (Cd), Barium (Ba), Aluminum (Al), and Copper (Cu) (Al Osman et al. 2019). Other authors include Arsenic (As), Nickel (Ni), Silver (Ag), and Zinc (Zn) to this group of metals and metalloids (Li et al. 2019, Srivastava et al. 2021).

These elements belong to the class of inorganic pollutants and are widely known and reported for their potential risks to public health and the environment due to their characteristics of environmental persistence, toxicity, and bioaccumulation (Ali and Khan 2018, Ali et al. 2019, Rigueto et al. 2021a, Srivastava et al. 2021, Yin et al. 2019). Some of these metals still have the ability to biomagnify along the trophic chain, increasing their deleterious potential to organisms (Ali et al. 2019). In general, metal with a specific gravity of 5.0 and above and those with atomic number above 20 are considered as heavy metals (Luckey et al. 1975).

The Agency for Toxic Substances and Disease Registry (ATSDR) and the United States Environmental Protection Agency (U.S. EPA) have listed substances that are most found at facilities on the National Priorities List and that have the most significant threat potential to human health, where As, Pb, Hg, Cd occupy the first, second, third and seventh positions in this list (ATSDR 2019). Based on these aspects, toxic metals are common among the main pollutants investigated among the scientific community and environmental agencies (Ali et al. 2019).

Regarding the origin of these metals, natural processes and anthropic activities are reported as the main sources of pollution by toxic metals (Ali et al. 2019, Bozorg-Haddad et al. 2021, Edelstein and Ben-Hur 2018, Li et al. 2019, Yin et al. 2019). More specifically, weathering of metal-bearing rocks and volcanic eruptions are the main natural sources of toxic metals, while mining and different industrial and agro-industrial activities comprise the anthropic sources (Ali et al. 2019, Edelstein and Ben-Hur 2018). In the case of anthropogenic sources, there has been an exponential increase in the generation of contaminants since the Industrial Revolution (Rai et al. 2019).

Toxic metals can enter the human body through different exposure routes, such as ingestion, inhalation, and dermal absorption of water, food, and air contaminated with these elements (Ali et al. 2019, Fu and Xi 2019, Al Osman et al. 2019). Some factors influence the retention of these substances in a living organism, including the metal speciation, the physiological mechanisms of this organism, as well as aspects related to homeostasis and detoxification (Ali et al. 2019).

According to Fu and Xi (2019), drinking water contaminated with toxic metals is the main form of exposure for humans, and the concentration of these elements in drinking water has exceeded recommended limits in recent years. Exposure to toxic metals through ingestion and inhalation is reported to present the greatest toxicological risk to human health (Srivastava et al. 2021).

Toxic metals are capable of causing several pathologies to humans and other living organisms, which varies according to different aspects, such as level, form, and duration of exposure to these elements, in addition to comorbidities, age, and other conditions of the contaminated (Al Osman et al. 2019, Walton et al. 2011). This exposure to these contaminants, especially via ingestion of contaminated water, can result in deleterious effects on human metabolism (Fu and Xi 2019). In general, these effects result from the production of reactive oxygen species, which can damage proteins and DNA, as well as have mutagenic, teratogenic, and carcinogenic potential (Fu and Xi 2019, Leong and Chang 2020). As a result, these elements are reported to cause high morbidity (e.g., retardation, cancers, behavioral disorders, respiratory problems, immunological, endocrinal, and neurological effects, among others) and even mortality (Edelstein and Ben-Hur 2018, Fu and Xi 2019, Al Osman et al. 2019, Rai et al. 2019).

Table 1 presents the Minimal Risk Levels (MRL) and/or the guideline value for drinking water for the main toxic metals, according to Agency for Toxic Substances and Disease Registry and World Health Organization (WHO), respectively.

According to Rai et al. (2019), toxic metals can also drastically affect the soil. In addition, the authors emphasize that these contaminants negatively influence enzymes and other compounds related to the germination process of different crops, which can affect plant physiology at different stages of its growth. Consequently, high concentrations of these inorganic compounds in the soil result in lower crop yields (Edelstein and Ben-Hur 2018).

Thus, it is necessary to reduce the generation of solid waste containing toxic metals. Chromium-tanned leather waste, for example, is usually sent to landfills, as it is faster and more economical; however, it is not an ecologically correct alternative, as it can leach and contaminate the soil and water resources (Rigueto et al. 2020).

Table 1: Minimal risk levels and guideline value for drinking water for several toxic metals.

Elements	Route	Duration	Minimal Risk Levels (MRL)	Provisional guideline value for drinking water
Al	Oral	Acute / Chronic	1 mg/kg/day	No guideline value
As	Oral	Acute	0.005 mg/kg/day	0.01 mg/L
		Chronic	0.0003 mg/kg/day	
Cd	Inhalation	Acute	0.00003 mg/m^3	0.003 mg/L
	Inhalation	Chronic	0.00001 mg/m^3	
	Oral	Intermediate	0.0005 mg/kg/day	
	Oral	Chronic	0.00001 mg/kg/day	
Cr(VI)	Oral	Intermediate	0.005 mg/kg/day	0.05 mg/L (Total chromium)
		Chronic	0.0009 mg/kg/day	
Cu	Oral	Acute / Intermediate	0.01 mg/kg/day	2 mg/L
Hg	Inhalation	Chronic	0.0002 mg/m^3	0.006 mg/L
Ni	Inhalation	Intermediate	0.0002 mg/m^3	0.07 mg/L
	Inhalation	Chronic	0.00009 mg/m^3	
Pb	-	-	-	0.01 mg/L
Zn	Oral	Intermediate / Chronic	0.3 mg/kg/day	No guideline value

Source: ATSDR (2019), WHO (2017).

In this sense, studies have aimed at the use of solid waste containing toxic metals to obtain useful materials; for example, the use of chromium-tanned leather wastes as an adsorbent (Piccin et al. 2012), fertilizers, and vermicompost (Nunes et al. 2018, Nogueira et al. 2010), extraction of gelatin from chromium-tanned leather wastes for the production of beads applied in the removal of emerging contaminants (Rigueto et al. 2021b) and polymeric films for application as ground cover (Rosseto et al. 2021), reuse of hazardous spent fluorescent lamps glass waste as supplementary cementitious material (Pitarch et al. 2021), among other materials of commercial and/or industrial interest (Rigueto et al. 2020, Dettmer et al. 2010).

2. Toxic metals' removal from aqueous matrices and soil

The technologies for water and wastewater treatment containing toxic metals have shown advantages that allow their practical development, such as electrolysis, ion exchange, chemical precipitation, membrane separation, adsorption, substitution coprecipitation, TiO_2 photocatalysis, Fenton oxidation, and use of ionic liquids. However, the choice of the appropriate technique depends on factors such as economic and environmental impacts (Alessandretti et al. 2021, Rajadurai and Anguraj 2021, Zhu et al. 2019), in this context, Table 2 presents the main advantages and disadvantages of toxic metals treatment techniques.

Regarding toxic metals' removal from soil, different remediation technologies can be applied to remove toxic metals from the soil, including physical, chemical, and biological approaches (Li et al. 2019). As mentioned earlier, the importance of carrying out the remediation of these inorganic pollutants is evident (Yin et al. 2019). Among the techniques, the use of biochar for soil decontamination is a promising and efficient alternative for this purpose (Rai et al. 2019). Biochar can act in the process of alleviating toxic metal contamination in the soil by immobilizing, stabilizing, and reducing the bioavailability of these contaminants (Rempel et al. 2021).

Other more eco-friendly approaches to the problem of toxic metals include phytoremediation and the use of different microorganisms since they are efficient and cost-effective (Li et al. 2019, Yin et al. 2019). In the case of vegetables and fruits, there are different ways to reduce the accumulation of inorganic contaminants in these foods, such as grafting, transgenic plants, and microorganisms (Edelstein and Ben-Hur 2018).

The use of different techniques is fundamental to reduce the concentration or bioavailability of toxic metals in different environmental compartments (water, soil and, air) and matrices (drinking water and food). Such approaches can favor the achievement of several Sustainable Development Goals (SDGs), especially those related to human health (SDG 3—Good Health and Well-being), maintenance of water quality (SDG 6—Clean Water and Sanitation; SDG 14—Life below Water),

Table 2: An overview of main advantages and disadvantages of toxic metals treatment techniques.

Technique	Advantage	Disadvantage	References
Electrolysis	• Convenient; • Easy to operate; • Effective with high concentration wastewater and having a large processing capacity	• Expensive approach (electrical energy required and the consumption of soluble anode materials); • Electrodes are easily passivated (inappropriate for the treatment of low-concentration complex toxic metal wastewater); • Secondary pollution can occur due to the production of residual organic ligands.	Zhu et al. (2019)
Ion exchange	• Recovery of metal value; • Selectivity; • Less sludge volume produced; • Meeting of strict discharge specifications	• Cannot handle concentrated metal solution (the matrix gets easily fouled by organics and other solids in the wastewater); • Nonselective and is highly sensitive to the pH of the solution.	Zewail and Yousef (2015), Barakat et al. (2011)
Chemical precipitation	• Simplicity of the process; • Inexpensive equipment requirement; • Convenient and safe operations.	• Requires a large amount of chemicals to reduce metals to an acceptable level for discharge; • Excessive sludge production that requires further treatment; • Slow metal precipitation; • Poor settling; • Aggregation of metal precipitates; • Long-term environmental impacts of sludge disposal.	Barakat et al. (2011), Aziz et al. (2008)
Membrane separation	• High efficiency and selectivity; • Effective with a wide range of metals; • Low energy requirements; • Easy operation.	• High membrane preparation costs; • Ease of membrane blocking; • Inability to stretch membranes; • High requirement for additional complex agents.	Zhu et al. (2019)

Table 2 contd. ...

...Table 2 contd.

Technique	Advantage	Disadvantage	References
Adsorption	• Low-cost technique; • Simple operation; • High adsorption capacity; • Effective for a variety of metals; • Have no secondary pollution implications; • Biodegradable biosorbent materials of residual industrial, agro-industrial or natural origin can be used; • Possibility to reuse the material in the adsorption for more than one cycle.	• Difficulty in the regeneration of materials; • Requiring a limited pH range; • Costs associated with the large-scale production (depending on the adsorbent material used).	De Rossi et al. (2020), Zhu et al. (2019), Renu et al. (2017)
Substitution coprecipitation	• Easy to operate; • High efficiency for toxic metals treatment; • Makes full use of iron-based materials.	• High rate of raw chemical consumption; • Narrow working pH range; • Agglomeration and passivation of the iron matrix; • Secondary pollution from released organic chelates.	Shan et al. (2018), Zhu et al. (2019)
TiO_2 photocatalysis	• Environmentally-friendly technique; • Ability to achieve complete mineralization of organic ligands under ambient temperature and atmospheric pressure conditions; • Generate low-toxicity by-products; • Highly-efficient, rapid and economical treatment method; • Has no secondary pollution implications.	• Electrons and holes can recombine easily, limiting the practical use of TiO_2.	Chowdhury et al. (2014), Zhu et al. (2019)
Fenton oxidation	• The use of strong oxidant (radical $^{\bullet}OH$) provides high activity, fast reaction rates, and mild reaction condition requirements; • Improve the degradation efficiency; • Application potential for the treatment of refractory pollutants and effectively meeting discharge effluent water quality requirements.	• Narrow pH range; • Generate large amounts of iron residue; • Have high operational costs; • Low water volume treatment capacity.	Zhu et al. (2019)

Table 2 contd. ...

...Table 2 contd.

Technique	Advantage	Disadvantage	References
Ionic liquids	• Fast reaction rate; • Recoverability with ease; • Higher selectivity; • Better reliability; • Reuse of spent solvent.	• Leaching of some ionic liquids into the aqueous phase, which may cause additional environmental risks; • Some ionic liquids have high viscosity, slower phase separation and toxicity.	Yudaev and Chistyakov (2022), Rajadurai and Anguraj (2021), Leyma et al. (2016)

soil and biodiversity (SDG 15—Life on Land) and food security (SDG 2—Zero Hunger).

3. Final considerations

This introductory chapter discusses an overview of toxic metals, including main sources, exposure routes, adverse effects, and treatment approaches. We emphasize that the continuity of studies aimed at the development or improvement of techniques for toxic metals removal from various environmental compartments is essential to make it possible to reduce the concentration or bioavailability of this class of environmental contaminants, as well as research aimed at the environmental monitoring, and the public health impact associated with human exposure to toxic metals.

The next chapters of this book will discuss topics about groundwater quality, some water treatment techniques for toxic metal removal, the generation, disposal, and recovery of solid wastes containing toxic metals, and genotoxicity and mutagenicity of metal-based nanomaterials.

References

Al Osman, M., Yang, F., and Massey, I.Y. 2019. Exposure routes and health effects of heavy metals on children. Biometals, 32: 563–573. doi: 10.1007/s10534-019-00193-5.

Alessandretti, I., Rigueto, C.V.T., Nazari, M.T., Rosseto, M., and Dettmer, A. 2021. Removal of diclofenac from wastewater: A comprehensive review of detection, characteristics and tertiary treatment techniques. J. Environ. Chem. Eng., 9(6): 106743. doi: 10.1016/j.jece.2021.106743.

Ali, H., and Khan, E. 2018. What are heavy metals? Long-standing controversy over the scientific use of the term 'heavy metals'—proposal of a comprehensive definition. Toxicol. Environ. Chem., 100(1): 6–19. doi: 10.1080/02772248.2017.1413652.

Ali, H., Khan, E., and Ilahi, I. 2019. Environmental chemistry and ecotoxicology of hazardous heavy metals: environmental persistence, toxicity, and bioaccumulation. J. Chem., 6730305. doi: 10.1155/2019/6730305.

ATSDR. Agency for Toxic Substances and Disease Registry. 2019. ATSDR's Substance Priority List. Available via https://www.atsdr.cdc.gov/spl/index.html#2019spl Acessed 15 Jan 2022.

Aziz, H.A., Adlan, M.N., and Ariffin, K.S. 2008. Heavy metals (Cd, Pb, Zn, Ni, Cu and Cr (III)) removal from water in Malaysia: post treatment by high quality limestone. Bioresour. Technol., 99(6): 1578–1583. doi: 10.1016/j.biortech.2007.04.007.

Barakat, M.A. 2011. New trends in removing heavy metals from industrial wastewater. Arab. J. Chem., 4(4): 361–377. doi: 10.1016/j.arabjc.2010.07.019.

Bozorg-Haddad, O., Delpasand, M., and Loáiciga, H.A. 2021. Water quality, hygiene, and health. pp. 217–257. *In*: Bozorg-Haddad, O. [ed.]. Economical, Political, and Social Issues in Water Resources, Elsevier.

Chowdhury, P., Elkamel, A., and Ray, A.K. 2014. Photocatalytic processes for the removal of toxic metal ions. *In*: Sharma, S. [ed.]. Heavy Metals in Water: Presence, Removal and Safety, Royal Society of Chemistry, 25–43. doi: 10.1039/9781782620174-00025.

De Rossi, A., Rigueto, C.V.T., Dettmer, A., Colla, L.M., and Piccin, J.S. 2020. Synthesis, characterization, and application of Saccharomyces cerevisiae/alginate composites beads for adsorption of heavy metals. J. Environ. Chem. Eng., 8(4): 104009. doi: 10.1016/j.jece.2020.104009.

Dettmer, A., Nunes, K.G.P., Gutterres, M., and Marcílio, N.R. 2010. Obtaining sodium chromate from ash produced by thermal treatment of leather wastes. Chem. Eng. J., 160(1): 8–12. doi: 10.1016/j.cej.2010.02.018.

Duffus, J.H. 2002. "Heavy metals" a meaningless term? (IUPAC Technical Report). Pure Appl. Chem., 74(5): 793–807. doi: 10.1351/pac200274050793.

Edelstein, M., and Ben-Hur, M. 2018. Heavy metals and metalloids: Sources, risks and strategies to reduce their accumulation in horticultural crops. Scientia Horticulturae, 234: 431–444. doi: 10.1016/j.scienta.2017.12.039.

Fu, Z., and Xi, S. 2020. The effects of heavy metals on human metabolism. Toxicol. Mech. Methods, 30(3): 167–176. doi: 10.1080/15376516.2019.1701594.

Leong, Y.K., and Chang, J.S. 2020. Bioremediation of heavy metals using microalgae: recent advances and mechanisms. Bioresour. Technol., 303: 122886. doi: 10.1016/j.biortech.2020.122886.

Leyma, R., Platzer, S., Jirsa, F., Kandioller, W., Krachler, R., and Keppler, B.K. 2016. Novel thiosalicylate-based ionic liquids for heavy metal extractions. J. Hazard. Mater., 314: 164–171. doi: 10.1016/j.jhazmat.2016.04.038.

Li, C., Zhou, K., Qin, W., Tian, C., Qi, M., Yan, X., and Han, W. 2019. A review on heavy metals contamination in soil: effects, sources, and remediation techniques. Soil Sediment Contam., 28(4): 380–394. doi: 10.1080/15320383.2019.1592108.

Luckey, T.D., B. Venugopal, and Hutcheson, D. 1975. Heavy metal toxicity safety and Hormology, Supplement Vol. I. Academic Press, New York, 121p.

Nogueira, F.G., Prado, N.T., Oliveira, L.C., Bastos, A.R., Lopes, J.H., and Carvalho, J.G. 2010. Incorporation of mineral phosphorus and potassium on leather waste (collagen): A new NcollagenPK-fertilizer with slow liberation. J. Hazard. Mater., 176(1-3): 374–380. doi: 10.1016/j.jhazmat.2009.11.040.

Nunes, R.R., Pigatin, L.B.F., Oliveira, T.S., Bontempi, R.M., and Rezende, M.O.O. 2018. Vermicomposted tannery wastes in the organic cultivation of sweet pepper: growth, nutritive value and production. Int. J. Recycl. Org. Waste Agric., 7(4): 313–324. doi: 10.1007/s40093-018-0217-7.

Piccin, J.S., Gomes, C.S., Feris, L.A., and Gutterres, M. 2012. Kinetics and isotherms of leather dye adsorption by tannery solid waste. Chem. Eng. J., 183: 30–38. doi: 10.1016/j.cej.2011.12.013.

Pitarch, A.M., Reig, L., Gallardo, A., Soriano, L., Borrachero, M.V., and Rochina, S. 2021. Reutilisation of hazardous spent fluorescent lamps glass waste as supplementary cementitious material. Constr. Build Mater., 292: 123424. 10.1016/j.conbuildmat.2021.123424.

Rai, P.K., Lee, S.S., Zhang, M., Tsang, Y.F., and Kim, K.H. 2019. Heavy metals in food crops: Health risks, fate, mechanisms, and management. Environ. Int., 125: 365–385. doi: 10.1016/j.envint.2019.01.067.

Rajadurai, V., and Anguraj, B.L. 2021. Ionic liquids to remove toxic metal pollution. Environ. Chem. Lett., 19(2): 1173–1203. doi: 10.1007/s10311-020-01115-5.

Rempel, A., Nazari, M.T., Braun, J.C.A., Kreling, N.E., Treichel, H., and Colla, L.M. 2021. Application of Biochar for soil remediation. pp. 403–425. *In*: Kapoor, R.T., Treichel, H., and Shah, M.P. [eds.]. Biochar and its Application in Bioremediation, Springer, Singapore.

Renu, Agarwal, M., and Singh, K. 2017. Heavy metal removal from wastewater using various adsorbents: a review. J. Water Reuse Desalin., 7(4): 387–419. doi: 10.2166/wrd.2016.104.

Rigueto, C.V.T., Rosseto, M., Krein, D.D.C., Ostwald, B.E.P., Massuda, L.A., Zanella, B.B., and Dettmer, A. 2020. Alternative uses for tannery wastes: a review of environmental, sustainability, and science. J. Leather Sci. Eng., 2(1): 1–20. doi: 10.1186/s42825-020-00034-z.

Rigueto, C.V.T., Nazari, M.T., Massuda, L.Á., Ostwald, B.E.P, Piccin, J.S., and Dettmer, A. 2021a. Production and environmental applications of gelatin-based composite adsorbents for contaminants removal: a review. Environ. Chem. Lett. 19: 2465–2486. doi: 10.1007/s10311-021-01184-0.

Rigueto, C.V.T., Nazari, M.T., Rosseto, M., Massuda, L.A., Alessandretti, I., Piccin, J.S., and Dettmer, A. 2021b. Emerging contaminants adsorption by beads from chromium (III) tanned leather waste recovered gelatin. J. Mol. Liq., 330: 115638. doi: 10.1016/j.molliq.2021.115638.

Rosseto, M., Rigueto, C.V.T., Krein, D.D.C., Massuda, L.A., Balbé, N.P., Colla, L.M., and Dettmer, A. 2021. Combined effect of transglutaminase and phenolic extract of S pirulina platensis in films based on starch and gelatin recovered from chrome III tanned leather waste. Biofuel Bioprod. Biorefin., 15(5): 1406–1420. doi: 10.1002/bbb.2244.

Shan, C., Xu, Z., Zhang, X., Xu, Y., Gao, G., and Pan, B. 2018. Efficient removal of EDTA-complexed Cu (II) by a combined Fe (III)/UV/alkaline precipitation process: Performance and role of Fe (II). Chemosphere, 193: 1235–1242. doi: 10.1016/j.chemosphere.2017.10.119.

Srivastava, A., Dutta, S., Ahuja, S., and Sharma, R.K. 2021. Green chemistry: key to reducing waste and improving water quality. pp. 359–407. *In*: Ahuja, S. [ed.]. Handbook of Water Purity and Quality 2nd ed., Academic Press.

Stojanovic, A., and Keppler, B.K. 2012. Ionic liquids as extracting agents for heavy metals. Sep. Sci. Technol., 47(2): 189–203. doi: 10.1080/01496395.2011.620587.

Walton, J.R. 2011. Bioavailable aluminum: Its effects on human health. pp. 331–342. *In*: Nriagu, J.O. [ed.]. Encyclopedia of Environmental Health, Elsevier Science.

WHO. World Health Organization. 2017. Guidelines for drinking-water quality (4th ed.). Available via https://www.who.int/publications/i/item/9789241549950. Acessed 25 Jan 2022.

Yin, K., Wang, Q., Lv, M., and Chen, L. 2019. Microorganism remediation strategies towards heavy metals. Chem. Eng. J., 360: 1553–1563. doi: 10.1016/j.cej.2018.10.226.

Yudaev, P.A., and Chistyakov, E.M. 2022. Ionic liquids as components of systems for metal extraction. ChemEngineering, 6(1): 6. doi: 10.3390/chemengineering6010006.

Zewail, T.M., and Yousef, N.S. 2015. Kinetic study of heavy metal ions removal by ion exchange in batch conical air spouted bed. Alex. Eng. J., 54(1): 83–90. doi: 10.1016/j.aej.2014.11.008.

Zhu, Y., Fan, W., Zhou, T., and Li, X. 2019. Removal of chelated heavy metals from aqueous solution: A review of current methods and mechanisms. Sci. Total Environ., 678: 253–266. doi: 10.1016/j.scitotenv.2019.04.416.

2

Groundwater Quality Indexing for Drinking Purpose from Arsenic Prone Areas, West Bengal
A Health Risk Assessment Study

Madhurima Joardar, Nilanjana Roy Chowdhury, Antara Das, Deepanjan Mridha and *Tarit Roychowdhury**

1. Introduction

Arsenic (As) has been categorized as group I human carcinogen (ATSDR 2014) among the toxic substances. The United States Agency for Toxic Substances and Disease Registry (ATSDR 2007) has marked arsenic under group I category, depending on its level of health hazard and toxicity. Groundwater arsenic-contamination is considered as one of the most serious emerging problem in today's world (Mohana et al. 2020, Landrigan et al. 2018). In different regions of the world, considering past decades, groundwater arsenic-contamination is one of the major community and environmental health concern (Sharma et al. 2017, Tabassum et al. 2019). The GMB plain is undertaken as the menace due to arsenic-contamination in groundwater (Chakraborti et al. 2018, Das et al. 2020, Shakoor et al. 2016). West Bengal under GMB plain (world's largest delta) is rich with arsenic in soil (Chakraborti et al. 2013, Chowdhury et al. 2018a). In West Bengal, Gaighata and Deganga are the two well-known arsenic affected blocks in North 24 Parganas district (Chowdhury et al. 2018a, b, Joardar et al. 2021a, b, Roychowdhury 2010). The use of groundwater is increasing due to rapid increase of population. Groundwater, the most abundant natural resource, is

School of Environmental Studies, Jadavpur University, Kolkata-700032, India
* Corresponding author: rctarit@yahoo.com, tarit.roychowdhury@jadavpuruniversity.in

mainly used for both domestic and irrigational purposes (Chakraborti et al. 2011). Arsenic in groundwater varies depending on several factors like depth of the tube-well, source and geographical location (Rabanni et al. 2017, Shahid et al. 2017). In Bengal delta, the inhabitants residing in arsenic-contaminated areas are using groundwater from both the sources like domestic shallow tube-well (STW) and deep tube-well (DTW) installed by the local government on a daily basis for household purposes (especially for drinking and cooking purposes). The depth of domestic shallow tube-wells ($\leq$ 140 ft) is lesser compared to deep tube-wells ($\geq$ 500 ft) (Goel et al. 2019, van Geen et al. 2003). The easy installation and use of STW naturally propagates consequent arsenic withdrawal leading to harmful human health risks, which includes cancer (skin, lung and bladder) (Argos et al. 2010, Mohana et al. 2020, Naujokas et al. 2013). A large population worldwide is facing carcinogenic health risk due to acute and chronic toxicity of arsenic through consumption of groundwater, which is documented as a severe environmental health concern (Celik et al. 2008, Chakraborti et al. 2013, Smith et al. 2009). Exposure to arsenic toxicity causes numerous health hazards in humans like pigmentation, cancer (skin, lung, and bladder), melanosis, keratosis, etc. (Abbas et al. 2018, Chakraborti et al. 2018, Roychowdhury 2010). Considering the toxic potentiality of arsenic, World Health Organization (WHO) has suggested the recommended value of arsenic as 0.01 mg/l in drinking water (Shakoor et al. 2016, WHO 2011).

The main objective of our study is to highlight the groundwater quality of two different sources (domestic shallow and deep tube-well) mainly used for drinking purposes from two arsenic-affected blocks of West Bengal. The suitability of groundwater is determined mainly depending on the physico-chemical parameters along with the Water Quality Index (WQI). A statistical interpretation has been performed to evaluate the relation between arsenic and iron present in groundwater. Principal Component Analysis has been carried out to comprehend the inter-dependence of the groundwater quality parameters, considering the depth as observational levels. The study investigates the quality of groundwater from domestic shallow and deep tube-well for human consumption, based on the human health risk assessment (cancer and non-cancer). As a result, highlighting the human health impact due to groundwater arsenic-contamination will lead to awareness among the exposed populations and suitable mitigation strategies. From environmental health perception, it is of paramount important to determine groundwater (especially for drinking purposes) arsenic concentration along with various other water quality parameters and subsequent human health risks. To the best of our knowledge, this is a first comparative study carried out on domestic shallow and deep tube-well groundwater quality from two arsenic-affected blocks in Bengal delta to evaluate the arsenic level and associated human health risks.

2. Materials and methods

2.1 Study area

The study area includes two severely arsenic-affected blocks located in North 24 Parganas district, West Bengal which is a part in the Gangetic delta, lying east of the Hooghly River. The two arsenic-affected blocks are namely: Gaighata (Latitude: 22°

55′ 48″ N and Longitude: 88° 43′ 48″ N) and Deganga (Latitude: 22° 41′ 36″ N and Longitude: 88° 40′ 41″ N). Equally, Deganga and Gaighata block have been already reported as severely arsenic-contaminated sites in West Bengal, India (Chowdhury et al. 2018a, Joardar et al. 2021a).

2.2 Sample collection and preservation

Groundwater samples were collected from two different sources (deep and domestic shallow tube-wells) of the study areas which are usually used for drinking purposes among the rural population. The Global Positioning System (GPS Model: GARMIN Etrex 30x) information with proper latitudes and longitudes has been taken for precise detection of the water samples. Groundwater samples were collected in duplicates and stored in pre-washed polyethylene bottles of 30 ml and 250 ml, respectively. The 30 ml water samples were preserved by adding 0.1% (v/v) concentrated nitric acid for evaluation of arsenic and iron. Likewise, the water samples (250 ml) were stored without addition of any preservative for estimation of different drinking water quality parameters. The polyethylene bottles were transported to the laboratory by placing it in an ice-cool box and stored at 4°C prior to analysis. The details of sample collection and preservation have been mentioned earlier (Das et al. 2020, Ghosh et al. 2019).

2.3 Sample analysis

Different physico-chemical parameters namely pH, temperature, Electrical Conductivity (EC), Total Dissolved Solids (TDS), Total Hardness (TH), Total Alkalinity (TA), Chloride (Cl^-), Sulphate (SO_4^{2-}), Iron (Fe), Nitrate (NO_3^-), Fluoride (F^-), and Arsenic (As) present in the groundwater samples were quantified. The physical water quality parameters were analyzed using digital instruments like pH (TOSHCON, pH meter CL 46+), and EC (SYSTRONICS, Conductivity Meter 306) and TDS were estimated using HANNA, HI 98194 digital multi-parameter waterproof instrument. Analysis of total arsenic in water samples was performed using HG-AAS ((Hydride Generation-Atomic Absorption Spectrophotometric) method (Atomic Absorption Spectrophotometer (Model: Varian AA140, USA)) coupled with Vapor Generation Accessory (VGA-77) of the software version 5.1 (Chowdhury et al. 2020, Joardar et al. 2021a). Using Spectrophotometric and Nephelometric turbidity method, iron and sulphate content was measured in UV–Vis spectrophotometer (Thermo-scientific, Orion Aquamate, 8000, made in USA) (APHA 1998, Fries et al. 1977). The parameters like nitrate and fluoride were determined using Ion Selective Electrode method (Model: Thermo Scientific Orion Star A214). The parameters, namely Total Hardness (TH), Total Alkalinity (TA) and Chloride ions, were calculated using Complexometric, Acid and Argentometric titration method, respectively (APHA 2005).

In the water samples, Mg^{2+} and Ca^{2+} ion concentrations were calculated using the following equations, respectively:

$$[Mg^{2+}] = (Magnesium\ hardness/4.11) \tag{1}$$

$$[Ca^{2+}] = (Calcium\ hardness/2.50) \tag{2}$$

The conversion factors 4.11 and 2.50 are obtained from the total hardness of the water that contributed both Mg and Ca hardness which is considered as $CaCO_3$ (Molecular Weight = 100) (Das and Nag 2015, Todd 1980). The details of information on the analytical methodologies used for estimation of the different cations and anions in the groundwater samples have been mentioned earlier (Das et al. 2020).

2.4 Chemicals and reagents

During the experimental analysis of the water samples, double-distilled water was used thoroughly. All chemicals and reagents of analytical grade were used at the time of experimental work. For total arsenic estimation, a standard stock solution of arsenate (1000 mg/l) from Merck, Darmstact, Germany was used. For HG-AAS method, 0.6% $NaBH_4$ mixed with 0.5% NaOH, and 5–10 M HCL were used during analysis. An aqueous solution of 10% of KI and concentrated HCl (8%) were added to the water samples during sample preparation and kept for 45 min prior to estimation of arsenic (Chowdhury et al. 2020, Joardar et al. 2021a). Standard stock solutions of 1000 mg/l for iron (NIST, $Fe(NO_3)_3$ in HNO_3 0.5 mol/l), fluoride (Orion 940907, Thermo Scientific, USA), and nitrate (as nitrogen, Orion 920707, Thermo Scientific, USA) have been thoroughly used during the analytical experiments. The details of the methodology, instruments and chemicals used to estimate each water quality parameter have been described in our earlier publication (Das et al. 2020).

2.5 Quality control and quality assurance

The quality assurance of the observed analytical data was validated in the course of standardization, routine blank measurements and analysis of spiked samples. Quality control analysis was performed by analyzing blank, duplicates samples, and calculating recovery of spiked digested samples. Inter-laboratory tests were also performed (Das et al. 2020, Rahman et al. 2002).

2.6 Statistical analysis

Pearson correlation matrix of the water samples were performed to find the inter-relation among the water quality parameters using Excel 2017 (Microsoft Office). The 2D Kernel density plot is non-parametric method for probability density functions, which is preferred to recognize the distribution of groundwater concentrations through a colour density representation of scatter plot (Duong 2007). Multivariate analysis through Principal Component Analysis (PCA) has been performed to comprehend the inter-dependence of the water quality parameters, considering the depth as observational levels (Fang et al. 2012). Origin 2018 software has been used to perform the 2D Kernel density and Principal Component Analysis (PCA).

2.7 Data analysis

2.7.1 Water quality index (WQI)

Water quality index (WQI) is an evaluation scale that highlights the overall quality of water for drinking purposes in the studied areas (Gupta and Misra 2018, Lumb et al. 2011, Meng et al. 2016, Xiao et al. 2014). WQI reflects the composite influence of different water quality parameters in this study. According to the assigned 'weight' of the parameters, the 'relative weight' (W_i) was evaluated.

The 'relative weight' (W_i) was calculated using the following equation:

$$W_i \text{ (Relative Weight)} = w_i/(\textstyle\sum_{i=1}^{n} (w_i))$$

where, w_i = Weight of each parameter; n= Number of parameters.
Quality rating scale (q_i) for each parameter is obtained by the following equation:

$$q_i \text{ (Quality Rating)} = (C_i / S_i) *100$$

where, C_i = Concentration of each chemical parameter in each water sample (mg/l); S_i = Indian drinking water standard for each chemical parameter according to the guidelines of the BIS (2012).
WQI was evaluated using the following equation:

$$SI_i \text{ (sub index of } i^{th} \text{ parameter)} = W_i * q_i$$

$$WQI = \textstyle\sum SI_i$$

where,
q_i = Rating based on concentrate of ith parameter (Das et al. 2020).
The computed WQI values are categorized to determine the quality of water as: "excellent" (value < 50), "good" (range = 50–100), "poor" (range = 100–200), "very poor" (range = 200–300), and "inappropriate for drinking" (value > 300).

2.7.2 Health risk assessment

In humans, the health risk is caused due to consumption of toxic elements through daily dietary intakes. In our study, the Health Risk Analysis was performed on the exposed populations caused due to ingestion of toxic elements through groundwater present in both deep (DTW) and domestic shallow (STW) tube-wells. Health risk analysis was performed using the recommended models/formulas adapted from USEPA (United States Environmental Protection Agency) policy.

Carcinogenic risk analysis has been performed for arsenic, which is well-known as a group I carcinogen (IARC 2012, Waqas et al. 2017). Carcinogenic (As) and non-carcinogenic (As, NO_3^- and F^-) risk has been measured for the exposed populations (Adimalla et al. 2018, Narsimha and Rajitha 2018, USEPA 2006). For the exposed population, the health risk is evaluated based on chronic daily intake (CDI) of toxic elements, carcinogenic risk (CR) and hazard quotient (HQ).

The chronic daily intake (CDI) of toxic elements was calculated (Das et al. 2020, Kumar et al. 2016) using the following equation:

$$\text{CDI} = \frac{\textbf{(Concentration * Ingestion Rate * Exposure Frequency * Exposure Duration)}}{\textbf{(Body Weight *Average Lifetime)}}$$

where,

C   =   Concentration of As in DTW or STW water (μg/l)

IR   =   Ingestion rate of drinking water (2.5 l/day) (USEPA 2014)

EF   =   Exposure Frequency (365 days/year)

ED   =   Exposure duration (age) (65 years for an adult) (WHO 2011)

BW   =   Standard Body Weight (70 kg for an adult) (Goswami et al. 2020)

AT   =   Average Lifetime (365 * 65 = 23725 days)

For arsenic intake, Carcinogenic risk (CR) was determined using the following equation:

CR = CDI * CSF

Cancer Slope Factor is considered as 1.5 per mg/kg/day for arsenic (USEPA 2005).

The acceptable or tolerable range for carcinogenic health risk by the USEPA is from 10^{-6} to 10^{-4} (Kazi et al. 2016, USEPA 2011).

For As, NO_3^- and F^- intake, non-carcinogenic risk (Hazard Quotient, i.e., HQ) was calculated using the formula:

HQ = CDI/RfD

'Reference Dose' (RfD) value represents the chronic oral exposure dose for the toxic elements as follows, i.e., arsenic (0.0003 mg/kg/day), nitrate (1.6 mg/kg/day) and fluoride (0.06 mg/kg/day), respectively, as mentioned in several studies (Rasool et al. 2015, Su et al. 2013, Narsimha and Rajitha 2018). The potential health risk for non-cancerous effect is indicated when HQ value > 1 (Kazi et al. 2016).

3. Results and discussion

3.1 *Water quality parameters of domestic shallow and deep tube-wells*

The overall groundwater quality of the domestic shallow and deep tube-well has been evaluated through analysis of the physical and chemical parameters. The statistical presentation of all the water quality parameters from the studied two blocks has been shown in Table 1.

In Gaighata block, the respective mean value of pH for domestic shallow (n = 14) and deep tube-well (n = 14) groundwater samples are 7.28 and 7.41, respectively, which signifies that the groundwater is alkaline in character. The alkaline nature is confirmed observing the mean value of total alkalinity for the groundwater samples, i.e., 468 and 464 mg/l, respectively, which is higher than the Indian Standard value of drinking water (BIS 2012). The mean value of temperature observed for the groundwater samples of Gaighata block varies between 25.1–27.6°C. The mean values of conductance and TDS of the domestic shallow and deep tube-well groundwater samples are 0.89 mS/cm, 164 mg/l and 0.97 mS/cm, 60.7 mg/l, respectively. The TDS observed in the drinking water samples is within the permissible limit

Table 1: Different water quality parameters of domestic shallow and deep tube-wells from two studied blocks (a) Gaighata and (b) Deganga.

a) Gaighata block		Statistical parameters				
Category	Water quality parameters	Mean	SD	Median	Min	Max
Domestic shallow tube-well (n = 14)	pH	7.28	0.33	7.25	6.9	7.9
	Temperature (°C)	26.3	0.45	26.2	25.9	27.6
	Conductance (ms/cm)	0.89	0.19	0.85	0.64	1.4
	TDS (mg/l)	164	79.5	150	50	300
	Total hardness (mg/l)	464	126	480	250	680
	Alkalinity (mg/l)	468	56.5	460	400	560
	Sulphate (mg/l)	51.4	114	11.8	0.1	433
	Chloride (mg/l)	88.8	24.8	79.9	49.7	134
	Arsenic (mg/l)	0.042	0.02	0.04	0.01	0.08
	Iron (mg/l)	7.65	2.39	8.22	2.61	11.3
	Fluoride (mg/l)	0.13	0.047	0.13	0.07	0.23
	Nitrate (mg/l)	4.52	5.34	2.48	0.49	20.1
	Calcium (mg/l)	122	36.5	12.5	60	179
	Magnesium (mg/l)	41.8	30.4	36	13.6	128
Deep tube-well (n = 14)	pH	7.41	0.19	7.4	7.1	7.8
	Temperature (°C)	25.3	0.32	25.1	25.1	26.1
	Conductance (ms/cm)	0.97	0.08	0.95	0.87	1.09
	TDS (mg/l)	60.7	21.3	50	50	100
	Total hardness (mg/l)	391	54.1	400	300	510
	Alkalinity (mg/l)	464	56.1	460	370	600
	Sulphate (mg/l)	1.82	2.72	0.1	0.1	7.54
	Chloride (mg/l)	82.2	30.8	81.7	28.4	142
	Arsenic (mg/l)	0.008	0.01	0.003	0.003	0.04
	Iron (mg/l)	1.48	0.75	1.745	0.19	2.82
	Fluoride (mg/l)	0.30	0.06	0.325	0.15	0.38
	Nitrate (mg/l)	1.39	0.43	1.22	1.018	2.44
	Calcium (mg/l)	75.8	18.4	79.2	43.2	101
	Magnesium (mg/l)	49.1	14.7	46.7	27.7	80.8

Table 1 contd. ...

...Table 1 contd.

(b) Deganga block	Statistical parameters					
Category	Water quality parameters	Mean	SD	Median	Min	Max
Domestic shallow tube-well (n = 20)	pH	7.35	0.097	7.36	7.18	7.52
	Temperature (°C)	31.8	0.16	31.8	31.6	32.1
	Conductance (ms/cm)	0.11	0.02	0.1	0.083	0.152
	TDS (mg/l)	3050	3605	1500	200	14000
	Total hardness (mg/l)	369	81.2	346	268	600
	Alkalinity (mg/l)	473	43.1	469	377	614
	Sulphate (mg/l)	9.03	9.31	6.2	0.4	39.1
	Chloride (mg/l)	61.4	29.3	58.9	20.8	135
	Arsenic (mg/l)	0.661	0.73	0.39	0.07	2.84
	Iron (mg/l)	5.79	3.04	6.09	0.41	11.4
	Fluoride (mg/l)	0.18	0.06	0.17	0.11	0.37
	Nitrate (mg/l)	1.98	1.34	1.41	0.87	5.58
	Calcium (mg/l)	90.5	17.7	88	60.8	128
	Magnesium (mg/l)	34.9	18.7	34.1	4.8	92.5
Deep tube-well (n = 20)	pH	7.66	0.14	7.66	7.49	8.05
	Temperature (°C)	31.8	0.12	31.8	31.6	32.1
	Conductance (ms/cm)	0.10	0.06	0.09	0.07	0.36
	TDS (mg/l)	1250	1606	600	200	6800
	Total hardness (mg/l)	314	55.8	322	140	380
	Alkalinity (mg/l)	443	62.7	443	316	525
	Sulphate (mg/l)	5.03	2.62	4.7	1.3	10.9
	Chloride (mg/l)	52.6	135	24.2	6.9	626.5
	Arsenic (mg/l)	0.023	0.04	0.008	0.003	0.153
	Iron (mg/l)	3.08	4.04	1.5	0.75	16.9
	Fluoride (mg/l)	0.21	0.04	0.215	0.079	0.28
	Nitrate (mg/l)	1.98	1.30	1.97	0.69	6.63
	Calcium (mg/l)	76.5	17.5	76.8	20.8	114
	Magnesium (mg/l)	29.9	6.1	29.4	18.5	40.9

(500 mg/l), which signifies that the presence of inorganic salts and minerals is lesser in this studied area. The mean chloride value observed for groundwater samples of domestic shallow (88.8 mg/l) and deep (82.2 mg/l) is within the recommended value (250 mg/l). The presence of chloride ions in water samples is influenced by TDS, which means that the taste (salinity) of groundwater is suitable for drinking purpose (Balakrishnan et al. 2011). The mean arsenic concentrations observed for groundwater samples from domestic shallow and deep tube-well are 0.042 mg/l (range = 0.01–0.08 mg/l) and 0.008 mg/l (range = 0.003–0.04 mg/l), respectively. The observed mean iron concentrations for domestic shallow (7.65 mg/l) and deep tube-well (1.48 mg/l) water samples are higher than the recommended value (0.3 mg/l), which indicated that the groundwater is enriched with iron. The mean value of Total Hardness (TH) measured for both the domestic shallow and deep tube-well drinking water samples is higher than its permissible value (200 mg/l) in drinking water. The higher mean concentration of TH in water samples signifies that the feature of groundwater is relatively hard. The presence of calcium, magnesium, and iron ions denotes the hardness of drinking water. The mean calcium ion concentration observed for both domestic shallow (122 mg/l) and deep (75.8 mg/l) is higher than mean magnesium ion of domestic shallow (41.8 mg/l) and deep (49.1 mg/l), respectively. For domestic shallow water samples, the observed mean concentrations of nitrate, fluoride and sulphate ions are 4.52 mg/l (range = 0.49–20.1 mg/l), 0.13 mg/l (range=0.07-0.23 mg/l) and 51.4 mg/l (range = 0.1–433 mg/l), respectively. Likewise, for the deep water samples, the mean concentrations observed are as follows for nitrate (mean = 1.39 mg/l; range = 1.02–2.44 mg/l), fluoride (mean = 0.30 mg/l; range = 0.15–0.38 mg/l), and sulphate (mean = 1.82 mg/l; range = 0.1–7.54 mg/l). These observed concentrations of nitrate, fluoride and sulphate ions for both domestic shallow and deep water samples of the studied Gaighata block are lower than the recommended values 45, 1.5 and 200 mg/l, respectively. Overall, the mean anionic concentration of the water samples from the studied block showed the order of the ions as chloride > sulphate > nitrate which justifies the water quality of Gaighata block as basic in nature.

In a similar way, for Deganga block, the respective mean value of pH for domestic shallow (n = 20) and deep (n = 20) tube-well groundwater samples are 7.35 and 7.66, respectively, which signifies that the groundwater is alkaline in character. The alkaline nature is confirmed observing the mean value of total alkalinity for the groundwater samples, i.e., 473 and 443 mg/l, respectively, which is higher than the recommended Indian Standard value. The mean value of temperature observed for the groundwater samples of Deganga block varies from 31.6–32.1°C. The mean values of conductance and TDS of the domestic shallow and deep tube-well groundwater samples are 0.11 mS/cm, 3050 mg/l and 0.10 mS/cm, 1250 mg/l, respectively. The TDS observed in the drinking water samples is much higher than the permissible limit which signifies the presence of inorganic salts and minerals in the studied area. The mean chloride value observed for groundwater samples of domestic shallow (61.4 mg/l) and deep (52.6 mg/l) is within the recommended value. The mean arsenic concentration observed for groundwater samples from domestic shallow and deep tube-wells are 0.661 mg/l (range: 0.07–2.84 mg/l), and 0.023 mg/l (range: 0.003–1.53 mg/l), respectively. The observed mean iron concentration for domestic shallow

(5.79 mg/l) and deep (3.08 mg/l) water samples is higher than the recommended value, indicating that the groundwater is iron enriched. The mean value of Total Hardness (TH) measured for both the domestic shallow (369 mg/l) and deep (314 mg/l) tube-well water samples is higher than the permissible limit. The higher mean concentration of TH in water samples signifies that the feature of groundwater is relatively hard. The presence of calcium, magnesium, and iron ions denotes the hardness of drinking water. The mean calcium ion concentration observed for both domestic shallow (90.5 mg/l), deep (76.5 mg/l) is higher than mean magnesium ion of domestic shallow (34.9 mg/l) and deep (29.9 mg/l), respectively. For domestic shallow water samples, the observed mean concentrations of nitrate, fluoride and sulphate ions are 1.98 mg/l (range = 0.87–5.58 mg/l), 0.18 mg/l (range = 0.11–0.37 mg/l) and 9.03 mg/l (range = 0.4–39.1 mg/l), respectively. Likewise, for the deep water samples, the mean concentrations observed are as follows for nitrate (mean = 1.98 mg/l; range = 0.69–6.63 mg/l), fluoride (mean = 0.21 mg/l; range = 0.079–0.28 mg/l), and sulphate (mean = 5.03 mg/l; range = 1.3–10.9 mg/l). These observed concentrations of nitrate, fluoride and sulphate ions for both domestic shallow and deep water samples of the studied Deganga block are lower than the recommended values 45, 1.5 and 200 mg/l, respectively. Overall, the mean anionic concentration of the water samples from the studied block showed the order of the ions as chloride > sulphate > nitrate, which justifies that the water quality of Deganga block is basic in nature.

3.2 Distribution of arsenic and iron in groundwater with depth

To understand the distributions of both the arsenic and iron concentrations in groundwater collected from domestic shallow and deep tube-wells, 2D-Kernel Density plot has been shown with depth range variation from the two studied blocks (Fig. 1).

In Gaighata block, major presence of arsenic in domestic shallow and deep tube-well groundwater is confined at a depth range of 50–100 ft and 580–780 ft, respectively (Fig. 1a, b). According to the density plot, the iron concentration greater than 0.3 mg/l is scattered proportionally with the depth range of both domestic shallow and deep tube-wells. Likewise, in Deganga block, major presence of arsenic as well as iron in domestic shallow and deep tube-well groundwater is confined at a depth range of 80–130 ft and 500–800 ft, respectively (Fig. 1c, d).

Approximately 78.5 and 70% of the deep tube-well groundwater samples from the respective Gaighata (n = 14, depth = 580–780 ft) and Deganga (n = 20, depth = 500–800 ft) block has been observed with arsenic less than the WHO recommended value in drinking water, i.e., 0.01 mg/l (WHO 2011). Groundwater from the domestic shallow tube-wells (depth = 50–130 ft) in both the studied blocks shows high arsenic concentration compared to the deep aquifer. About 28% of the groundwater samples in the depth range of < 110 ft were found as arsenic-contaminated above 50 μg/l in West Bengal, India (Chowdhury et al. 1999). As a result, consumption of arsenic through groundwater from domestic shallow tube-wells poses severe health risk to the populations residing in the arsenic-affected blocks.

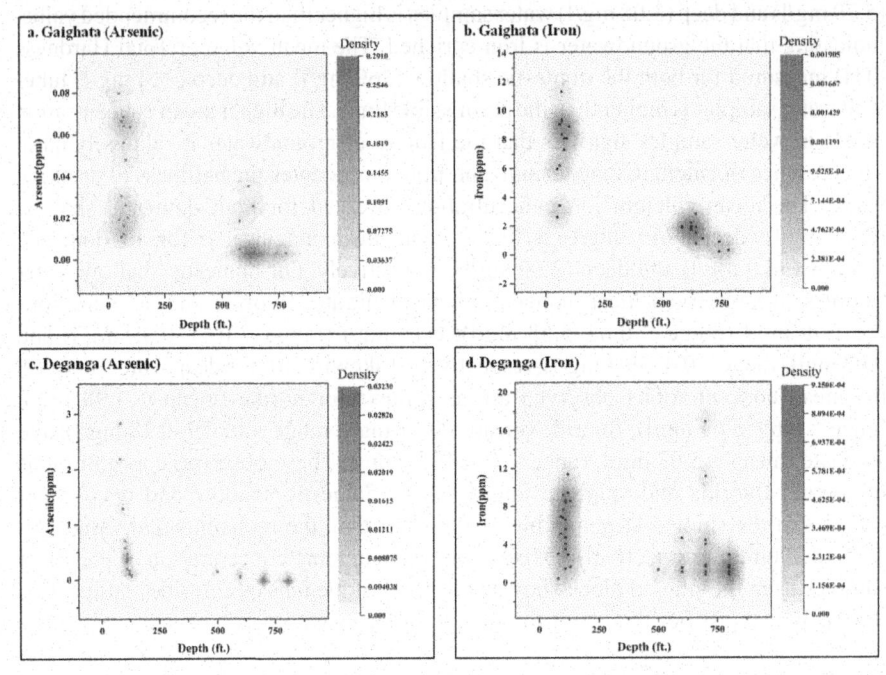

Fig. 1: Kernel Density plot of arsenic and iron from the two studied blocks.

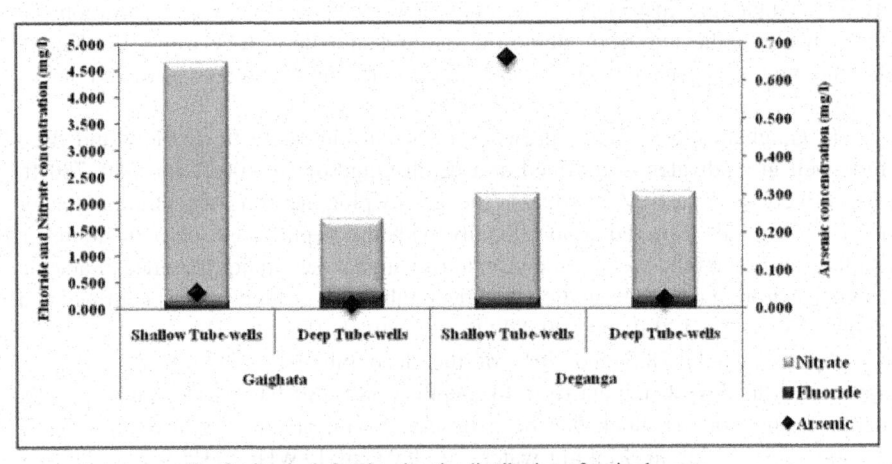

Fig. 2: Stacked plot showing the distribution of toxic elements.

3.3 Statistical distribution of toxic elements (As, NO$_3^-$, and F$^-$)

Arsenic, nitrate and fluoride are categorized as toxic elements present in groundwater. The statistical distribution of the three toxic elements in the studied blocks has been shown Fig. 2. The mean arsenic concentration in domestic shallow tube-well groundwater of the respective Gaighata and Deganga block is 4.2 and 66.1 fold higher than the WHO recommended limit. The deep tube-well groundwater of Gaighata block is within the permissible limit, whereas in Deganga block it is 2 fold higher. The mean arsenic concentration of domestic shallow tube-wells is 5.25 and 28.7 times higher than the mean arsenic of deep tube-well groundwater of Gaighata and Deganga block, respectively. Exposure to high arsenic concentration in drinking water leads to severe human health risks like arsenicosis, skin pigmentation, cancer (skin, lung and bladder), etc. (Abdul et al. 2015, Bhowmick et al. 2018, Roychowdhury 2008, 2010).

The fluoride as well as nitrate concentration in the domestic shallow and deep tube-wells groundwater from the studied blocks is within the permissible limit of drinking water. The recommended value of fluoride and nitrate in drinking water is 1.5 and 45 mg/l, respectively (BIS 2012, WHO 2011). Consumption of high fluoride containing drinking water results in decaying of bones, dental and skeletal fluorosis in children and adults (Narsimha and Rajitha 2018). Globally, nitrate is the most considerable chemical toxin in groundwater. Exposure of nitrate through drinking water causes health risk in infants (Fan and Steinberg 1996).

3.4 Correlation and Principal Component Analysis of the water quality parameters

The correlation among the water quality parameters has been depicted in Table 2. In Gaighata block, for both domestic shallow and deep tube-well, a moderately positive relation has been observed between pH and alkalinity (0.48, 0.59) compared to the Deganga block groundwater samples. This suggests that the water quality of Gaighata and Deganga block is alkaline in nature, which is confirmed by the mean concentration of pH and alkalinity in groundwater samples. The TDS and EC for the groundwater samples showed no significant variations. A strong positive correlation between EC and chloride was observed for both the types of groundwater samples. The interdependence of the water quality parameters depicts that the quality of domestic shallow groundwater is hard in nature compared to deep tube-well groundwater.

The different water quality parameters from the two studied blocks have been plotted through Principal Component Analysis (PCA) (Fig. 3). PCA provides information on the water quality parameters, which depict the entire data elucidation, and reduction followed by summarizing the statistical correlation among them with least loss of original information (Helena et al. 2000). The biplot was plotted for both domestic shallow and deep tube-well groundwater with depth as observational levels and eleven variables (pH, conductance, TDS, total hardness, alkalinity, sulphate, chloride, arsenic, iron, fluoride, and nitrate).

In Gaighata block, for domestic shallow tube-well water, the two Principal Components cumulatively showed 52.6% of the total variation, where PC1 and PC2 contributed 26.98% and 25.62%, respectively. The Group 1 in PCA biplot was formed between arsenic, nitrate, pH, iron and alkalinity. Among the distinctly placed parameters (total hardness, TDS, conductance, chloride, sulphate and fluoride), conductance has strong relation with chloride. Similarly, for deep tube-well groundwater, the cumulative two Principal Components showed 51.17% of the total variation, where PC1 and PC2 contributed 30.71% and 20.46%, respectively. The PCA biplot showed the dependence among the parameters as Group 1 (arsenic

Table 2: Correlation matrixes of the physico-chemical parameters of domestic shallow and deep tube-wells from the studied blocks.

Gaighata	pH	EC	TDS	TH	Alkalinity	Sulphate	Chloride	Arsenic	Iron	Fluoride	Nitrate
a) Domestic shallow tube-wells											
pH	1										
EC	-0.22	1									
TDS	-0.60	0.17	1.00								
TH	-0.59	0.81	0.35	1							
Alkalinity	0.48	0.33	-0.62	0.04	1						
Sulphate	0.03	-0.02	-0.20	0.21	0.25	1					
Chloride	0.03	0.70	0.04	0.43	0.36	-0.49	1				
Arsenic	0.22	-0.34	-0.09	-0.46	-0.13	-0.31	-0.06	1			
Iron	0.08	0.28	-0.37	-0.07	0.51	-0.05	0.15	0.31	1		
Fluoride	-0.16	-0.26	0.22	-0.06	-0.24	0.61	-0.68	-0.32	-0.16	1	
Nitrate	0.15	-0.15	-0.19	-0.22	0.15	-0.06	-0.18	0.28	0.24	-0.05	1

Gaighata	pH	EC	TDS	TH	Alkalinity	Sulphate	Chloride	Arsenic	Iron	Fluoride	Nitrate
b) Deep tube-wells											
pH	1										
EC	0.35	1									
TDS	0.26	-0.10	1								
TH	0.00	-0.09	-0.34	1							
Alkalinity	0.59	0.19	-0.20	0.32	1						
Sulphate	-0.08	0.21	0.44	-0.18	-0.11	1					
Chloride	0.26	0.86	0.12	-0.11	0.13	0.35	1				
Arsenic	-0.68	-0.52	-0.27	0.06	-0.34	-0.28	-0.49	1			
Iron	-0.18	-0.37	-0.30	0.22	-0.35	-0.03	-0.56	0.12	1		
Fluoride	0.23	0.25	0.12	0.05	-0.04	0.34	0.23	-0.79	0.26	1	
Nitrate	0.60	-0.02	-0.01	0.24	0.37	-0.20	-0.18	-0.14	0.26	0.00	1

Deganga
c) Domestic shallow tube-wells

	pH	EC	TDS	TH	Alkalinity	Sulphate	Chloride	Arsenic	Iron	Fluoride	Nitrate
pH	1										
EC	-0.10	1									
TDS	0.21	-0.28	1								
TH	0.02	0.48	-0.09	1							
Alkalinity	0.01	0.79	-0.05	0.69	1						
Sulphate	0.10	0.36	-0.16	-0.13	0.20	1					
Chloride	-0.09	0.82	0.04	0.50	0.57	0.33	1				
Arsenic	0.42	-0.25	0.16	-0.12	-0.09	-0.06	-0.35	1			
Iron	-0.34	-0.04	-0.12	-0.09	0.06	0.40	-0.16	-0.41	1		
Fluoride	0.20	0.10	0.09	0.21	0.15	-0.32	0.10	-0.17	-0.37	1	
Nitrate	0.01	0.15	0.10	0.73	0.33	-0.10	0.38	-0.27	0.00	-0.06	1

Deganga
d) Deep tube-wells

	pH	EC	TDS	TH	Alkalinity	Sulphate	Chloride	Arsenic	Iron	Fluoride	Nitrate
pH	1										
EC	0.66	1									
TDS	-0.06	0.02	1								
TH	-0.62	-0.62	0.05	1							
Alkalinity	0.27	0.47	0.07	0.28	1						
Sulphate	0.08	0.20	-0.41	0.06	0.28	1					
Chloride	0.65	0.98	0.04	-0.75	0.29	0.13	1				
Arsenic	-0.31	-0.20	-0.09	-0.19	-0.55	-0.15	-0.08	1			
Iron	-0.09	-0.10	-0.15	0.42	0.29	-0.01	-0.13	-0.11	1		
Fluoride	0.02	0.11	0.01	0.26	0.55	0.20	-0.02	-0.78	-0.12	1	
Nitrate	0.70	0.89	-0.04	-0.42	0.53	0.19	0.85	-0.25	0.23	0.02	1

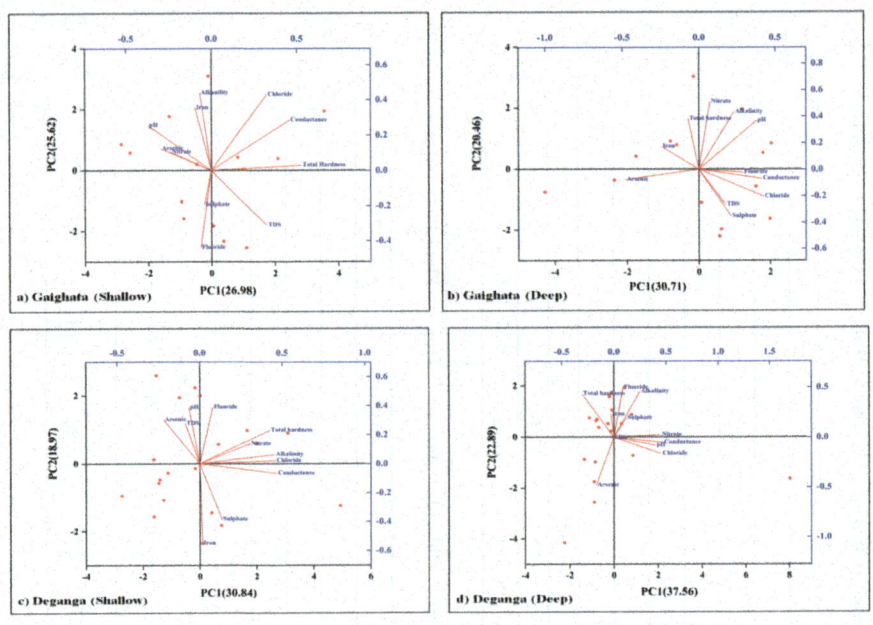

Fig. 3: PCA showing the scores of the first two principal components (PC1 and PC2) of the total variance.

and iron), Group 2 (total hardness, nitrate, pH and alkalinity) and Group 3 (TDS, sulphate, chloride, conductance, and fluoride).

Likewise, in Deganga block, domestic shallow tube-well water showed two Principal Components cumulatively as 49.81% of the total variation, where PC1 and PC2 contributed 30.84% and 18.97%, respectively. The interdependence relation among the water parameters has been observed as Group 1 (iron and sulphate), Group 2 (alkalinity, chloride, conductance), Group 3 (arsenic, pH, TDS, fluoride) and Group 4 (total hardness and nitrate). In a similar way, the deep tube-well groundwater showed 60.45% of the total variation as cumulative two Principal Components, where PC1 and PC2 contributed 37.56% and 22.89%, respectively. The PCA biplot showed inter-relation among the parameters as Group 1 (alkalinity, chloride, conductance, TDS, nitrate and pH), and Group 2 (sulphate, fluoride) and total hardness as well as arsenic was placed distinctly.

Overall, in the domestic shallow aquifer samples of Gaighata block, the high ionic electrical conductivity resembles the high chloride concentration. The presence of total hardness has caused the high value of alkalinity, whereas the Deganga domestic shallow water samples showed a good correlation between electrical conductivity and chloride, alkalinity, and total hardness. The water quality parameters of the domestic shallow water samples are distributed throughout the hyper-plane of the PC1 and PC2. In case of deep tube-wells, the water quality parameters are mainly distributed throughout the hyper-plane of the PC2, except arsenic and iron (Gaighata block) and only arsenic (Deganga block) falls under PC1. The varied distribution pattern of the parameters might be due to different geographical position of the aquifers from the studied blocks.

3.5 Assessment of Water Quality Index (WQI)

Ten significant physico-chemical parameters were considered according to their significant weight (w_i) with respect to the overall water quality for drinking purposes. For the toxic element, i.e., arsenic, the highest weight of 5 has been considered depending on its potential harmful nature. In case of fluoride and nitrate, a weight of 4 was taken as it plays an important role in groundwater pollution. The physico-chemical parameters along with their respective calculated relative weights and Indian Standards in drinking water has been shown in Table 3 (mentioned in Section 2.7.1). The Water Quality Index (WQI) has been evaluated for both the domestic shallow and deep tube-wells from the two studied blocks using the equations as described in Section 2.7.1. The respective range of WQI categorizes the water quality for drinking purpose.

Water Quality classification of domestic shallow and deep groundwater based on WQI assessment has been shown in Fig. 4. About 14% and 86% of the domestic shallow tube-well groundwater (n = 14) samples from Gaighata block are estimated as 'poor', and 'unsuitable for drinking', respectively, whereas 100% of the domestic shallow tube-well groundwater (n = 20) samples from Deganga block are considered as 'unsuitable for drinking'. In case of deep tube-well groundwater (n = 14) samples collected from Gaighata block, about 36% and 64% are categorized under 'good' and 'poor' drinking water, respectively. Likewise, in Deganga block, it was observed about 25, 45, 10, and 20% of the deep tube-well groundwater (n = 20) samples are under the categories 'good', 'poor', 'very poor', and 'unsuitable for drinking' respectively. Hence, Water Quality Index (WQI) showed that approximately 36% and 25% of deep tube-well water samples are categorized under 'good water quality', whereas 86% and 100% of the domestic shallow tube-well water samples are not recommended for drinking purpose in Gaighata and Deganga, respectively. As a whole, the domestic shallow tube-well water quality is mainly 'unsuitable for drinking' from both the

Table 3: Relative weight of physico-chemical parameters in groundwater.

Parameters	Weight (w_i)	Relative weight (W_i)	Indian Standards (mg/l, except pH) (BIS 2012, WHO 2011)
Arsenic	5	0.167	0.01
Iron	3	0.1	0.3
Ph	3	0.1	6.5–8.5
TDS	2	0.067	500–2000
Total Hardness	2	0.067	200–600
Alkalinity	2	0.067	200–600
Fluoride	4	0.133	1.0–1.5
Nitrate	4	0.133	45
Sulphate	2	0.067	200–400
Chloride	3	0.1	250–1000
	$\Sigma = 30$	$\Sigma = 1$	

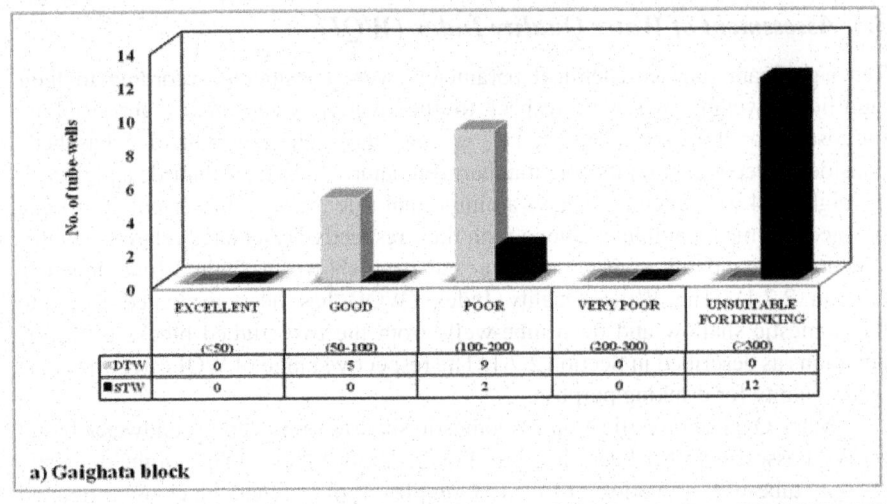

a) Gaighata block

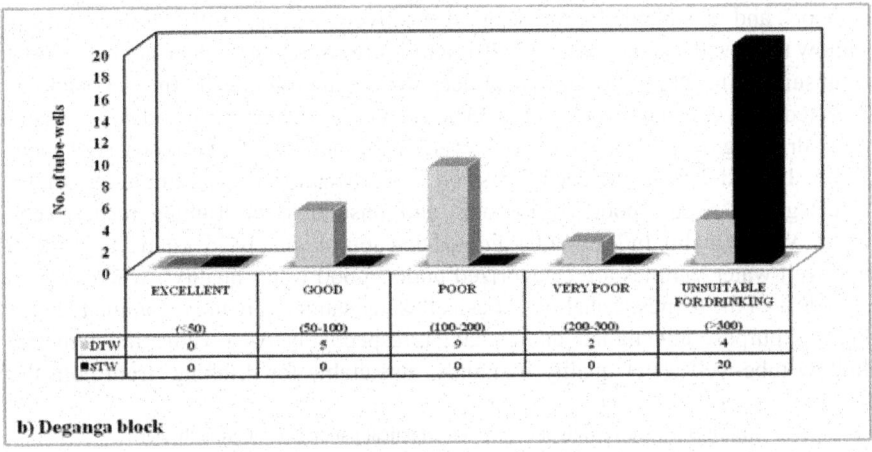

b) Deganga block

Fig. 4: Water Quality classification based on WQI assessment.

studied blocks. On the other hand, the deep tube-well groundwater from the studied areas is safer compared to shallow level; however, it needs to evaluate the quality before use for domestic purposes.

3.6 Human health risk assessment through consumption of toxic contaminants present in groundwater (As, NO_3^-, F^-)

The human health carcinogenic and non-carcinogenic risk has been evaluated based on the equations as mentioned in the *Section 2.7.2*. Cancer as well as non-cancer risk assessment has been performed for arsenic, a group I toxic and carcinogenic pollutant (ATSDR 2007, IARC 2012, USEPA 2006). Only non-cancer risk assessment (Hazard Quotient) has been performed for nitrate and fluoride (Adimalla et al. 2018, Das et al. 2020, Narsimha and Rajitha 2018). The health risk assessment (both cancer and non-cancer) of the populations through these toxic contaminants present in groundwater has been shown in Table 4.

Table 4: Health risk assessment (cancer and non-cancer) for the studied populations through consumption of the toxic elements.

Studied block	Sources (Tube-well)	Toxic contaminants	Cancer risk Mean	Cancer risk Range	Non-cancer risk Mean	Non-cancer risk Range
Gaighata	Deep	Arsenic	$4.38*10^{-4}$	$1*10^{-4} - 1.8*10^{-3}$	0.97	0.36–4.19
		Fluoride	-	-	0.005	0.001-0.021
		Nitrate	-	-	0.0002	0.00006–0.0007
	Domestic shallow	Arsenic	$2.24*10^{-3}$	$5*10^{-4} - 4*10^{-3}$	4.97	1.31–9.51
		Fluoride	-	-	0.024	0.006–0.047
		Nitrate	-	-	0.0009	0.0002–0.002
Deganga	Deep	Arsenic	$1.2*10^{-3}$	$2*10^{-4} - 8.2*10^{-3}$	2.68	0.35–18.2
		Fluoride	-	-	0.013	0.002–0.09
		Nitrate	-	-	0.0005	0.00006–0.003
	Domestic shallow	Arsenic	$3.54*10^{-2}$	$4*10^{-3} - 0.152$	78.7	8.09–337
		Fluoride	-	-	0.39	0.04–1.68
		Nitrate	-	-	0.15	0.002–0.06

The potential carcinogenic and non-carcinogenic risk due to arsenic toxicity has been observed higher among the populations of Deganga through domestic shallow tube-wells (3.54*10–2 and 78.6) and deep tube-wells (1.21*10–3 and 2.68), compared to the Gaighata populations (domestic shallow tue-wells: 2.24*10–3 and 4.97) and (deep tube-wells: 4.38*10–4 and 0.97), respectively. The standard level for lifetime carcinogenic risk by arsenic is $1*10^{-6}$ (USEPA 2005). The observed range of HQ value greater than 1 for arsenic is a serious matter of concern for the studied two populations who have been consuming arsenic-contaminated groundwater for a prolonged time, which might be responsible for severe health hazards in near future. The range of cancer risk through intake of domestic shallow and deep tube-well groundwater is much higher than its threshold level. For both the studied blocks, the risk through domestic shallow tube-well groundwater is higher than the deep tube-well groundwater. This study indicates that the exposure of drinking water from domestic shallow tube-well has a greater potential to cause serious health problems including lung, liver, urinary or skin cancer (Alam et al. 2016). The studied populations from the two blocks have been already exposed to arsenic toxicity through consumption of arsenic-contaminated drinking water and foodstuffs (rice grains and vegetables) being cultivated by using arsenic-contaminated irrigation water and soil (Chowdhury et al. 2018a, b, Roychowdhury 2010). Rice grain is the main staple crop, which contributes a considerable amount of inorganic arsenic through daily diet that poses severe health risk, due to arsenic toxicity, for the population (Roychowdhury 2008, 2010). Therefore, it is important to take necessary steps for improvement of drinking water quality as it is related with the health issues of the local population. The calculated mean value of non-cancer risk (HQ) for nitrate and fluoride found in Gaighata block are 0.0009, 0.024 for domestic shallow and 0.0002, 0.005 for deep tube-well groundwater samples, respectively. Likewise, Deganga block showed nitrate and fluoride HQ value of 0.015, 0.39 through domestic shallow and 0.0005, 0.013 for deep tube-well groundwater samples, respectively. The acceptable level of risk (non-cancerous diseases) for each element is unity, i.e., when the HQ value is greater than 1, then an unacceptable non-carcinogenic risk exists (Das et al. 2020, USEPA 2005). The evaluated non-cancer risk (HQ value) for both nitrate and fluoride present in groundwater is less than the tolerable level of risk for non-cancerous diseases. As a result, there is no human health risk from nitrate and fluoride at present through intake of groundwater from both the studied Gaighata and Deganga blocks.

4. Conclusion

Consumption of groundwater for drinking purpose in arsenic-prone areas is of great concern. The present study focuses on the overall groundwater quality of the domestic shallow and deep tube-wells from two arsenic-affected blocks through evaluation of different physico-chemical parameters. The groundwater has been categorized as alkaline and hard in nature. About 78.5 and 70% of the deep tube-well groundwater samples from Gaighata (depth = 580–780 ft) and Deganga (depth = 500–800 ft) block, respectively, have been identified with arsenic concentration less than 0.01 mg/l. The mean arsenic concentration of domestic shallow tube-well is 5.25 and 28.7 fold higher than the mean arsenic of deep tube-well from Gaighata and Deganga block, respectively. Presence of other toxic elements like nitrate and fluoride in groundwater does not play any significant role, as the values are within the permissible limit. So, it can be concluded that there is no human health risk from nitrate and fluoride through intake of groundwater from both the studied blocks. The groundwater quality index (WQI value) showed that 36% and 25% of deep tube-well water samples are characterized under 'good water quality', recommended for drinking purpose in Gaighata and Deganga, respectively. The domestic shallow tube-well water quality indexing clearly indicates that the water is 'unsuitable for drinking' from both the studied blocks. As a result, the groundwater from deep aquifer is comparatively safer with respect to shallow aquifer for drinking, cooking and other household purposes; however, continuous monitoring of water quality is required. The potential carcinogenic and non-carcinogenic risk due to severe arsenic exposure through drinking water is high from domestic shallow tube-well compared to deep tube-well among the studied blocks. This study will assist the exposed population and authorities to be aware of the potential health risks posed due to intake of arsenic-contaminated water. Further studies need to be carried out on the species distribution of arsenic in the different sources of groundwater. It is vital to acquire significant actions for improvement of the drinking water quality as it is directly related with the health of the local population. Supplementing of safe drinking water through proper surface water treatment plants, groundwater arsenic removal plants and rainwater harvesting and promoting awareness among the populations is strongly recommended in the arsenic-exposed areas. A standard monitoring of the deep tube-wells is suggested to maintain the water quality index.

All these recommended pathways lead to prevent the severe health risks caused due to devastating calamity of arsenic.

References

Abbas, G., Murtaza, B., Bibi, I., Shahid, M., Niazi, N.K., Khan, M.I., Amjad, M., and Hussain, M. 2018. Arsenic uptake, toxicity, detoxification, and speciation in plants: physiological, biochemical, and molecular aspects. International Journal of Environmental Research and Public Health, 15(1): 59.

Abdul, K.S.M., Jayasinghe, S.S., Chandana, E.P., Jayasumana, C., and De Silva, P.M.C. 2015. Arsenic and human health effects: A review. Environmental Toxicology and Pharmacology, 40(3): 828–846.

Adimalla, N., Li, P., and Qian, H. 2018. Evaluation of groundwater contamination for fluoride and nitrate in semi-arid region of Nirmal Province, South India: a special emphasis on human health risk assessment (HHRA). Human and Ecological Risk Assessment: An International Journal, 25(5): 1107–1124. https://doi.org/10.1080/10807039.2018.1460579.

Alam, M.O., Shaikh, W.A., Chakraborty, S., Avishek, K., and Bhattacharya, T. 2016. Groundwater arsenic contamination and potential health risk assessment of Gangetic Plains of Jharkhand, India. Exposure and Health, 8(1): 125–142.

APHA. 1998. Standard Methods for the Examination of Water and Wastewater, twentieth ed. American Public Health Association, American Water Works Association and Water Environmental Federation, Washington DC.

APHA. 2005. Standard Methods for the Examination of Water and Wastewater, twentyfirst ed. American Public Health Association/American Water Works Association/ Water Environment Federation, Washington DC.

Argos, M., Kalra, T., Rathouz, P.J., Chen, Y., Pierce, B., Parvez, F., Islam, T., Ahmed, A., Rakibuz-Zaman, M., Hasan, R., and Sarwar, G. 2010. Arsenic exposure from drinking water, and all-cause and chronic-disease mortalities in Bangladesh (HEALS): a prospective cohort study. The Lancet, 376(9737): 252–258.

ATSDR. 2007. Toxicological Profile for Arsenic. Agency for Toxic Substances and Disease Registry, Division of Toxicology, Atlanta, GA.

ATSDR. 2014. Toxicological profile for Arsenic. Agency for Toxic Substances and Disease Registry, Division of Toxicology, Atlanta, GA.

Balakrishnan, P., Saleem, A., and Mallikarjun, N.D. 2011. Groundwater quality mapping using geographic information system (GIS): a case study of Gulbarga City, Karnataka, India. Afr. J. Environmental Science Technology, 5(12): 1069–1084. https://doi.org/ 10.5897/AJEST11.134.

Bhowmick, S., Pramanik, S., Singh, P., Mondal, P., Chatterjee, D., and Nriagu, J. 2018. Arsenic in groundwater of West Bengal, India: a review of human health risks and assessment of possible intervention options. Science of the Total Environment, 612: 148–169.

BIS. 2012. Indian Standard Specifications for Drinking Water. IS:10500. Bureau of Indian Standards, New Delhi, India.

Celik, I., Gallicchio, L., Boyd, K., Lam, T.K., Matanoski, G., Tao, X., Shiels, M., Hammond, E., Chen, L., Robinson, K.A., and Caulfield, L.E. 2008. Arsenic in drinking water and lung cancer: a systematic review. Environmental Research, 108(1): 48–55.

Chakraborti, D., Das, B., and Murrill, M.T. 2011. Examining India's groundwater quality management. Environmental Science Technology, 45: 27–33.

Chakraborti, D., Rahman, M.M., Das, B., Nayak, B., Pal, A., Sengupta, M.K., Hossain, M.A., Ahamed, S., Sahu, M., Saha, K.C., Mukherjee, S.C., Pati, S., Dutta, R.N., and Quamruzzaman, Q. 2013. Groundwater arsenic contamination in Ganga–Meghna–Brahmaputra plain, its health effects and an approach for mitigation. Environmental Earth Sciences, 70(5): 1993–2008. https://doi.org/10.1007/s12665- 013-2699-y.

Chakraborti, D., Singh, S.K., Rahman, M.M., Dutta, R.N., Mukherjee, S.C., Pati, S., and Kar, P.B. 2018. Groundwater arsenic contamination in the Ganga River Basin: a future health danger. International Journal of Environmental Research and Public Health, 15(2): 180–198. https://doi.org/10.3390/ijerph15020180.

Chowdhury, N.R., Das, R., Joardar, M., Ghosh, S., Bhowmick, S., and Roychowdhury, T. 2018a. Arsenic accumulation in paddy plants at different phases of pre-monsoon cultivation. Chemosphere 210: 987–997.

Chowdhury, N.R., Ghosh, S., Joardar, M., Kar, D., and Roychowdhury, T. 2018b. Impact of arsenic contaminated groundwater used during domestic scale post harvesting of paddy crop in West Bengal: Arsenic partitioning in raw and parboiled whole grain. Chemosphere, 211: 173–184.

Chowdhury, N.R., Das, A., Mukherjee, M., Swain, S., Joardar, M., De, A., Mridha, D., and Roychowdhury, T. 2020. Monsoonal paddy cultivation with phase-wise arsenic distribution inexposed and control sites of West Bengal, alongside its assimilation in rice grain. Journal of Hazardous Materials, 400: 123206.

Chowdhury, T.R., Basu, G.K., Mandal, B.K., Biswas, B.K., Samanta, G., Chowdhury, U.K., Chanda, C.R., Lodh, D., Roy, S.L., Saha, K.C., and Roy, S. 1999. Arsenic poisoning in the Ganges delta. Nature, 401: 545–546.

Das, A., Das, S.S., Chowdhury, N.R., Joardar, M., Ghosh, B., and Roychowdhury, T. 2020. Quality and health risk evaluation for groundwater in Nadia district, West Bengal: An approach on its suitability for drinking and domestic purpose. Groundwater for Sustainable Development, 10: 100351. https://doi.org/10.1016/j.gsd.2020.100351.

Das, S., and Nag, S.K. 2015. Deciphering groundwater quality for irrigation and domestic purposes—a case study in Suri I and II blocks, Birbhum District, West Bengal, India. Journal of Earth System Science, 124(5): 965–992.

Duong, Tarn. 2007. Ks: Kernel Density Estimation and Kernel Discriminant Analysis for Multivariate Data in R. Journal of Statistical Software, 21(7): 10.18637/jss.v021.i07.

Fan, A.M., and Steinberg, V.E. 1996. Health implications of nitrate and nitrite in drinking water: an update on methemoglobinemia occurrence and reproductive and developmental toxicity. Regulatory Toxicology and Pharmacology, 23(1): 35–43. https://doi.org/ 10.1006/rtph.1996.0006.

Fang, S.M., Jia, X.B., Yang, X.Y., Li Y.D., and An, S.Q. 2012. A method of identifying priority spatial patterns for the management of potential ecological risks posed by heavy metals. Journal of Hazardous Materials, 237-238: 290–298.

Fries, J., Getrost, H., and Merck, D.E. 1977. Organic reagents trace analysis. E. Merck, Darmstadt, Germany, 975.

Ghosh, S., Majumder, S., and Roychowdhury, T. 2019. Assessment of the effect of urban pollution on surface water-groundwater system of Adi Ganga, a historical outlet of river Ganga. Chemosphere, 237: 124507.

Goel, V., Islam, M.S., Yunus, M., Ali, M.T., Khan, A.F., Alam, N., Faruque, A.S.G., Bell, G., Sobsey, M. and Emch, M. 2019. Deep tubewell microbial water quality and access in arsenic mitigation programs in rural Bangladesh. Science of the Total Environment, 659: 1577–1584.

Goswami, R., Kumar, M., Biyani, N., and Shea, P.J. 2020. Arsenic exposure and perception of health risk due to groundwater contamination in Majuli (river island), Assam, India. Environmental Geochemistry and Health, 42(2): 443–460.

Gupta, R., and Misra, A.K. 2018. Groundwater quality analysis of quaternary aquifers in Jhajjar District, Haryana, India: Focus on groundwater fluoride and health implications. Alexandria Engineering Journal, 57(1): 375–381.

Helena, B., Pardo, R., Vega, M., Barrado, E., Fernandez, J.M., and Fernandez, L. 2000. Temporal evolution of groundwater composition in an alluvial aquifer (Pisuergariver, Spain) by principal component analysis. Water Research, 34: 807–816.

International Agency for Research on Cancer (IARC). 2012. Arsenic, Metals, Fibres, and Dusts, Volume 100C.A Review of Human Carcinogens.http://monographs.iarc.fr/ENG/Monographs/vol100C/mono100C.pdf.

Joardar, M., Das, A., Mridha, D., De, A., Chowdhury, N.R., and Roychowdhury, T. 2021a. Evaluation of acute and chronic arsenic exposure on school children from exposed and apparently control areas of West Bengal, India. Exposure and Health, 13: 33–50.

Joardar, M., Das, A., Chowdhury, N. R., Mridha, D., De, A., Majumdar, K. K., and Roychowdhury, T. 2021b. Health effect and risk assessment of the populations exposed to different arsenic levels in drinking water and foodstuffs from four villages in arsenic endemic Gaighata block, West Bengal, India. Environmental Geochemistry and Health, 43(8): 3027–3053.

Kazi, T.G., Brahman, K.D., Afridi, H.I., Arain, M.B., Talpur, F.N., and Akhtar, A. 2016. The effects of arsenic contaminated drinking water of livestock on its total levels in milk samples of different cattle: risk assessment in children. Chemosphere, 165: 427–433.

Kumar, M., Rahman, M.M., Ramanathan, A., and Naidu, R. 2016. Arsenic and other elements in drinking water and dietary components from the middle Gangetic plain of Bihar, India: health risk index. Science of the Total Environment, 539: 125–134.

Landrigan, P.J., Fuller, R., Acosta, N.J., Adeyi, O., Arnold, R., Baldé, A.B., Bertollini, R., Bose-O'Reilly, S., Boufford, J.I., Breysse, P.N., and Chiles, T. 2018. The Lancet Commission on pollution and health. The Lancet, 391(10119): 462–512.

Lumb, A., Sharma, T.C., and Bibeault, J.F. 2011. A review of genesis and evolution of Water Quality Index (WQI) and some future directions. Water Quality Exposure and Health, 3: 11–24.

Meng, Q., Zhang, J., Zhang, Z., and Wu, T. 2016. Geochemistry of dissolved trace elements and heavy metals in the Dan River Drainage (China): distribution, sources, and water quality assessment. Environmental Science and Pollution Research, 23(8): 8091–8103.

Mohana, A.A., Rahman, M.A., and Islam, M.R. 2020. Deep and shallow tube-well water from an arsenic-contaminated area in rural Bangladesh: risk-based status. International Journal of Energy and Water Resources, 4(2): 163–179.

Narsimha, A., and Rajitha, S. 2018. Spatial distribution and seasonal variation in fluoride enrichment in groundwater and its associated human health risk assessment in Telangana State, South India. Human and Ecological Risk Assessment: An International Journal, 24(8): 2119–2132.

Naujokas, M.F., Anderson, B., Ahsan, H., Aposhian, H.V., Graziano, J.H., Thompson, C., and Suk, W.A. 2013. The broad scope of health effects from chronic arsenic exposure: update on a worldwide public health problem. Environmental Health Perspectives, 121(3): 295–302.

Rabbani, U., Mahar, G., Siddique, A., and Fatmi, Z. 2017. Risk assessment for arsenic-contaminated groundwater along River Indus in Pakistan. Environmental Geochemistry and Health, 39: 179–190.

Rahman, M.M., Mukherjee, D., Sengupta, M.K., Chowdhury, U.K., Lodh, D., Chanda, C.R., Roy, S., Selim, M., Quamruzzaman, Q., Milton, A.H., Shahidullah, S.M., Rahman, M.T., and Chakraborti, D. 2002. Effectiveness and reliability of arsenic field testing kits: are the million dollar screening projects effective or not? Environmental Science and Technology, 36(24): 5385–5394. https://doi.org/10.1021/es020591o.

Rasool, A., Xiao, T., Baig, Z.T., Masood, S., Mostofa, K.M., and Iqbal, M. 2015. Co-occurrence of arsenic and fluoride in the groundwater of Punjab, Pakistan: source discrimination and health risk assessment. Environmental Science and Pollution Research, 22(24): 19729–19746.

Roychowdhury, T. 2008. Impact of sedimentary arsenic through irrigated groundwater on soil, plant, crops and human continuum from Bengal delta: special reference to raw and cooked rice. Food and Chemical Toxicology, 46(8): 2856–2864. https://doi.org/ 10.1016/j.fct.2008.05.019.

Roychowdhury, T. 2010. Groundwater arsenic contamination in one of the 107 arsenic-affected blocks in West Bengal, India: Status, distribution, health effects and factors responsible for arsenic poisoning. International Journal of Hygiene and Environmental Health, 213(6): 414–427.

Shahid, M., Khalid, M., Dumat, C., Khalid, S., Niazi, N.K., Imran, M., Bibi, I., Ahmad, I., Hammad, M., and Tabassum, R.A. 2017. Arsenic level and risk assessment of groundwater in Vehari, Punjab Province, Pakistan. Exposure and Health. https://doi.org/10.1007/s12403-017-0257-7.

Shakoor, M.B., Niazi, N.K., Bibi, I., Murtaza, G., Kunhikrishnan, A., Seshadri, B., Shahid, M., Ali, S., Bolan, N.S., and Ok, Y.S. 2016. Remediation of arseniccontaminated water using agricultural wastes as biosorbents. Critical Review. Environmental Science and Technology, 46: 467–499.

Sharma, S., Kaur, I., and Nagpal, A.K. 2017. Assessment of arsenic content in soil, rice grains and groundwater and associated health risks in human population from Ropar wetland, India, and its vicinity. Environmental Science and Pollution Research, 1–13.

Smith, A.H., Ercumen, A., Yuan, Y., and Steinmaus, C.M. 2009. Increased lung cancer risks are similar whether arsenic is ingested or inhaled. Journal of Exposure Science and Environmental Epidemiology, 19(4): 343.

Su, X., Wang, H., and Zhang, Y. 2013. Health risk assessment of nitrate contamination in groundwater: a case study of an agricultural area in Northeast China. Water Resources Management, 27(8): 3025–3034.

Tabassum, R.A., Shahid, M., Dumat, C., Niazi, N.K., Khalid, S., Shah, N.S., Imran, M., and Khalid, S., 2019. Health risk assessment of drinking arsenic-containing groundwater in Hasilpur, Pakistan: effect of sampling area, depth, and source. Environmental Science and Pollution Research, 26(20): 20018–20029.

Todd, D.K. 1980. Groundwater Hydrology, 2nd Edition. Wiley, New York, 527.

USEPA. 2005. Guidelines for Carcinogen Risk Assessment. Risk Assessment Forum. United States Environmental Protection Agency, Washington, C. EPA/630/P-03/ 001F.

USEPA. 2006. USEPA Region III Risk-Based Concentration Table: Technical Background Information. United States Environmental Protection Agency, Washington, DC.

USEPA. 2011. Screening Level (RSL) for Chemical Contaminant at Superfound Sites. U.S. Environmental Protection Agency,Washington, DC.

USEPA. 2014. Human Health Evaluation Manual, Supplemental Guidance: Update of Standard Default Exposure Factors-OSWER Directive 9200.1-120, 6.

Van Geen, A., Ahmed, K.M., Seddique, A.A., and Shamsudduha, M. 2003. Community wells to mitigate the arsenic crisis in Bangladesh. Bulletin of the World Health Organization, 81: 632–638.

Waqas, H., Shan, A., Khan, Y.G., Nawaz, R., Rizwan, M., Saif-Ur-Rehman, M., Shakoor, M.B., Ahmed, W., and Jabeen, M. 2017. Human health risk assessment of arsenic in groundwater aquifers of Lahore, Pakistan. Human Ecological Risk Assessment: An International Journal, 23(4): 836–850.

WHO. 2011. Guidelines for drinking-water quality. WHO Chronicle, 38(4): 104–108.

Xiao, J., Jin, Z., and Wang, J. 2014. Geochemistry of trace elements and water quality assessment of natural water within the Tarim River Basin in the extreme arid region, NW China. Journal of Geochemical Exploration, 136: 118–126.

3

Genotoxicity and Mutagenicity of Metal-based Nanomaterials, with an Emphasis on using *Drosophila*

Mohamed Alaraby,[1] *Doaa Abass*[1] and *Ricard Marcos*[2,*]

1. Introduction

The small size scale of nanomaterials (NMs) is the most important factor underlining their chemico-biological activity/reactivity. This is mainly due to their large exposed surface area, in comparison with bulk materials. Due to their physicochemical properties, NMs are involved in many applications. Thus, metal oxide nanoparticles (NPs) have been introduced in electronics (Chavali and Nikolova 2019), catalysis (Akbari et al. 2018), medical applications (Augustine et al. 2017), electrochemical sensing and biosensing (George et al. 2018), optical sensor (Maruthupandy et al. 2017), cosmetic products (Subramaniam et al. 2019), and in many other fields. Thus, NMs production is a fertile area for investment, and it has been estimated that the USD 4.1 billion for the global nanomaterials market in 2015 would reach USD 22.88 billion by 2027 (Global Nanomaterials Market Size Report 2020).

The extended and increased use of NMs supposes their growing presence in the environment. Consequently, all organisms including humans are exposed to these materials, whether directly or indirectly via transferring NMs through the food chain (Judy et al. 2011). The tiny size of NMs contributes to the facility of crossing

[1] Zoology Department, Faculty of Sciences, Sohag University (82524), Sohag, Egypt.
[2] Group of Mutagenesis, Department of Genetics and Microbiology, Faculty of Biosciences, Universitat Autònoma de Barcelona, Campus of Bellaterra, 08193 Cerdanyola del Vallès (Barcelona), Spain.
* Corresponding author: ricard.marcos@uab.cat

biological barriers, transferring them through body fluids, and reaching the different body tissues (Alaraby et al. 2015a). Inside cells, NMs can cause chemical and physical stress that initiate a biological response to counteract the harsh impact of NMs. The NMs' action involves discharging harmful ions or interacting with cellular organelles and other cell components such as proteins, lipids, and nuclear acids (Karlsson et al. 2013). So, oxidative stress, DNA damage, and mutagenicity are potential events associated with NMs exposure, when the cellular antioxidant capacity and repairing machinery of the cell fail to restore homeostasis. Accordingly, it is of high importance to detect the potential biological impacts of NMs using different *in vivo* and *in vitro* models, mainly focusing on genotoxicity and mutagenesis as early events of more complex health impacts, such as carcinogenesis.

2. Metal-based nanomaterials (MBNMs)

Metal-based NMs (MBNMs) are considered as one specific category of nanomaterials as they have special physicochemical properties (optical, electrical, chemical, and physical) derived from their metallic nature (Mortezaee et al. 2019). Metals are electropositive elements, lustrous, hard, malleable, ductile, good conductors of heat and electricity, and with relatively low ionization energies. Due to their malleability and ductility, metals can be easily converted into a huge number of structural geometries (like thin sheets and small wires) at the nanoscale that largely changes their chemical and physical properties (Bhattacharya and Mukherjee 2008). Metal nanomaterials are prepared from different metals such as silver, copper, iron, alumina, zinc, titanium, and silica, among others, and using different methodological approaches (Siril and Türk 2020). Metals easily react with oxygen to form a metal oxide, and as nanoparticles, they are very useful due to their semi-conductive properties and catalytic activity (Narayan et al. 2019). Thus, MBNMs are characterized by their small band gaps, which dramatically elevated their chemical and conductivity properties; nonetheless, they keep similar crystal structures like their bulk materials (Golbamaki et al. 2015).

3. Toxicity of MBNMs

The wide use of MBNMs, in addition to their important chemically reactivity, imposes many questions regarding their potential health harmful impacts. Early researchers tried to determine the potential biological interactions of MBNMs using different models, as an approach to predict the associated risk in case of human exposure. Toxicity and cytotoxicity evaluation has been the simplest way to determine the risk associated with potential MBNMs exposure. Many reviews are highlighting the general toxicity of different metal oxide nanoparticles (Sarkar et al. 2014, Sengul and Asmatulu 2020), or establishing comparisons between different metal oxide nanomaterials like TiO_2, ZnO, Fe_3O_4, and Al_2O_3 with particle sizes ranging from 30 to 45 nm (Jeng and Swanson 2006). In those studies, authors used different biomarkers of toxicity such as cellular morphology, mitochondrial function, membrane leakage of lactate dehydrogenase (LDH), the permeability of plasma membrane, and apoptosis, to measure toxic effects in Neuro-2A cells. The reported results indicated that only ZnONPs were able to significantly affect different toxicity biomarkers at low concentrations.

The use of established cell lines has been the most normal approach applied to determine the potentially toxic effects of MBNMs. Nevertheless, *in vitro* effects have been considered to be far away from the effects that MBNMs can produce *in vivo*. In this way, the results obtained *in vitro* are difficult to extrapolate for human exposure scenarios, and not very powerful for risk assessment purposes. Although in terms of risk assessment, the results obtained using mammalian models are the best option, there is a strong pressure to reduce, refine, and replace the use of laboratory animals (Hartung and Sabbioni 2011). In such a scenario, the use of lower eukaryotic *in vivo* models, such as the fruit fly (*Drosophila melanogaster*), is a good alternative. *D. melanogaster* is one of the most reliable models, and it has been proposed as a useful tool to detect potential harmful effects of many different types of nanomaterials, including MBNMs (Ong et al. 2015, Alaraby et al. 2016a). Low cost, easy manipulation, short life cycle ~ 10 days (see Fig. 1), completely sequenced genome (13,600 genes), reliable results, and genetic similarity to humans, in addition to ethical considerations, are some of the relevant characteristics of *Drosophila* as a suitable model, explaining their wide use in many fields including the assessing of nanomaterials. We have already explored the advantages of using *Drosophila* to diagnose the impacts of several MBNMs such as CeO_2NPs (Alaraby et al. 2015a), ZnONPs (Alaraby et al. 2015b), CdSe quantum dots (Alaraby et al. 2015c), CuONPs (Alaraby et al. 2016b, 2017), NiONPs (Alaraby et al. 2018), AgNPs (Alaraby et al. 2019), CoNPs (Alaraby et al. 2020), and TiO_2NPs (Alaraby et al. 2021). In such studies, toxicity was determined as the decrease in the egg-to-adult viability, although changes in the expression levels of genes involved in the general stress response, such as the heat shock protein genes *Hsp70* and *Hsp83*, have also been used (Alaraby et al. 2016b).

4. Oxidative stress induced by MBNMs

The particulate nature of nanomaterials as well as their ability to release harmful ions contributes to increasing both physical and chemical stress in exposed cells that ultimately cause oxidative stress. Oxidative stress conditions result from the disequilibrium between the pro-oxidant and antioxidant status inside the cell/ organism. Thus, oxidative stress disturbs the cellular homeostasis and, if cells fail to maintain normal physiological redox conditions, a cascade of events take place inducing cytotoxicity, DNA damage, apoptosis, and cancer initiation (Liou et al. 2017). These drastic effects remark the relevance of evaluating oxidative stress, among the different potential harmful effects associated with NMs exposure. Several researchers have used different biological models to report oxidative stress induction resulting from different MBNMs exposures such as ZnONPs (Xia et al. 2008), CoONPs (Fahmy and Cormier 2009), AgNPs (Kim et al. 2009), NiONPs (Siddiqui et al. 2012), and TiO_2NPs (Alinovi et al. 2015). *Drosophila* has also been used to check the potential induction of oxidative stress by different MBNMs, as already reviewed (Alaraby et al. 2016a).

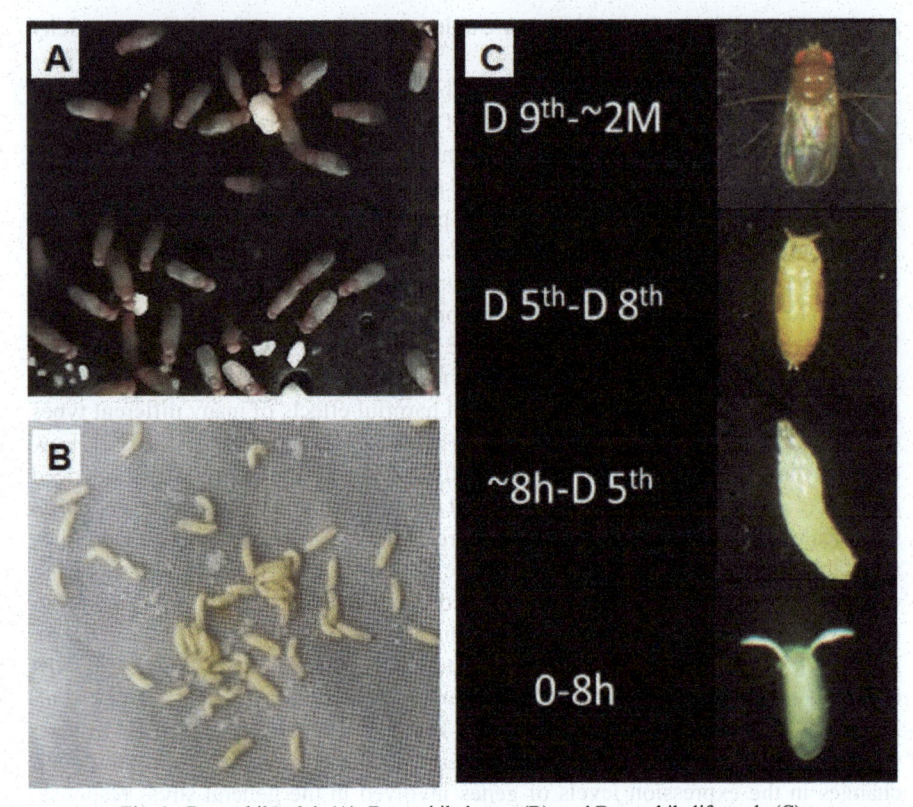

Fig. 1: Drosophila adult (A), Drosophila larvae (B), and Drosophila life cycle (C).

The physicochemical properties of NMs like particle diameter, charge, aspect ratio, and chemical composition are key factors for reactive oxygen species (ROS) production (Shvedova et al. 2012). ROS production can result in different forms including hydrogen peroxide (H_2O_2), hydroxyl radical (OH·), hydroxyl ion (OH⁻), superoxide anion ($O_2^{·-}$), and singlet oxygen ($^1O_2$). Mitochondria are very sensitive organelles and their damage is considered as the main source of oxidative stress (Ott et al. 2007). Interestingly, abnormal shape of mitochondria in midgut enterocytes of *Drosophila* has been reported to be associated with CeO_2NPs exposure (Alaraby et al. 2015c). Besides, it has recently been observed that nickel nanowires can internalize mitochondria (Alaraby et al. 2020b). Hemocytes are a suitable cell target in *Drosophila*, to evaluate oxidative stress. It must be remembered that *Drosophila* has an open circulatory system termed hemocoel that is analogous to human blood. Inside the hemocoel, five types of hemocytes are found, which are analogous to human lymphocytes (see Fig. 2). Once nanomaterials succeeded in passing the intestinal barriers, they directly distributed into hemocoel contacting with hemolymph components, especially with hemocytes. Thus, hemocytes have been proposed as a suitable cell-model in *Drosophila* to detect MBNMs' oxidative stress (Alaraby et al. 2015c). Earlier in 2012, Galeone and co-authors suggested the direct interaction of

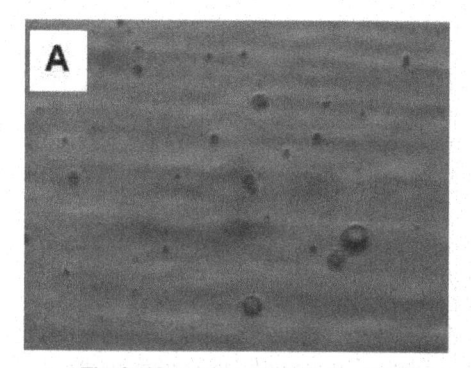 

Fig. 2: Hemocytes. Normal view (A), fluorescent hemocytes showing oxidative stress (B).

NMs with hemocytes, since they found that the hemocytes of CdTeQDs-exposed *Drosophila* produced fluorescent signals (Galeone et al. 2012). Further studies have used hemocytes to detect the intracellular levels of ROS resulting from MBNMs. Thus, NiONPs, AgNPs, and TiO$_2$NPs exposure were able to produce ROS in a dose-dependent manner (Alaraby et al. 2018, 2019, 2020), although others such as ZnONPs, CuONPs, and CeO$_2$NPs did not induce intracellular ROS (Alaraby et al. 2015a, b, c). Such negative effects might be attributed to (i) the retaining of most of NMs particles by the peritrophic membrane (Alaraby et al. 2021), or (ii) the intrinsic phagocytic properties of some types of hemocytes, and their proved antioxidant capacity (Micheal and Subramanyam 2013).

5. Antioxidant response and MBNMs exposure

The cells of eukaryotes possess a preserving antioxidant capacity against oxidative stress to restore the internal hemostasis. The antioxidant defense mechanisms could be either enzymatic, via different antioxidant enzymes, or non-enzymatic via glutathione and ascorbic acid. The antioxidant defense of *Drosophila* is very sensitive, showing different type of responses according to the NM they are exposed to. MBNMs not only elevated ROS production but also deregulated several antioxidant genes. Regardless of its shapes, CuONPs and NiONPs exposures induced depletion effects in their expression (Alaraby et al. 2016b, 2018). But, on the other hand, ZnONPs exposure did not show any observable effects on their pro-oxidant capacity (Alaraby et al. 2015a), and the exposure to Zn ions significantly down-regulated *Sod* expression. All this indicates the sensitivity of the antioxidant system of *Drosophila* and the ability to detect the antioxidant fluctuation when flies are exposed to MBNMs or their salts. Interestingly, changes in the exposure doses modulate the gene expression levels. Thus, low doses of titanium nanowires failed to induce a change in the levels of antioxidant enzymes but, on the contrary, high doses mediated significant deregulation in the expression of *Sod2* and *Cat* (Alaraby et al. 2021). Furthermore, the *Cat* gene was over-expressed after CoNPs exposures (Alaraby et al. 2020). Overall, all these data ensure the effectiveness of the antioxidant systems of *Drosophila* to combat the oxidative events associated with MBNMs exposures.

6. Genotoxicity of MBNMs

Since MBNMs are very active in mediating oxidative stress induction, there is a general concern about their potential to induce genotoxic or mutagenic effects. Genotoxicity is regarded as the ability of any agent to interact/damage DNA, while mutagenicity refers to the fixed changes (at the level of gene or chromosome) once the repair mechanisms have acted via error-prone. MBNMs can cause DNA damage by different mechanisms: (i) via direct interaction with DNA, or (ii) via interactions between the NPs and proteins such as those participating in cell division, or (iii) via oxidative stress induction (Ansari et al. 2019). Many studies are reporting the genotoxic impact of MBNMs (Xie et al. 2011, Karlsson et al. 2014, Rajiv et al. 2016) using different methodological approaches. The use of *Drosophila* not only permits the diagnosis of the genotoxicity of different types of MBNMs but also decipher the mechanisms of action. At the molecular level, changes in the expression levels of two important gene markers such as *p53* and *Ogg1* have been used. The *p53* gene, known as the cell guardian, precludes genotoxicity via promoting cell cycle checkpoints, repairing DNA, or initiating apoptosis cascades to maintain genomic integrity (Ra et al. 2009). Thus, the expression of *p53* was over-expressed when *Drosophila* larvae are exposed to AgNPs (Ahamed et al. 2010, Alaraby et al. 2019), AuNPs (Vecchio et al. 2012), or CdQDs (Brunetti et al. 2013, Alaraby et al. 2015b). Moreover, the expression of p53 mRNA is dose-dependent (Alaraby et al. 2015c) indicating the sensitivity of *p53* as a biomarker of genotoxicity. The *Ogg1* (8-oxoguanine glycosylase) gene is another biomarker of genotoxicity used in *Drosophila*. It is important to remember that the Ogg1 enzyme is involved in eliminating oxidized DNA bases resulting from oxidative stress conditions (Okumura et al. 2019). In this way, changes in their expression mean that oxidative DNA damage has been induced. Using this marker, CuONPs failed to induce changes in the expression of the *Ogg1* gene, but nevertheless reduced the *Ogg1* expression induced by potassium dichromate (Alaraby et al. 2017). Changes in the expression levels of *Ogg1* were observed after exposure to Ni compounds (whether nanowires or nanospheres) (Alaraby et al. 2018), to CoNPs (Alaraby et al. 2020), and to TiO$_2$NPs (including nanowires, nanorods, and nanospheres) (Alaraby et al. 2021).

Another way to demonstrate DNA damage induction is by using the comet assay. This assay measures the induction of DNA breaks as well as oxidatively damaged DNA bases, by using single-cell gel electrophoresis, and it is applicable to test the effects of NMs in any type of eukaryotic cell (García-Rodríguez et al. 2019). This assay has also been widely used in *Drosophila* (Marcos and Carmona 2019). As we have already indicated, the Drosophila hemocytes are suitable for studying different types of effects including DNA damage measured by the comet assay. Using *Drosophila*, most of the studied MBNMs like cobalt, silver, nickel, Cd quantum dots, and titanium NM can induce DNA breaks. On the contrary, MBNMs like zinc and copper NPs failed to increase the basal level of DNA breaks in hemocytes. Regarding the genotoxic effects induced by ZnONPs, although no effects were initially reported (Alaraby et al. 2015a), a further study showed slight but significant genotoxic effects (Carmona et al. 2016). This would mean that ZnONPs can be considered as non-genotoxic or having weak genotoxic effects. The observed discrepancies can be due to the great ability of ZnONPs to dissolve in different media releasing Zn ions. On the contrary, AgNPs induced significant DNA damage in *Drosophila* hemocytes in a dose-dependent manner (Alaraby et al. 2019).

Cadmium, as a usual component of the quantum-dots (QD) core, has been evaluated to determine its genotoxic potential. The genotoxic potential of CdQD is associated with the release of Cd^{2+} ions, as QD are easily biodegraded and releasing ions (Kwon et al. 2012). When *Drosophila* larvae received CdQDs orally, this NM transits throughout the gastric tract being exposed to different pH conditions, which increases Cd^{2+} ions discharge. When the comet assay was used in hemocytes, significant genotoxic effects were observed (Alaraby et al. 2015c), which would agree with previously reported data, although genotoxicity was detected using the tunnel assay instead of the comet assay (Galeone et al. 2012). Genotoxic induction was also detected using CoNPs, where their effects were greater than those mediated by cobalt ions (Alaraby et al. 2020). These results are similar to those obtained in human bronchial BEAS-2B cells where CoNPs induced more genotoxic impact than Co ions (Uboldi et al. 2015). According to that, the direct contact of CoNPs with the nucleus may be the main factor involved in their genotoxicity (Thongkumkoon et al. 2014). In a recent study, TiO_2NPs were observed inside the nucleus of enterocytes of *Drosophila* larvae, associated with elevated levels of DNA damage (Alaraby et al. 2021). Thus, direct interaction of TiO_2NPs with the nucleus would induce DNA breaks, as suggested in human intestinal Caco-2 cells (García-Rodríguez et al. 2018). Interestingly, the observed effects were modulated by both length and shape; thus, nanowires of titanium mediated more DNA damage levels than nanorods or nanosphere (Alaraby et al. 2021). Regarding the modulator effects of length and shape, studies with NiONPs indicate that regardless of their form (nanowires, nanospheres, or even ions), all of them induced DNA breaks (Alaraby et al. 2018). So, the genotoxicity of MBNMs is not only due to ion release since many other factors, including chemical nature and high aspect ratio, can modulate their genotoxic potential.

7. Mutagenicity associated with MBNMs exposure

As previously indicated, it is important to differentiate between two interfering terms, "genotoxicity and mutagenicity". Genotoxicity is the window for mutagenicity, where the genotoxic materials disturb genetic information, leading to the formation of mutation and, lately, to cancer process initiation. In this regard, we can say that genotoxic materials may or may not be mutagenic but all mutagenic agents without exception are genotoxic, so mutagenicity is the next stage of genotoxicity if the genetic disturbance is large to the degree the repair system fails to restore the normal conditions (Toyokuni 1998). In such context, it must be indicated that although most of the MBNMs act as genotoxicants, increasing the levels of DNA damage, few of them act as mutagens (or even acting as antimutagens) (Alaraby et al. 2015c, 2017).

Fortunately, there is a useful classical assay in *Drosophila*, namely the wing-spots assay, having the ability to detect different types of mutation through a simple and easy assay (Marcos and Carmona 2019). At the phenotypic level, the induction of mutations is detected by the presence of mutant wing-spots in individuals who are thansheterozygous for two genes modulating the morphology of the wing hairs, namely multiple-wing hair (*mwh*) and flare (*flr*). The loss of heterozygosity in one of the two gene markers is responsible for the appearance of a mutant clone (*mwh* or *flr*) with a varied number of mutated cells in the wing-blade (Fig. 3). Drosophila wing cells reach 24,400 cells (de Andrade et al. 2004). Normally, each wing cell of *Drosophila* has one spiny hair, but a mutation in the *mwh* gene leads to the appearance of 3-to-5 hairs. If the mutation occurs in the *flr* gene, the hair morphology changes presenting a deformed like burned hair. Due to the facility to detect the wing mutant spots (microscopically), as well as its reliability, the assay has been used to evaluate many agents, including MBNMs.

It should be indicated that the appearance of mutant clones can be due to different mutational events including point mutations and deletion. Nevertheless, mutant clones can appear resulting from somatic recombination. In this regard, it must be highlighted that somatic recombination is a very relevant mechanism involved in carcinogenesis processes (Luo et al. 2000). Thus, the wing-spot using *Drosophila* is

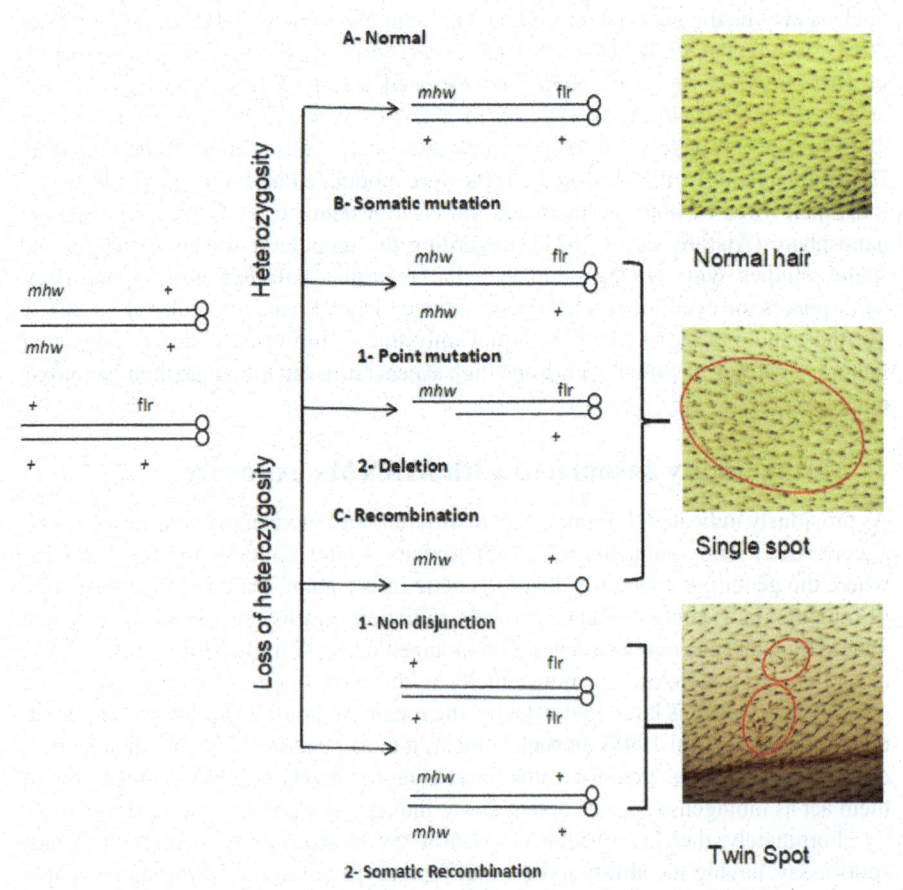

Fig. 3: Scheme showing how the wing-spot test works. Left: mechanism involved. Right: pictures showing the resulting mutant clones.

the only one *in vivo* assay detecting the induction of somatic recombination (Marcos and Carmona 2019).

Regarding the reported studies evaluating the mutagenicity of MBNMs in the wing spot assay, a study evaluating the effects of AgNPs demonstrated that small but significant increases in the frequency of total spots were obtained. Interestingly, most of the mutant clones induced were associated with the induction of somatic recombination (Demir et al. 2011). Although one initial study with TiO_2NPs produced negative results (Carmona et al. 2015), a further study combining the use of normal strains and high-bioactivation strains was able to demonstrate that a positive induction of mutant clones was observed when the high-bioactivation strain was used (Reis et al. 2016). Similarly, when ZnONPs were evaluated, negative results were obtained in two different studies (Alaraby et al. 2015b, Carmona et al. 2016); nevertheless, when the high bioactivation strains of *Drosophila* were used a positive effect was observed (Reis et al. 2015). These discrepancies can be attributed to the weak mutagenic effects observed. In this way, small changes in the protocol, like the use of strain with a high metabolic activity, can move from one negative result to a positive one. Nickel nanoparticles have also been evaluated in the wing-sot assay. In this case, positive effects were observed in the standard cross with the five concentrations tested; nevertheless, only the highest concentration tested was mutagenic when the high-bioactivation cross was used (De Carli et al. 2018). When the mutagenic effects of CoNPs were evaluated, a clear positive induction of mutant clones was observed (Vales et al. 2013, Ertugrul et al. 2020). Interestingly, in the most recent study, it was observed that the co-treatment with melatonin was able to significantly reduce the effects induced by CoNPs (Ertugrul et al. 2020). Regarding the mutagenic effects of CuONPs, positive induction was observed in the wing-spot test (Carmona et al. 2015). Interestingly, in another study, its antimutagenic effects against potassium dichromate (PD) and ethylmethanesulfonate (EMS) were evaluated. Although no effects were observed in the co-treatment with EMS, CuONP were able to reduce the mutagenicity induced by PD. Interestingly, that antimutagenic potential may be explained by the ability of CuONPs to partially restore the expression levels of the repair gene *Ogg1*, and the antioxidant gene *Cu,ZnSod*, both of which were inhibited by PD treatment (Alaraby et al. 2017).

All these studies demonstrated that some MBNMs can induce positive mutagenic effects in *Drosophila*, although in those cases where the origin of the mutant clones was evaluated, the induction of somatic recombination was the mechanism of action usually found.

8. Antigenotoxic potential of MBNMs

Not all nanomaterials are intrinsically harmful since some nanomaterials, especially MBNMs, can be incorporated in several medical applications as contrast agents, diagnostic imaging and probes, cell manipulations, drugs, and therapeutic carriers (Kwon et al. 2018). By using *Drosophila*, it has been observed that some NMs like nickel nanowires, as high aspect ratio nanomaterials, can result in no harm from the genotoxic point of view when introduced in medical applications, as effective contrast agent tools (Alaraby et al. 2018). On the other hand, MBNMs can also be involved in the treatment of different diseases, especially those especially dangerous like cancer, killing tumor cells via thermotherapy (van Landeghem et al. 2009), or via oxidation (Siddiqui et al. 2013). Interestingly, some MBNMs have proved to have antioxidant intrinsic features (Vinardell and Mitjans 2015). Thus, CeO_2NPs show a selective action according to the cell type, protecting normal cells and killing cancer cells (Wason et al. 2013), which was attributed to the intrinsic antioxidant properties of cerium (Shcherbakov et al. 2015). Cerium would act as an antioxidant enzyme scavenging a variety of radicals in cells (Nelson et al. 2016). This behavior has also been observed in Drosophila, where CeO_2NPs reduced the oxidative stress-induced in *Drosophila* hemocytes by pro-oxidant agents (Alaraby et al. 2015c). This behavior would explain why CeO_2NPs failed to induce mutagenicity in *Drosophila* via using the wing-spot assay but, on the contrary, reduced the mutagenicity of a well-known genotoxic agent (potassium dichromate). Copper is another metal with reported antigenotoxic potential; thus, although it is an essential and biocompatible element (Johnson et al. 1992), its behavior in a nanoparticulated form significantly differs from its bulky form. In this regard, although several studies have reported cytotoxic potential (Ingle et al. 2014, Adeyemi et al. 2020), CoNPs were able to inhibit the genotoxicity of potassium dichromate (Alaraby et al. 2017).

Taking into account the relevance of the potential antigenotoxicity of MBNMs, especially those synthesized from biocompatible elements, extensive research is needed to discover those MBNMs posing such abilities (Barabadi et al. 2019).

9. General conclusions

MBNMs production is increasing and, consequently, they represent an important health/environmental concern due to its wide global spread. Although metal exposure is classically considered as a real carcinogenic risk, due to their well demonstrated genotoxic potential, their potential effects at the nanoscale require further work. According to the well-known relationship between genotoxicity, mutagenicity, and carcinogenicity, determining the DNA damaging potential of MBNMs compounds represent an important challenge to evaluate their potential carcinogenic risk.

In this context, to have robust tools is essential. Thus, *Drosophila* appears as a useful model to evaluate the genotoxic/mutagenic potential of MBNMs. Although the comet assay is a widely used assay to detect genotoxicity in any eukaryotic organism, and that can be also used in *Drosophila*, there is an important lack of powerful *in vivo* assays detecting mutagenicity. Although originally *Drosophila* was used to detect germinal induced mutations, the standard assay (the sex-linked recessive lethal assay) used was too long and tedious. Nevertheless, the proposal of a new assay (SMART, somatic mutation and recombination test) significantly changed the way to carry out mutagenicity studies, mainly due to its simplicity. It should be pointed out that, in addition to the induction of the somatic mutation, the assay is also able to determine the induction of somatic recombination, which is a mechanism strongly linked to the cancer induction steps. As previously reported, the use of the SMART assay that detects somatic mutations in the wing cells has been successfully used in the evaluation of MBNMs. In general, the observed effects are lower than those obtained using genotoxicity assays. This only means that some of the genotoxic effects observed are properly repaired and, consequently, they are not fixed as mutations. In this context, and terms of risk, we can remark the advantages of the use of mutagenicity tests, in comparison with genotoxicity tests, to determine the potential risk posed by MBNMs exposure. Thus, a good proposal is that genotoxicity and mutagenicity tests should be part of one battery to be used in the evaluation of MBNMs. Genotoxicity would measure the potential interaction with DNA, while mutagenicity would test the ability to fix such genetic damage. It must be remembered that one proved mechanism of action of MBNMs is the induction of oxidative damage and this type of lesion is easily recognized and repaired by the cells and, consequently, most of these lesions are not ending as mutational events.

Another aspect to be highlighted is that several MBNMs have a dual behavior sharing both genotoxic and antigenotoxic potential. Thus, solving under which conditions MBNMs behave in one or another way is crucial. As we have indicated, *Drosophila* can be used in such type of studies, which reinforce its potential as a model to evaluate the risks associated with MBNMs exposures.

References

Adeyemi, J.A., Machado, A.R.T., Ogunjimi, A.T., Alberici, L.C., Antunes, L.M.G., and Barbosa, F. Jr. 2020. Cytotoxicity, mutagenicity, oxidative stress and mitochondrial impairment in human hepatoma (HepG2) cells exposed to copper oxide, copper-iron oxide and carbon nanoparticles. Ecotoxicol. Environ. Saf., 189: 109982.

Ahamed, M., Posgai, R., Gorey, T.J., Nielsen, M., Hussain, S.M., and Rowe, J.J. 2010. Silver nanoparticles induced heat shock protein 70, oxidative stress and apoptosis in *Drosophila melanogaster*. Toxicol. Appl. Pharmacol., 242: 263–269.

Akbari, A., Amini, M., Tarassoli, A., Eftekhari-Sis, B., Ghasemian, N., and Jabbari, E. 2018. Transition metal oxide nanoparticles as efficient catalysts in oxidation reactions. Nano-Structures & Nano-Objects, 14: 19–48.

Alaraby, M., Hernández, A., Annangi, B., Demir, E., Bach, J., Rubio, L. et al. 2015a. Antioxidant and antigenotoxic properties of CeO$_2$ NPs and cerium sulphate: Studies with *Drosophila melanogaster* as a promising *in vivo* model. Nanotoxicology, 9: 749–759.

Alaraby, M., Annangi, B., Hernández, A., Creus, A., and Marcos, R. 2015b. A comprehensive study of the harmful effects of ZnO nanoparticles using *Drosophila melanogaster* as an *in vivo* model. J. Hazard. Mater. 296: 166–174.

Alaraby, M., Demir, E., Hernández, A., and Marcos, R. 2015c. Assessing potential harmful effects of CdSe quantum dots by using *Drosophila melanogaster* as *in vivo* model. Sci. Total Environ., 530: 66–75.

Alaraby, M., Annangi, B., Marcos, R. and Hernandez, A. 2016a. *Drosophila melanogaster* as a suitable *in vivo* model to determine potential side effects of nanomaterials: A review. J. Toxicol. Environ. Health, Part B, 19: 65–104.

Alaraby, M., Hernández, A., and Marcos, R. 2016b. New insights in the acute toxic/genotoxic effects of CuO nanoparticles in the *in vivo Drosophila* model. Nanotoxicology, 10: 749–760.

Alaraby, M., Hernández, A., and Marcos, R. 2017. Copper oxide nanoparticles and copper sulphate act as antigenotoxic agents in *Drosophila melanogaster*. Environ. Mol. Mutagen., 58: 46–55.

Alaraby, M., Hernández, A., and Marcos, R. 2018. Systematic *in vivo* study of NiO nanowires and nanospheres: biodegradation, uptake and biological impacts. Nanotoxicology, 12: 1027–1044.

Alaraby, M., Romero, S., Hernández, A., and Marcos, R. 2019. Toxic and genotoxic effects of silver nanoparticles in *Drosophila*. Environ. Mol. Mutagen., 60: 277–285.

Alaraby, M., Demir, E., Domenech, J., Velázquez, A., Hernández, A., and Marcos, R. 2020. *In vivo* evaluation of the toxic and genotoxic effects of exposure to cobalt nanoparticles using *Drosophila melanogaster*. Environ. Sci.: Nano., 7: 610–622.

Alaraby, M., Hernández, A., and Marcos, R. 2021. Novel insights into biodegradation, interaction, internalization and impacts of high-aspect-ratio TiO$_2$ nanomaterials: A systematic *in vivo* study using *Drosophila melanogaster*. J. Hazard. Mater. 124474. Online ahead of print. doi: 10.1016/j.jhazmat.2020.124474.

Alinovi, R., Goldoni, M., Pinelli, S., Campanini, M., Aliatis, I., Bersani, D. et al. 2015. Oxidative and pro-inflammatory effects of cobalt and titanium oxide nanoparticles on aortic and venous endothelial cells. Toxicol. *in Vitro*, 29: 426–437.

Ansari, M.O., Parveen, N., Ahmad, M.F., Afrin, S., Rahman, Y., Jameel, S., et al. 2019. Evaluation of DNA interaction, genotoxicity and oxidative stress induced by iron oxide nanoparticles both *in vitro* and *in vivo*: attenuation by thymoquinone. Sci. Rep., 9: 1–14.

Augustine, R., Mathew, A.P., and Sosnik, A. 2017. Metal oxide nanoparticles as versatile therapeutic agents modulating cell signaling pathways: linking nanotechnology with molecular medicine. Appl. Mater. Today, 7: 91–103.

Barabadi, H., Najafi, M., Samadian, H., Azarnezhad, A., Vahidi, H., Mahjoub, M.A. et al. 2019. A systematic review of the genotoxicity and antigenotoxicity of biologically synthesized metallic nanomaterials: are green nanoparticles safe enough for clinical marketing? Medicina (Kaunas), 55: 439.

Bhattacharya, R., and Mukherjee, P. 2008. Biological properties of "naked" metal nanoparticles. Adv. Drug Deliv. Rev., 60: 1289–1306.

Brunetti, V., Chibli, H., Fiammengo, R., Galeone, A., Malvindi, M.A., and Vecchio, G. 2013. InP/ZnS as a safer alternative to CdSe/ZnS core/shell quantum dots: *in vitro* and *in vivo* toxicity assessment. Nanoscale, 5: 307–317.

Carmona, E.R., Inostroza-Blancheteau, C., Obando, V., Rubio, L., and Marcos, R. 2015. Genotoxicity of copper oxide nanoparticles in *Drosophila melanogaster*. Mutat. Res. Genet. Toxicol. Environ. Mutagen., 791: 1–11.

Carmona, E.R., Inostroza-Blancheteau, C., Rubio, L., and Marcos, R. 2016. Genotoxic and oxidative stress potential of nanosized and bulk zinc oxide particles in *Drosophila melanogaster.* Toxicol. Ind. Health, 32: 1987–2001.

Chavali, M.S., and Nikolova, M.P. 2019. Metal oxide nanoparticles and their applications in nanotechnology. SN Appl. Sci., 1: 607.

De Carli, R.F., Chaves, D.D.S., Cardozo, T.R., de Souza, A.P., Seeber, A., Flores, W.H. et al. 2018. Evaluation of the genotoxic properties of nickel oxide nanoparticles *in vitro* and *in vivo*. Mutat. Res. Genet. Toxicol. Environ. Mutagen., 836: 47–53.

Demir, E., Vales, G., Kaya, B., Creus, A., and Marcos, R. 2011. Genotoxic analysis of silver nanoparticles in *Drosophila*. Nanotoxicology, 5: 417–424.

Ertuğrul, H., Yalçın, B., Güneş, M., and Kaya, B. 2020. Ameliorative effects of melatonin against nano and ionic cobalt induced genotoxicity in two *in vivo Drosophila* assays. Drug Chem. Toxicol., 43: 279–286.

Fahmy, B., and Cormier, S.A. 2009. Copper oxide nanoparticles induce oxidative stress and cytotoxicity in airway epithelial cells. Toxicol. *in Vitro*, 23: 1365–1371.

Galeone, A., Vecchio, G., Malvindi, M.A., Brunetti, V., Cingolani, R., and Pompa, P.P. 2012. *In vivo* assessment of CdSe-ZnS quantum dots: coating dependent bioaccumulation and genotoxicity. Nanoscale, 4: 6401e7.

García-Rodríguez, A., Vila, L., Cortés, C., Hernández, A., and Marcos, R. 2018. Effects of differently shaped TiO_2NPs (nanospheres, nanorods and nanowires) on the *in vitro* model (Caco-2/HT29) of the intestinal barrier. Part. Fibre Toxicol., 15: 1–16.

García-Rodríguez, A., Rubio, L., Vila, L., Xamena, N., Velázquez, A., Marcos, R. et al. 2019. The comet assay as a tool to detect the genotoxic potential of nanomaterials. nanomaterials (Basel), 9: 1385.

George, J.M., Antony, A., and Mathew, B. 2018. Metal oxide nanoparticles in electrochemical sensing and biosensing: a review. Microchim. Acta, 185: 358.

Global Nanomaterials Market Size Report, 2020. ID: GVR-4-68038-565-6, pages 1-113.

Golbamaki, N., Rasulev, B., Cassano, A., Robinson, R.L.M., Benfenati, E., Leszczynski, J. et al. 2015. Genotoxicity of metal oxide nanomaterials: review of recent data and discussion of possible mechanisms. Nanoscale, 7: 2154–2198.

Hartung, T., and Sabbioni, E. 2011. Alternative *in vitro* assays in nanomaterial toxicology. Wiley Interdiscip. Rev. Nanomed. Nanobiotechnol., 3: 545–573.

Ingle, A.P., Duran, N., and Rai, M. 2014. Bioactivity, mechanism of action, and cytotoxicity of copper-based nanoparticles: a review. Appl. Microbiol. Biotechnol., 98: 1001–1009.

Jeng, H.A., and Swanson, J. 2006. Toxicity of metal oxide nanoparticles in mammalian cells. J. Environ. Sci. Health P. A, 41: 2699–2711.

Johnson, M.A., Fischer, J.G., and Kays, S.E. 1992. Is copper an antioxidant nutrient? Crit. Rev. Food Sci. Nutrit., 32: 1–31.

Judy, J.D., Unrine, J.M., and Bertsch, P.M. 2011. Evidence for biomagnification of gold nanoparticles within a terrestrial food chain. Environ. Sci. Technol., 45: 776–781.

Karlsson, H.L., Cronholm, P., Hedberg, Y., Tornberg, M., De Battice, L., Svedhem, S. et al. 2013. Cell membrane damage and protein interaction induced by copper containing nanoparticles -Importance of the metal release process. Toxicology, 313: 59–69.

Karlsson, H.L., Gliga, A.R., Calléja, F.M., Gonçalves, C.S., Wallinder, I.O., Vrieling, H. et al. 2014. Mechanism-based genotoxicity screening of metal oxide nanoparticles using the ToxTracker panel of reporter cell lines. Part. Fibre Toxicol., 11: 41.

Kim, S., Choi, J.E., Choi, J., Chung, K.H., Park, K., Yi, J. et al. 2009. Oxidative stress-dependent toxicity of silver nanoparticles in human hepatoma cells. Toxicol. *in Vitro*, 23: 1076–1084.

Kwon, D., Kim, M.J., Park, C., Park, J., Choi, K., and Yoon, T.H. 2012. *In vivo* biodegradation of colloidal quantum dots by a freshwater invertebrate *Daphnia magna*. Aquat. Toxicol., 114: 217–222.

Kwon, H.J., Shin, K., Soh, M., Chang, H., Kim, J., Lee, J., et al. 2018. Large-scale synthesis and medical applications of uniform-sized metal oxide nanoparticles. Adv. Mater., 30: 1704290.

Liou, S.H., Wu, W.T., Liao, H.Y., Chen, C.Y., Tsai, C.Y., Jung, W.T. et al. 2017. Global DNA methylation and oxidative stress biomarkers in workers exposed to metal oxide nanoparticles. J. Hazard. Mater., 331: 329–335.

Luo, G., Santoro, I.M., McDaniel, L.D., Nishijima, I., Mills, M., Youssoufian, H. et al. 2000. Cancer predisposition caused by elevated mitotic recombination in Bloom mice. Nat. Genet., 26: 424–429.

Marcos, R., and Carmona, E.R. 2019. The wing-spot and the comet tests as useful assays for detecting genotoxicity in *Drosophila*. Methods Mol. Biol., 2031: 337–348.

Maruthupandy, M., Zuo, Y., Chen, J.S., Song, J.M., Niu, H.L., Mao, C.J. et al. 2017. Synthesis of metal oxide nanoparticles (CuO and ZnONPs) via biological template and their optical sensor applications. Appli. Surf. Sci., 397: 167–174.

Micheal, A.S., and Subramanyam, M.V.V. 2013. Antioxidant enzymes as defense mechanism against oxidative stress in midgut tissue and hemocytes of *Bombyx mori* larvae subjected to various stressors. Arch. Insect Biochem. Physiol., 84: 222–234.

Mortezaee, K., Najafi, M., Samadian, H., Barabadi, H., Azarnezhad, A., and Ahmadi, A. 2019. Redox interactions and genotoxicity of metal-based nanoparticles: A comprehensive review. Chem. Biol. Interact., 312: 108814.

Narayan, N., Meiyazhagan, A., and Vajtai, R. 2019. Metal nanoparticles as green catalysts. Materials, 12: 3602.

Nelson, B.C., Johnson, M.E., Walker, M.L., Riley, K.R., and Sims, C.M. 2016. Antioxidant cerium oxide nanoparticles in biology and medicine. Antioxidants, 5: 15.

Okumura, K., Nishihara, S., and Inoue, Y.H. 2019. Genetic identification and characterization of three genes that prevent accumulation of oxidative DNA damage in *Drosophila* adult tissues. DNA Repair (Amst), 78: 7–19.

Ong, C., Yung, L.Y.L., Cai, Y., Bay, B.H., and Baeg, G.H. 2015. *Drosophila melanogaster* as a model organism to study nanotoxicity. Nanotoxicology, 9: 396–403.

Ott, M., Gogvadze, V., Orrenius, S., and Zhivotovsky, B. 2007. Mitochondria oxidative stress and cell death. Apoptosis, 12: 913–922.

Ra, H., Kim, H.L., Lee, H.W., and Kim, Y.H. 2009. Essential role of *p53* in TPEN induced neuronal apoptosis. FEBS Lett., 583: 1516–1520.

Rajiv, S., Jerobin, J., Saranya, V., Nainawat, M., Sharma, A., Makwana, P. et al. 2016. Comparative cytotoxicity and genotoxicity of cobalt (II, III) oxide, iron (III) oxide, silicon dioxide, and aluminum oxide nanoparticles on human lymphocytes *in vitro*. Hum. Exp. Toxicol., 35: 170–183.

Reis, Éde. M., de Rezende, A.A., Santos, D.V., de Oliveria, P.F., Nicolella, H.D., Tavares, D.C. et al. 2015. Assessment of the genotoxic potential of two zinc oxide sources (amorphous and nanoparticles) using the *in vitro* micronucleus test and the *in vivo* wing somatic mutation and recombination test. Food Chem. Toxicol., 84: 55–63.

Reis, Éde. M., Rezende, A.A., Oliveira, P.F., Nicolella, H.D., Tavares, D.C., Silva, A.C. et al. 2016. Evaluation of titanium dioxide nanocrystal-induced genotoxicity by the cytokinesis-block micronucleus assay and the *Drosophila* wing spot test. Food Chem. Toxicol., 96: 309–319.

Sarkar, A., Ghosh, M., and Sil, P.C. 2014. Nanotoxicity: oxidative stress mediated toxicity of metal and metal oxide nanoparticles. J. Nanosci. Nanotechnol., 14: 730–743.

Sengul, A.B., and Asmatulu, E. 2020. Toxicity of metal and metal oxide nanoparticles: a review. Environ. Chem. Lett., 10: 1659–1683.

Shcherbakov, A.B., Zholobak, N.M., Baranchikov, A.E., Ryabova, A.V., and Ivanov, V.K. 2015. Cerium fluoride nanoparticles protect cells against oxidative stress. Mater. Sci. Eng. C Mater. Biol. Appl. 50: 151–159.

Shvedova, A.A., Pietroiusti, A., Fadeel, B., and Kagan, V.E. 2012. Mechanisms of carbon nanotube-induced toxicity: focus on oxidative stress. Toxicol. Appl. Pharmacol., 261: 121–133.

Siddiqui, M.A., Ahamed, M., Ahmad, J., Khan, M.M., Musarrat, J., Al-Khedhairy, A.A. et al. 2012. Nickel oxide nanoparticles induce cytotoxicity, oxidative stress and apoptosis in cultured human cells that is abrogated by the dietary antioxidant curcumin. Food Chem. Toxicol., 50: 641–647.

Siddiqui, M.A., Alhadlaq, H.A., Ahmad, J., Al-Khedhairy, A.A., Musarrat, J., and Ahamed, M. 2013. Copper oxide nanoparticles induced mitochondria mediated apoptosis in human hepatocarcinoma cells. PLoS One, 8: e69534.

Siril, P.F., and Türk, M. 2020. Synthesis of metal nanostructures using supercritical carbon dioxide: a green and upscalable process. Small, 16: e2001972.

Subramaniam, V.D., Prasad, S.V., Banerjee, A., Gopinath, M., Murugesan, R., Marotta, F. et al. 2019. Health hazards of nanoparticles: understanding the toxicity mechanism of nanosized ZnO in cosmetic products. Drug Chem. Toxicol., 42: 84–93.

Thongkumkoon, P., Sangwijit, K., Chaiwong, C., Thongtem, S., Singjai, P., and Yu, L.D. 2014. Direct nanomaterial-DNA contact effects on DNA and mutation induction. Toxicol. Lett., 226: 90–97.

Toyokuni, S. 1998. Oxidative stress and cancer: the role of redox regulation. Biotherapy, 11: 147–154.

Uboldi, C., Orsière, T., Darolles, C., Aloin, V., Tassistro, V., George, I. et al. 2015. Poorly soluble cobalt oxide particles trigger genotoxicity via multiple pathways. Part. Fibre Toxicol., 13: 5.

Vales, G., Demir, E., Kaya, B., Creus, A., and Marcos, R. 2013. Genotoxicity of cobalt nanoparticles and ions in *Drosophila*. Nanotoxicology, 7: 462–468.

van Landeghem, F.K., Maier-Hauff, K., Jordan, A., Hoffmann, K.T., Gneveckow, U., Scholz, R. et al. 2009. Post-mortem studies in glioblastoma patients treated with thermotherapy using magnetic nanoparticles. Biomaterials, 30: 52–57.

Vecchio, G., Galeone, A., Brunetti, V., Maiorano, G., Rizzello, L., Sabella, S. et al. 2012. Mutagenic effects of gold nanoparticles induce aberrant phenotypes in *Drosophila melanogaster*. Nanomedicine, 8: 1–7.

Vinardell, M.P., and Mitjans, M. 2015. Antitumor activities of metal oxide nanoparticles. Nanomaterials, 5: 1004–1021.

Wason, M.S., Colon, J., Das, S., Seal, S., Turkson, J., Zhao, J. et al. 2013. Sensitization of pancreatic cancer cells to radiation by cerium oxide nanoparticle-induced ROS production. Nanomedicine, 9: 558–569.

Xia, T., Kovochich, M., Liong, M., Madler, L., Gilbert, B., Shi, H. et al. 2008. Comparison of the mechanism of toxicity of zinc oxide and cerium oxide nanoparticles based on dissolution and oxidative stress properties. ACS Nano, 2: 2121–2134.

Xie, H., Mason, M.M., and Wise, J.P. 2011. Genotoxicity of metal nanoparticles. Rev. Environ. Health, 26: 251–268.

4

Alleviation of Arsenic Stress in Plants using Nanofertilizers and its Extent of Commercialization
A Systemic Review

Iravati Ray,[#] Deepanjan Mridha,[#] Madhurima Joardar, Antara Das, Nilanjana Roy Chowdhury, Ayan De and *Tarit Roychowdhury* [*]

1. Introduction

Green Revolution marked the beginning of a new era in agriculture when the production and productivity of crops took a quantum jump. One of the main reasons behind the success of Green Revolution is the application of technology in the field of agriculture. During the period 1958–1966, India witnessed the rapid rise in application of chemical fertilizers to the soils. This resulted in a change in colour of the plants from light green to dark green, indicating the adequate availability of nutrients leading to formation of more chlorophyll. From this, the term "Green Revolution" was born (Swaminathan 1969). But this uncontrolled application of chemical fertilizers was in a way futile and obviously unsustainable, which gave rise to the concept of Evergreen Revolution, proposed by M.S. Swaminathan, the "Father of Green Revolution in India". Excessive use of chemical fertilisers is hazardous to plants, soils, and, eventually, humans and other animals. Moreover, the fertilizers become unavailable to the plants due to leaching, photolysis, hydrolysis and

School of Environmental Studies, Jadavpur University, Kokata-700032, India.
* Corresponding author: rctarit@yahoo.com; tarit.roychowdhury@jadavpuruniversity.in
[#] The authors contributed equally to this work.

decomposition (Siddiqui et al. 2015). After application of the chemical fertilizers, about 40–70% of nitrogen, 80–90% of phosphorus and 50–90% of potassium are lost without even reaching to the plants (Pitambara et al. 2019).

Here comes the necessity of nanotechnology. Application of nanotechnology in agriculture is wide spread. It is used in crop protection, weed management, post-harvest care, diagnosing of pathogens, etc. Moreover, it is used as nano-pesticide, nano-sensors in agricultural tools, nano-nutrient and nanofertilizer (Sharma et al. 2019). Leaching can be prevented by nanofertilizers as they can be released in a controlled fashion (Akhtar et al. 2020). The detrimental effects of these fertilizers start right from their production and continue to exist even after their use. These adverse effects can only be reduced or removed by employing organic manure, biofertilizers or nanofertilizers (a comparatively new area, which needs more exploration) in the process of production of crops and propagate sustainable agriculture, which will give high yield with minimal environmental impacts. Although required in small quantity, nanofertilizers are much more bioavailable than their conventional counterparts. They can interact with other components present in the soil such as humic acid to increase their bioavailability (Pitambara et al. 2019). Nanofertilizers deliver nutrients to the plants by various mechanisms such as adsorption on nanoparticles, encapsulation in nanoparticulate shell, entrapment in polymeric nanoparticles and attachment of nanoparticles mediated ligands (Kaushik and Dijwanti 2019).

The world population is expected to hit 9.73 billion by 2050 (https://population.un.org/wpp/Graphs/DemographicProfiles/Line/900). We must switch to Climate Smart Agriculture (CSA) to increase agricultural productivity, ensure food security and development, reduce green-house emissions, etc. (Campbell et al. 2014). Nanofertilizers have the ability to boost agricultural output while decreasing pollution, increase plant height, root and shoot length, biomass, photosynthesis rate, and so on, as stated in numerous study studies (Ansari et al. 2020). The development of high-end nano equipment to increase the productivity and quality in the field of agriculture will reduce or even eliminate the reliance on conventional fertilizers (Sekhon 2014, Liu and Lal 2015). Nanofertilizers have also been reported to tolerate abiotic stresses like drought, flooding, and salinity as well as heavy metal contamination efficiently and can also work in association with microorganisms such as nanobiofertilizers (Zulfiqar et al. 2019). Sustainable agriculture can be practised by plugging together the benefits of genetic engineering and nanoparticles (Sangeetha et al. 2019). Arsenic (As) stress is one of the most important among these abiotic stresses. Quite a few researches have been conducted to overcome the adverse effects of several species of As with the help of nanofertilizers.

The problem of As toxicity is mainly predominant in South and South-East Asian countries, especially in Bangladesh and India (in the state of West Bengal) (Chatterjee et al. 1995, Chowdhury et al. 2000). It gets associated with the soil and groundwater with which irrigation of various crops is done. Rice plants and grains are known to extensively accumulate As due to this irrigation with contaminated water and is a potential source of toxicity when consumed (Roychowdhury et al. 2005). However, during the monsoon season, the toxicity decreases to some extent as rainwater dilutes the bioavailable As (Chowdhury et al. 2020). Presence of arsenic in drinking water is a major concern in several areas of West Bengal. Arsenic concentration in groundwater from Madhusudankati village of Gaighata block, North 24 Parganas district is 7 times higher than the permissible limit set by WHO (Joardar et al. 2020). Thus, simultaneous consumption of arsenic contaminated drinking water and rice grain poses a huge threat to the health of the people living in the exposed zones. Attempts to mitigate As toxicity in rice plants using various amendments are going on for the past few years (Mridha et al. 2021). Its toxicity and bioavailability depend on chemical forms. Inorganic As (As^{+3} and As^{+5}) compounds are more toxic than its organic counterparts and are considered as class-1 human carcinogen. Sedimentary rocks contain higher concentrations of arsenic than that of igneous (Bhumbla and Keefer 1994). Arsenic has two oxidation states, Arsenite (As^{+3}) and Arsenate (As^{+5}), of which As^{+3} is soluble and hence more toxic (WHO 1981) as As^{+5} cannot exhibit its toxic effect directly. As^{+3} is generally two to ten times more acutely toxic than As^{+5}. As^{+5} needs to get converted first to As^{+3} for showing its toxicity (Thomas et al. 2001). Arsenic binds to sulfhydryl groups present in keratinized tissue. When As^{+3} enters the human body system, it rapidly attacks the active site of the protein containing thiol group, thus making it inactive, which in turn leads to prevention of ATP formation and disruption of body enzymatic activity (Hughes 2002).

This review paper aims to focus on the different types of nanofertilizers used to alleviate As stress and their respective effects on the plants. It also aims to throw a light on the different processes of application of the nanofertilizers and their comparative efficacy. Depending on the efficacy, their potential of commercialization will also open up a scope for discussion. Nanomaterials have been applied on plants to alleviate stresses induced by other heavy metals. Application on nanosilica is known to alleviate Cd, Pb, Zn and Al toxicity in plants (Mathur and Roy 2020). Cadmium toxicity in plants has been mitigated by application of zinc oxide, iron oxide and titanium dioxide nanoparticles (Hussain et al. 2019, Ogunkunle et al. 2020, Priyanka et al. 2021, Rizwan et al. 2019). Human beings have developed nanoparticles from 4th century AD, evidences of which can be found in coloured stained-glass windows of cathedrals and many other things. Thus, developing a promising and sustainable nanofertilization technique is very much tangible in the near future and definitely its market will drastically grow within a few years. However, the biosafety aspects of nanofertilizers must be considered as well.

2. Nanofertilizers: A sustainable alternative

Unjudicial application of chemical fertilizers is going on since the green revolution. Despite increasing crop production, they disturb the mineral balance and in the long run destroy the soil fertility (Solanki et al. 2015). The detrimental effects of these fertilizers start right from their production and continue to exist even after their use. These adverse effects can only be reduced or removed by employing organic manure, biofertilizers or nanofertilizers in the process of production of crops. This will propagate sustainable agriculture, which will ensure high yield with minimal environmental impacts.

The use of organic, bio and nanofertilizers has been proven to be beneficial in many aspects. Nanomaterials have also been derived from waste materials and utilized for alleviation of As stress (Ray et al. 2021). In the past few years, the use of nanofertilizers as an alternative to chemical fertilizers has increased. Nanofertilizers can act as growth regulators, control nutrient and water release, etc. It also helps in producing improved quality agricultural products, retaining water and nutrient in soil, managing plant disease, remediation of heavy metal, etc. (Butt and Naseer 2020). In conventional fertilizers, solubility gets reduced upon mixing two or more fertilizers (Kant and Kafkafi 2013). But exactly the opposite happens in case of nanofertilizers as their particle size in much less than conventional chemical fertilizers. Solubility increases with decreasing particle size according to Ostwald-Freundlich equation:

$$\frac{S}{S_0} = exp\frac{\gamma V_m}{RTd}$$

Where, S is the solubility (mol/kg H_2O) at temperature T(K), V_m is molar volume (m^3/mol), γ is surface free energy (surface tension) (mJ/m^2), R is the universal gas constant (8314.5 mJ/mol K), S_0 is the solubility of the bulk material (mol/kg H_2O), and d is particle diameter (m).

From the above equation, it is evident that solubility increases as particle size decreases. For $\frac{S}{S_0} \ll 1$, the exponential term has to be much less than 1, which is only possible when the diameter of the particle is in the nano regime, provided the temperature is kept constant. Thus, nanoparticle takes an advantageous position in this regard (Sasson et al. 2007). These nanofertilizers upon application can easily enter through plants' cells when applied on surface, in turn increasing nutrient uptake efficiency. Due to this smaller particle size, the dissolution rate and the abundance of the nanoparticles in water/soil must be higher than their bulk counterparts (Liu and Lal 2015), while in the case of conventional fertilizers, mixing two fertilizers can also result in low nutrient availability and uptake by the plants. Nanofertilizers with high surface area improve nutrient uptake and nutrient use efficiency (Table 1). Nanofertilizers supply nutrient for extended duration and also minimize the loss of nutrients through leaching (Cui et al. 2010, Kant and Kafkafi 2013, Pitambara et al. 2019). However, excessive use of nanofertilizers, upon entering the food chain, may expose human and animal life to risks. ZnO nanofertilizer application above

Table 1: The positive and negative effects of different fertilizers.

Parameter	Chemical Fertilizer	Organic Fertilizer	Nanofertilizer	Biofertilizer	References
Nutrient Use Efficiency	–	+	+	+	Butt and Naseer 2020
Soil Fertility	–	+	+	+	Sharma et al. 2019
Cost Effectiveness	+	–	–	–	Dimkpa et al. 2014
Stress Mitigation	–	+	+	+	Pitambara et al. 2019
Easy Storage	+	–	+	–	Duhan et al. 2017
Environment Friendly	–	+	+	+	Solanki et al. 2015

'+' = more, and '–' = less

500 mg/kg has been proven to be detrimental for plants (Pullagurala et al. 2018). Nanoparticles may accumulate on surfaces of roots, thus decreasing pore size of cell wall. There are evidences of detrimental effects of nanofertilizers on algae, which plays an important role in maintaining the soil health (Kalwani et al. 2022). Metallic nanofertilizers release metal ions which cause cellular damages and reduce growth in plants. Iron oxide nanofertilizers clogs the root cells and reduces water and nutrient uptake of plants (Paramo et al. 2020, Zuverza-Mena et al. 2017). Nanofertilizer toxicity to plants is directly proportional to their concentration and exposure period, but inversely proportional to their size (Zuverza-Mena et al. 2017). Thus, to achieve the best results, it is critical to optimize the application concentration, exposure length, and size of nanofertilizers.

Compared to inorganic fertilizers, biodegradable organic fertilizers and manures are considered as safe options, having little or no harmful environmental and ecological effects (Table 1). They can provide necessary nutrients to the plant for growth and development. It is possible to manufacture them *in situ*. But it is very difficult to satisfy all the nutrient requirements of a crop using organic manure because the nutrient content of the manures generally do not match with the stoichiometric necessities of the crops (Connor and Loomis 1992). Compared to chemical fertilizers, the quantity of organic manure required is nearly double. Their nutrient release speed is very slow and they are also quite expensive.

Biofertilizers are very similar to nanofertilizers, but also have slight differences in their mechanism of action, way of application, effective rates required and also environmental effects (El-Ghamry et al. 2018). The main sources of biofertilizers are living organisms, decomposed organic wastes and decomposed dead organisms. Biofertilizers also come in the form of microbial strains that interacts with the rhizosphere and helps the plants in nutrient uptake. They are renewable sources of nutrients, which may increase crop yield and also improve soil property to maintain fertility. Biofertilizers are cheap and easier to produce by culturing microorganisms. But they can be disadvantageous also as they possess low shelf life, chances of contamination, and chances of becoming less effective. Their cell count decreases and they are highly temperature sensitive (Thomas and Singh 2019). Biofertilizers are slower than chemical fertilizers and difficult to store.

Another novel and cost-effective form of fertilizers is nanobiofertilizers. It has all the properties of biofertilizers as well as nanofertilizers and can be thought as a hybrid between nano and biofertilizers. In this type of fertilizers, the biofertilizers are coated with different nanomaterials, which are released slowly in a controlled fashion for a longer period of time. This helps in increasing the nutrient use efficiency of plants, in turn increasing the crop yield and productivity (Kumari et al. 2019). Thus, the conglomeration of bio and nanofertilizers results in a synergy, promoting sustainable agriculture. Quite a few research works have been conducted on the effects of biofertilizers on the production and yield of crops.

In this review, a compilation of the characteristics of different kind of fertilizers have been made. This may help in identifying the most effective and sustainable form of fertilizer, which can be commercialized to replace the conventional chemical fertilizers. The chemical fertilizers also have beneficial properties. They can be easily stored and are required in less quantity than organic fertilizers. However, compared to their detrimental effects, the benefits are nothing.

Keeping in mind about these advantages and disadvantages of the fertilizers, some practical applications can be studied. Plant physiology and morphology are parameters which can reflect the beneficial and harmful effects of the fertilizers used. Thus, to assess the most sustainable alternative to the conventional treatments, the physiology and morphology have to be monitored.

2.1 Comparative effects of nano, bio, organic and chemical fertilizers on plant morphology and physiology

The eternal hunger to increase the production and yield of crops has been fulfilled for several decades at the expense of polluting our mother Earth. Usage of chemical fertilizers makes agriculture the largest source of phosphorus and nitrogen pollution. They have the potential to contaminate groundwater systems (https://www.epa. gov/agriculture/agriculture-nutrient-management-and-fertilizer). So, now it is high time to eliminate these adverse effects. The use of organic and biofertilizers can maintain good water quality, prevent pollution and ensure sustainable agriculture. Agriculture must work with nature and not against it. Several irreversible changes have already occurred due to excessive agricultural pollution, but we must act now to stop more such changes from destroying the nature. Several research articles have been published, comparing the effects of chemical, organic, bio and nanofertilizers on plants. Application of fertilizers other than chemical fertilizers have often proven to be beneficial for the plants' morphology and physiology.

El-Henawy et al. (2018) studied the effects of application of different types of fertilizers (organic, mineral and nano) on red cabbage and broccoli. Their experiments revealed plant dependence on fertilizers for their vegetative growth. After first foliar application, the highest number of leaves per plants (both cabbage and broccoli) was obtained upon application of nano Se. Maximum plant height in case of broccoli was obtained upon application of nano Se, but in case of other morphological parameters NPK dominated. During second foliar application, number of leaves per plant was maximum in cabbage while the plant height was maximum in broccoli. But application of nano Cu did not yield the best of results compared to the other fertilizers in case of morphological parameters, except maximum head diameter was observed in broccoli after second foliar application. In case of chlorophyll content, the best results were shown by nano Cu for cabbage plants. However, the scenario was not the same for broccoli as compost tea yielded the highest amount of chlorophyll. Application of nano Cu also resulted in least infections in broccoli. Liu and Lal (2014) conducted a greenhouse experiment to assess the fertilizing effects of a new form of nanohydroxyapatite (nHA) on soybean compared to regular phosphorus fertilizer. They observed that upon application of the nanofertilizers, the growth rate increased by 32.6%, seed yield by 20.4%, biomass production above ground by 18.2% and that at below ground level by 41.2% compared to conventional phosphorus fertilizer. Ahmed et al. (2019) applied mineral fertilizers and nanofertilizers independently as well as by mixing them in different ratios on plots of maize intercropped with fodder crop legume, guar, cowpea and clitoria. The highest values of several parameters like plant height, stem diameter, yield was recorded when the ratio of nano is to mineral NPK fertilizer was 3:1. Similarly, the lowest values were observed during only mineral fertilizer application. Janmohammadi et al. (2016) compared the effects of nano-chelated complete fertilizer on potato with that of several other fertilizers and observed that nano-chelated complete fertilizers demonstrated the best results in almost every aspect, showing the maximum number of stems, highest main stem diameter, greatest number of leaves, maximum biological yield, highest number of tubers per plant, highest mean tuber weight, maximum mean tuber diameter, greatest tuber weight per plant and maximum tuber yield. Upon application of nano chelated Zn + B, the highest plant height was recorded. Highest nitrate content was recorded for conventionally fertilized plants. Najafi Disfani et al. (2017) compared the effects of nano Fe/SiO_2 and bulk Fe/SiO_2 to find that the nanofertilizers gave better results compared to their bulk counterpart. Germination time was decreased in both barley and maize using nanofertilizer. However, common Fe/SiO_2 treatment yielded the maximum seedling length in maize, whereas in barley, nano Fe/SiO_2 was more effective in terms of seedling length. Nano Fe/SiO_2 also managed to increase the dry weight more than common Fe/SiO_2 in both barley and maize. Babaei et al. (2017) drew comparison between nano Zn oxide, nano Fe oxide and nano Zn + Fe and three strains of biofertilizers: *Azotobacter*, *Azosperilium* and *Pseudomonas*. They found that under the highest salinity applied to wheat, the 3 biofertilizers increased grain yield by 2.6, 4.2 and 8.5%, while the nanofertilizers increased grain yield by 2.71, 2.17 and 17.39%, respectively. Rossi et al. (2018) studied a comparative test on coffee plants using Zn nanoparticles and bulk zinc sulphate in which Zn nanoparticles showed positive impact on fresh and dry weight of the plant roots, stems and leaves while $ZnSO_4$ had a negative impact. Mir et al. (2015) demonstrated synergistic effects of manure + biofertilizers, nanofertilizers + biofertilizers and manure + nanofertilizer. From variance analysis, they concluded that the synergies developed by manure + biofertilizer and nanofertilizer + biofertilizer were more pronounced than that of manure + nanofertilizer. Kah et al. (2018) undertook a comparative study on the efficacy of nanofertilizer and non-nanofertilizer. Their review inferred that the macronutrient, micronutrient and carrier nanofertilizers were about 19, 18 and 29% more efficient than their conventional counterpart, although some of them displayed lower efficacy, which may be due to toxicity from high concentrations. A few of the conducted studies have been complied in Table 2.

Nanofertilizers can enhance the nutrient effectiveness by three times, in turn reducing the required quantity of fertilizers. Apart from enhancing the quality of morphological and physiological characteristics of plant, nanoparticles are efficient in alleviating several abiotic stresses including heavy metal stress, thus, in turn making the plant morphologically superior. In this review, special emphasis has been given on mitigating arsenic stress through nanofertilizers.

3. Different nanofertilizers mitigating arsenic stress

In the past few decades, quite a few nanoparticles have evolved, application of which as nanofertilizers can mitigate arsenic stress and improve plant morphology. Many researchers have independently used nanofertilizers and many have compared the

Table 2: Results of application of nanoscale fertilizers to various plants with comparison to control or other fertilizers.

Nanofertilizer used	Compared with	Method of Application	Applied on	Results	References
Nano Se	Control NPK Compost Tea	Foliar	Cabbage Broccoli	Positive results were obtained in some aspects.	El-Henawy et al. 2018
Nano Cu	Control NPK Compost Tea	Foliar	Cabbage Broccoli	Mostly negative results were observed when compared to other fertilizers.	El-Henawy et al. 2018
Nano Hydroxyapatite	Conventional P Fertilizer	Fertilizer solution added to plant pots	Soybean	Positive results were observed.	Liu and Lal 2014
NPK	NPK mineral fertilizer	Soil Application	Maize intercropped with guar, cowpea and clitoria	Positive results were obtained when fertilizers were mixed in the proper ratio.	Ahmed et al. 2019
Nano chelated complete fertilizer (N, P, K, Fe, Zn, Ca, Mg, Mn, Cu, B, Mo)	1. Control 2. Conventional NPK 3. MOG Enzymatic Biofertilizer(Org C, N, K_2O, Fe, Cu, Enzymes) 4. Nano chelated Ca 5. Nano chelated + Zn Boron	Applied with irrigation water (Fertigation)	Potato	Positive results were obtained in every aspect.	Janmohammadi et al. 2016
Fe/SiO$_2$	Common Fe/SiO$_2$ Control	Soil Application	Barley Maize	Positive results were observed in most of the aspects.	Najafi Disfani et al. 2016
Nano Zn oxide Nano Fe oxide Nano Zn + Fe together	Three strains of biofertilizers: *Azotobacter, Azosperilium* and *Pseudomonas*	Foliar Application	Wheat	Mostly positive results were produced.	Babaei et al. 2017
Zn	Zinc Sulphate	Foliar Application	Coffee	All results were positive.	Rossi et al. 2018

effects of nanofertilizers and conventional bulk fertilizers in alleviating arsenic stress. But the number of comparative studies is far less than individual studies. Tripathi et al. (2016) compared the effects of Si NPs with that of bulk silicon and inferred that Si NPs were more efficient. Cui et al. (2019) applied different concentrations of silica nanoparticles on rice seedlings. Pre-treatment with NPs increased the number of viable cells. After the treatment, the survival percentage of the cells were also increased compared to control. Liu et al. (2014) conducted field, pot and hydroponic experiments of applying silica nanoparticles on rice foliage under As stress. They observed As mitigation upon treatment. Ma et al. (2020) compared the effects of soil application of ZnO NPs and Zn^{2+} ions on As stressed rice plants. Here the ions proved to be more efficient than the nanoparticles. Wang et al. (2019) performed a similar kind of experiment with CuO NPs and Cu^{2+} ions. Once again, the ions proved to be more efficient than the nanoparticles. Wu et al. (2020) also performed their experiment with ZnO NPs. They primed the rice seeds with ZnO NPs. Their treatment increased germination percentage, while decreasing the As content in roots and shoots. Huang et al. (2018) suspended two varieties of rice seedlings in various nanoparticle suspensions such as high quality graphene oxide, multiple layer graphene oxide, 20 nm and 40 nm hydroxyapatite, nano Fe_3O_4 and nano zero valent iron. They observed varied results with different concentrations of As. The effects of nanoparticles in As stress mitigation also varied across cultivars. Wang et al. (2018) performed their experiments with CeO_2 and ZnO nanoparticles and found positive result with ZnO NPs in terms of decreasing As concentration, while CeO_2 yielded negative results. Katiyar et al. (2020) suspended mung beans in green synthesized and chemically synthesized Titanium (Ti) nanoparticle suspensions separately. The green synthesized NPs decreased the As content quite efficiently, while increasing the growth parameters like dry/wet mass and radicle length, compared to the plants treated with As only. In case of chemically synthesized Ti NPs, decrease in As content was observed along with increase in mass and radicle length compared to the plants treated with only As. α-MnO_2 nanorods decreased the As concentration in rice husk and brown rice in both pot and field application. The leachability of As was also decreased (Li et al. 2019). A few of the conducted studies have been complied in Table 3.

4. Efficacy of nanofertilizers in mitigating arsenic stress with respect to concentration, application method and type of nanofertilizers

4.1 Efficacy vs. concentration

Data sets have been identified from various research articles to develop a relationship between nanofertilizer (NF) concentration, As concentration and % alleviation (Ahmad et al. 2020, Gil-Diaz et al. 2016, Katiyar et al. 2020, Liu et al. 2014, Li et al. 2019, Ma et al. 2020, Pan et al. 2019, Tripathi et al. 2016, Wang et al. 2018, 2019, Wu et al. 2020). The percentage alleviation is determined by the change in concentration of accumulated As in plant shoots before and after exposure to nanofertilizers. From the data, it is seen that different nanofertilizers behave differently. The data

Table 3: Arsenic alleviation capacity of different nanofertilizers compared to bulk fertilizers and control.

Nanofertilizer used	Compared with	Method of Application	Applied on	Arsenic Dose	Fertilizer Dose	Results	References
Silicon	Bulk Silicon	Seedlings were treated with SiNp and bulk Si solutions	Maize and Hybrid Maize	25 µM 50 µM	10 µM	Positive results were observed	Tripathi et al. 2016
SiO$_2$	Control	Cells were cultured in presence of Silica nanoparticles	Rice	10 µM 40 µM 80 µM	1.1 mM 1.0 mM	Positive results were observed	Cui et al. 2019
SiO$_2$	Control	Foliar application in field, pot and hydroponic experiments	Rice	50 µM 100 µM	5 mM	Positive results were observed	Liu et al. 2014
ZnO	Zn^{2+}	Soil Application	Rice	6.76 mg/kg	100 mg/kg	$\Pi_{Zn}^{2+} > \Pi_{ZnO}$	Ma et al. 2020
ZnO	Control	Seed Priming	Rice	2 mg/L	10 mg/L 20 mg/L 50 mg/L 100 mg/L 200 mg/L	Positive results were observed	Wu et al. 2020
CuO	Cu(II)	Seedlings were treated with nanofertilizer	Rice	1 mg/L	100 mg/L	$\Pi_{Cu}^{2+} > \Pi_{CuO}$	Wang et al. 2019
High quality graphene oxide	Control	Seedlings placed in nanoparticle and arsenic suspension	Rice cultivar T705 and Rice cultivar X24	0.8 mg/L 1.6 mg/L 3.2 mg/L 4.0 mg/L	0.1 g in 500 ml 1/10 Hoagland nutrient solution	Mixed results were observed	Huang et al. 2018
Multiple layer graphene oxide	Control	Seedlings placed in nanoparticle and arsenic suspension	Rice cultivar T705 and Rice cultivar X24	0.8 mg/L 1.6 mg/L 3.2 mg/L 4.0 mg/L	0.1 g in 500 ml 1/10 Hoagland nutrient solution	Mixed results were observed	Huang et al. 2018
20 nm Hydroxyapatite 40 nm Hydroxyapatite	Control	Seedlings placed in nanoparticle and arsenic suspension	Rice cultivar T705 and Rice cultivar X24	0.8 mg/L 1.6 mg/L 3.2 mg/L 4.0 mg/L	0.1 g in 500 ml 1/10 Hoagland nutrient solution	Mixed results were observed	Huang et al. 2018

Nano Fe_3O_4	Control	Seedlings placed in nanoparticle and arsenic suspension	Rice cultivar T705 and Rice cultivar X24	0.8 mg/L 1.6 mg/L 3.2 mg/L 4.0 mg/L	0.1 g in 500 ml 1/10 Hoagland nutrient solution	Mixed results were observed	Huang et al. 2018
Nano Zero valent Iron	Control	Seedlings placed in nanoparticle and arsenic suspension	Rice cultivar T705 and Rice cultivar X24	0.8 mg/L 1.6 mg/L 3.2 mg/L 4.0 mg/L	0.1 g in 500 ml 1/10 Hoagland nutrient solution	Positive results were observed	Huang et al. 2018
Nano Zero valent Iron	Control	Soil Application	Barley	> 5000 mg/kg	1% 10%	Positive results were observed	Gil-Diaz et al. 2016
CeO_2	Control	Seedlings placed in nanoparticle and arsenic suspension	Rice	1 mg/L	100 mg/L	Negative results were observed	Wang et al. 2018
ZnO	Control	Seedlings placed in nanoparticle and arsenic suspension	Rice	1 mg/L	100 mg/L	Positive results were observed	Wang et al. 2018
ZnO	Control	Foliar Application	Soybean	10 μM 20 μM	50 mg/L 100 mg/L	Positive results were observed	Ahmad et al. 2020
Green synthesized Titanium	Control	Seedlings placed in nanoparticle and arsenic suspension	Mung Bean	10 μM	0.001 g/mL	Positive results were observed	Katiyar et al. 2020
Chemically synthesized Titanium	Control	Seedlings placed in nanoparticle and arsenic suspension	Mung Bean	10 μM	0.001 g/mL	Positive results were observed	Katiyar et al. 2020
Silica	Control	Foliar Application	Brown Rice	22.6 mg/kg	5 mM	Positive results were observed	Pan et al. 2019
α-MnO_2 Rods	Control	Field Application Soil Application in Pots	Rice	3.06 mg/kg in husk 0.96 mg/kg in brown rice	0.2, 0.5, 1.0, 2.0 % of soil weight	Positive results were obtained	Li et al. 2019

Ŋ = Mitigation Efficiency

reported by researchers are insufficient to draw any comparison between the arsenic mitigation efficiency of nanofertilizers and other conventional fertilizers. Hence, there is scope of conducting further studies in this area. However, from the data we have determined the ratio of nanofertilizer and As concentration. Taking that value in the abscissa (x-axis), we have plotted the As alleviation percentage of nanofertilizers in the ordinate (y-axis). The data trend suggests that no definite relationship can be drawn between concentration and alleviation percentage as the data points are scattered throughout between 0 and 100th point of the X-axis (Fig. 1). Thus, the As alleviation does not only depend upon the nanofertilizer concentration but also on several other factors such as choice of nanofertilizer, the plant upon which it is acting, method of application, As speciation, etc.

But from the research conducted by Wu et al. (2020), it is quite evident that As alleviation increases with increasing nanofertilizer concentration in case of ZnO NF. But Ma et al. 2020 used the same nanoparticle on the same plant (rice) to obtain an alleviation efficiency higher (60.2%) than that of Wu et al. (2020). The difference between their experiments lies in the fact that Wu et al. (2020) studied on seed priming, while Ma et al. (2020) applied the nanofertilizer to soil. Thus, it can be emphasised that the method of application is important in conducting mitigation studies. Various research articles have also compared the different methods by studying their effects on plant physiology, morphology and even heavy metal remediation. Once again,

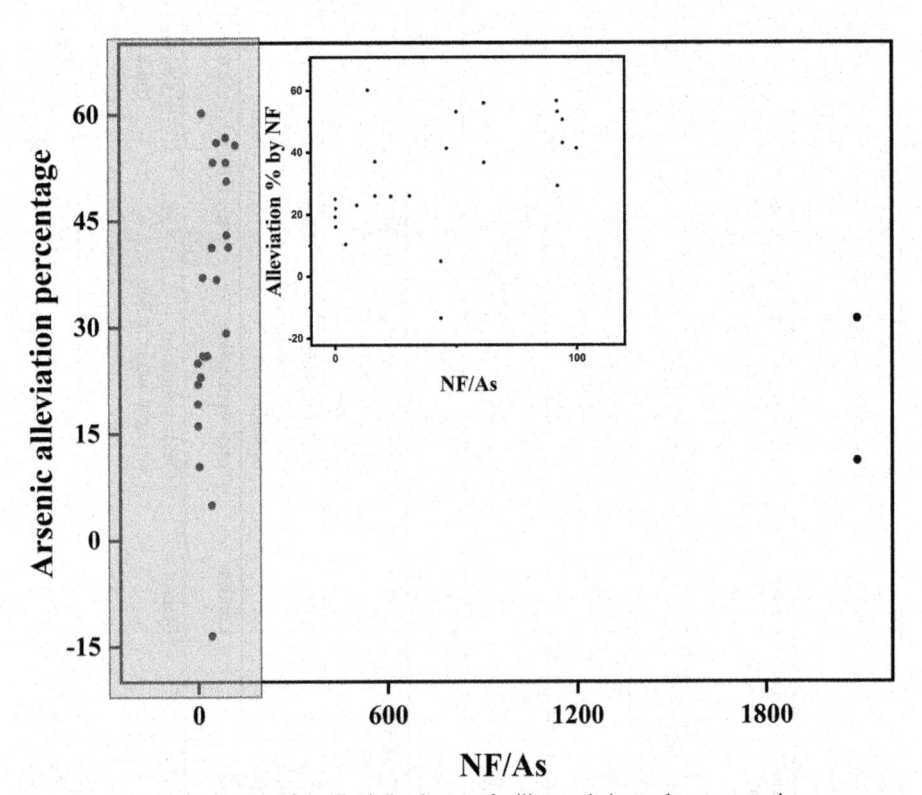

Fig. 1: Percentage of As alleviation by nanofertilizers relative to the concentration.

it may be mentioned as this review focuses on arsenic alleviation, there is a further scope of comparing the effectiveness of different application methods in alleviating As stress.

4.2 Efficacy vs. method of application

From the literatures surveyed, it was found that there are mainly three ways for application of nanofertilizers: (i) Foliar Application (ii) Seed Priming/Soaking (iii) Soil Application/Broadcasting (El-Ghamry et al. 2018). Different methods of nanofertilizer application and the gain in plant growth and yield under As stress, using nanofertilizer, were depicted in Fig. 2. In many cases during comparison between foliar application and other types of application, the former has emerged victorious in

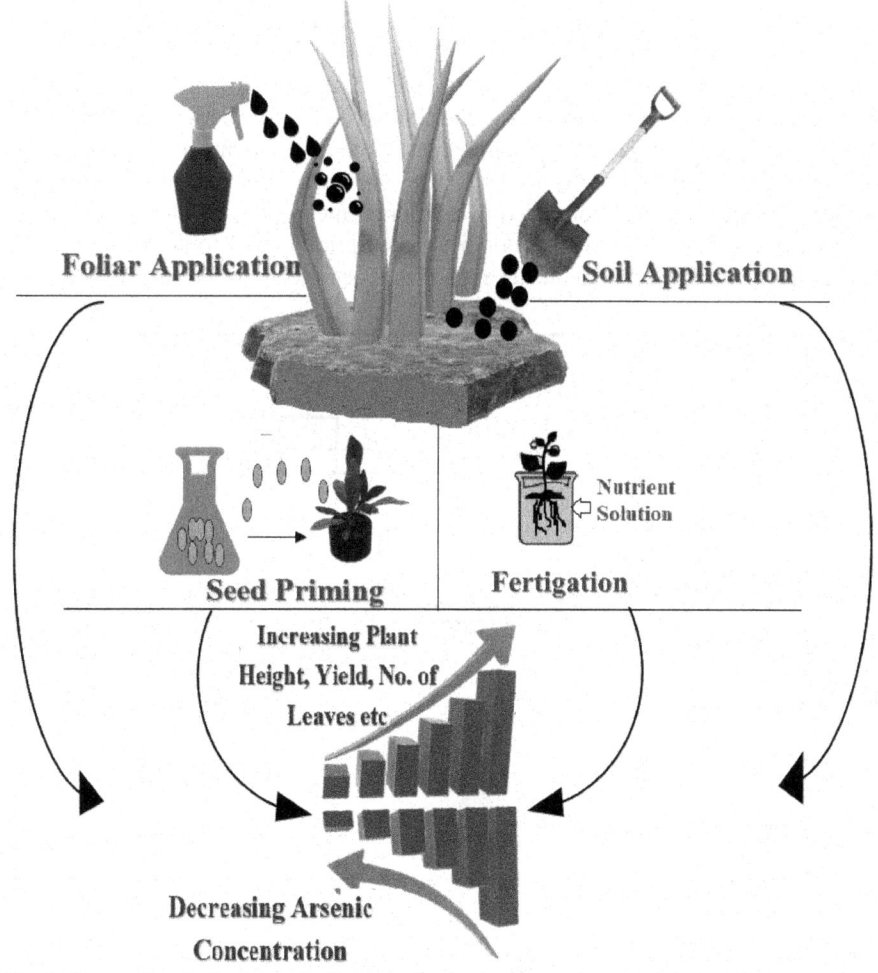

Fig. 2: Proposed methods of nanofertilizers application for better plant growth and yield under arsenic stress.

terms of giving the plant a better morphology. The percentage of As alleviated using different method of nanofertilizer application was presented in Fig. 3.

Again, based on nutrient demand of plants, nanofertilizers can be classified as: (i) macrofertilizer (ii) microfertilizer and (iii) nanoparticulate fertilizer (Chhipa 2017). There are mainly 14 macronutrients and micronutrients (N, P, S, K, Mg, Ca, Fe, Mn, Zn, Cu, B, Mo, Cl, Ni) along with few other beneficial elements (Na, Si, Co, I, V) for plant growth (Marschner 1995). These are irreplaceable mineral nutrients, without which a plant will fail to complete its life cycle, as they are directly involved in plant metabolism (Marschner 1995). With the help of nanotechnology, these nutrients are delivered by: (i) aeroponics, (ii) hydroponics, (fertigation, i.e., application of nanofertilizer with irrigation water) (iii) soil application, and (iv) foliar application (Solanki et al. 2015).

One of the most important macronutrients is nitrogen that causes 3-dimensional pollution by run off, leaching and volatilization (Cai et al. 2014). Cai et al. (2014) developed a process of controlling this 3-dimensional pollution by adding modified nanoclay to conventional fertilizer. It was named as loss control fertilizer (LCF). Comparison of leachate, volatilization and residue percentage between conventional fertilizer and LCF provided promising results. A similar experiment was carried out by Ma et al. (2019) by adding 6% LCU with 94% urea also yielding encouraging results. Aziz et al. (2019) compared two different methods of fertilizer application with nano chitosan and carbon nanotubes—seed priming and foliar application. It was found that foliar application was much more beneficial than seed priming.

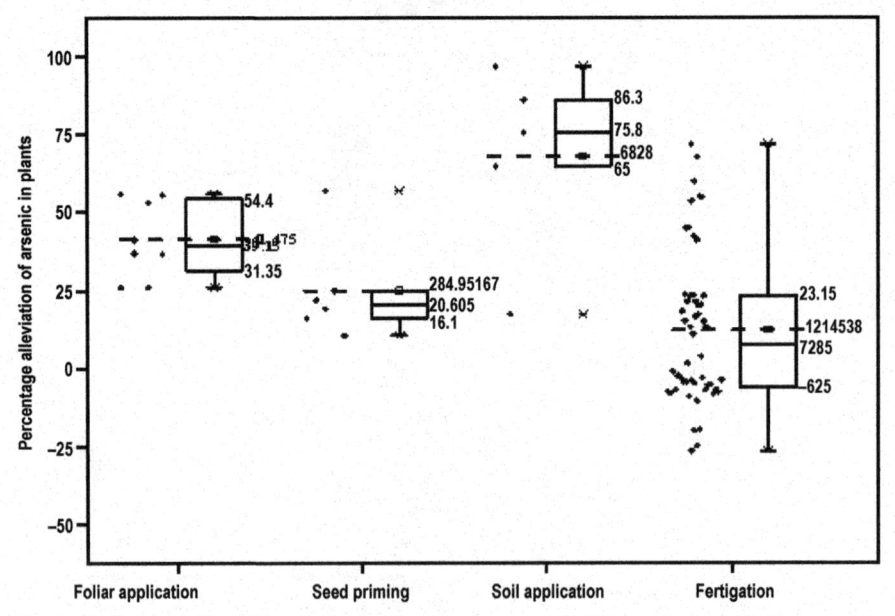

Fig. 3: Comparison of the different methods of application of nanofertilizers with respect to the percentage of As alleviation achieved. The lower and higher end of the boxes represent 25 and 75 percentiles, respectively. The solid lines within the boxes signify the median and the dashed lines signify the mean. The whiskers indicate 90 and 10 percentiles. The stars around the boxes demonstrate the data points

4.2.1 Foliar application

Liu et al. (2014) applied nanoscale silica sol on foliage to see a 22% increase in rice yield and a 53.2% and 41.3% decrease in As concentration compared to control exposed to 50 and 100 μM As, respectively. A field study with brown rice was conducted by Pan et al. (2019), which reported that the application of nanosilica sol (5 mM) on rice leaf at a dose of 750 L/ha resulted in 26–37% decrease in As concentration. Ahmad et al. (2020) investigated As alleviation by foliar application of ZnO nanoparticles on soybean plants at a concentration of 50 and 100 mg/kg. The As concentrations were 10 and 20 μM, respectively. At 10 μM As concentration, alleviation percentages of 36.7 and 55.6% were recorded in shoots upon application of 50 and 100 mg/kg of ZnO NPs. At 20 μM As concentration, alleviation percentages of 26 and 56% were achieved in shoots upon application of 50 and 100 mg/kg of ZnO NPs. Alleviation also occurred in roots of the plants.

4.2.2 Seed priming

Tripathi et al. (2016) investigated the alleviation of As^{+5} stress by applying Si nanoparticles on seeds and found out that growth of the maize plants and hybrid plants were reduced by 4 and 5% upon exposure to 25 μM As +Si NPs and by 12 and 16% upon exposure to 50 μM As + Si NPs, compared to 14 and 29% for 25 μM and 19 and 34% for 50 μM in control, respectively. Wu et al. (2020) found 10.4–57% of As decrease in shoots and 8.5-72.3% of As decrease in roots with application of ZnO nanoparticles on rice seedlings.

4.2.3 Soil application

Ma et al. (2020) treated the soil with ZnO nanoparticles and found MMA, DMA and As^{+5} to be lowered by 86.3, 75.8 and 17.3%, respectively, compared to the controls. Gil-Diaz et al. (2014) applied one commercially available nano zero valent iron (nZVI) at doses of 1 and 10% on soil to decrease the mobility and bioavailability of As. It was found that after the treatment, the amount of As in the residual fraction (not bioavailable) of the soil was significantly higher. Greater efficiency was achieved at a dose of 10%. In a similar experiment, Gil-Diaz et al. (2017) applied three commercially available nZVI slurries separately on two soil samples at doses of 0, 1, 5 and 10% (w/w). After treatment, the mobility and availability of As were investigated using Toxicity Characteristics Leaching Procedure (TCLP). It was inferred that all three treatments reduced As leachability. Gil-Diaz et al. (2016) applied nZVI at doses of 1 and 10% on pots containing seeds of barley sown in 130 g of soil. After treatment, As in the residual fraction increased, while decreasing trend was observed in other bioavailable fractions. Arsenic leaching was reduced more efficiently at the dose of 10% nZVI. Thus, the As uptake by barley plants was also decreased, especially in the treatments with 10% nZVI. Arsenic concentration in shoots was decreased by 65 and 97% in 1 and 10% treated soils, respectively. Baragano et al. (2019) compared the effectiveness of commercial nZVI and Goethite nanospheres at doses of 0.5, 2, 5, 10% in immobilizing As in polluted soil. After the treatment, TCLP test was conducted to evaluate the extent of immobilization. Both the nanoparticles reduced As leachability but Goethite showed a slight better result with immobilization percentages of 82.5, 99.3, 99.7, 99.8 compared to 41.6, 89.5,

96.2, 97.6 in nZVI at increasing doses of 0.5, 2, 5, and 10%, respectively. Gil-Diaz et al. (2019) conducted pilot scale study in two areas by applying 2.5% nZVI for a period of 32 months. The highest reduction was observed after 72 h of application. Li et al. (2019) applied α-MnO$_2$ nanorods on rice plants in field scale as well as in pots at doses of 0.2, 0.5, 1, and 2% of soil weight to see the positive results in all cases.

4.2.4 Fertigation

Huang et al. (2018) conducted hydroponic experiments by suspending two varieties of rice seedlings (T705 and X24) in various nanoparticle suspensions of high-quality graphene oxide, multiple layer graphene oxide, 20 nm and 40 nm hydroxyapatite, nano Fe$_3$O$_4$ and nZVI in Hoagland nutrient solutions. They found varied results in respect to As alleviation, plant growth and stress response. In some cases, As content in the above ground and underground part of rice cultivars were found to increase after treatment, while in some cases As alleviation was observed. The As alleviation efficiency differed from cultivar to cultivar. X24 variety only showed ameliorative effects at low As concentrations. Wang et al. (2018) found out little effects of As mitigation with CeO$_2$ nanoparticles but with ZnO nanoparticles, the total As in rice roots was found out to be 72 and 68% lower than the As^{+3} and As^{+5} treatment in controls. Wang et al. (2019) treated rice seedlings with Cu nanoparticles, which resulted in a decrease of 23% As^{+3} and 54% As^{+5} in roots and 45% As^{+3} and 55% As^{+5} in shoots compared to the respective controls.

A cumulative comparison has been drawn between the various methods of application of nanofertilizers in terms of their percentage As alleviation capability in plants (Ahmad et al. 2020, Baragano et al. 2019, Gil-Diaz et al. 2014, 2016, 2017, 2019, Liu et al. 2014, 2019, Ma et al. 2020, Tripathi et al. 2016, Wang et al. 2018, 2019, Wu et al. 2020). The medians for foliar, seed priming, soil application and fertigation lie at 39.15, 20.60 and 75.8 and 7.28%, respectively. Thus, in the conducted experiments, soil application has achieved greater levels of alleviation than the other methods. But since there have not been any perfect comparison between the methods, no conclusion can be made about their superiority. Thus, a new avenue can open up if all the methods are compared under the same conditions keeping all the parameters and constraints constant. In this way, the best method of alleviating As stress and consequently other heavy metal stress can be found out.

4.3 Efficacy vs. type of nanofertilizer

The extent of As alleviation also depends on the type of nanofertilizer used (Fig. 4). The use of ZnO in alleviating As stress in plants has been extensive (Ahmad et al. 2020, Ma et al. 2020, Wang et al. 2018, Wu et al. 2020). Nano zero valent iron, however, gave the better mitigation efficiency in barley plants (Gil-Diaz et al. 2016) (Fig. 4). nZVI has been extensively used by Gil-Diaz et al. (2014, 2016, 2017, 2019) for many field studies to investigate the immobilization of As in soil. Huang et al. (2018) also used Fe NPs, hydroxyapatite and graphene oxides with moderate As alleviation efficiency in rice seedlings. Arsenic alleviation percentage varied with varying concentrations and also the chosen rice cultivar. However, the

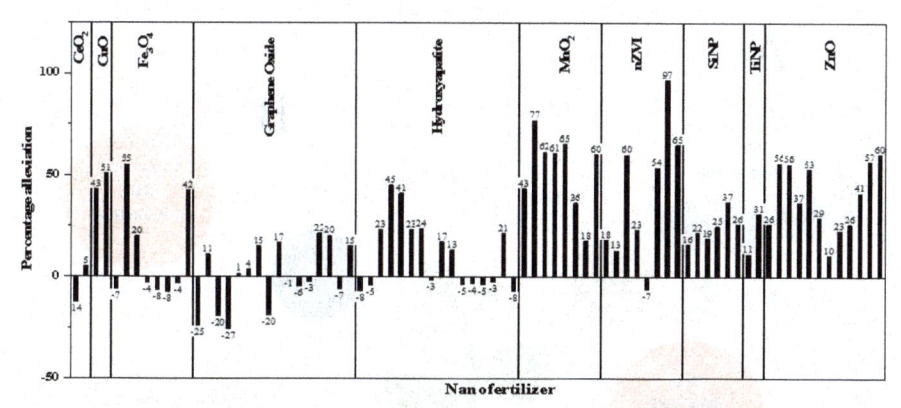

Fig. 4.   Percentage alleviation of As in plants by different types of nanofertilizers.

As concentration in the rice roots increased after the treatments. High As alleviation efficiency has also been recorded for CuO nanofertilizers (Wang et al. 2019). Still these fertilizers do not always show the same efficacy as they cannot alleviate all the As species similarly. Silicon and silica nanoparticles have also shown a steady mitigation capability at a range of 19–38% (Cui et al. 2019, Liu et al. 2014, Pan et al. 2019, Tripathi et al. 2016). However, to the best of our knowledge, no arsenic alleviation experiments have been conducted by silica or silicon nanoparticles by the method of soil application. So, there lies one more opportunity to venture into. In fact, all the nanofertilizers that are mentioned here have not been implemented by all the three methods of application. If such studies are conducted, then the best method in arsenic mitigation can be found out.

5. Mechanism of arsenic alleviation by nanofertilizers in plants

Different nanofertilizers follow different mechanisms in alleviating As from soil. Plants under As stress undergo oxidative deterioration of lipids due to generation of reactive oxygen species (ROS), which through a series of chain reactions carry out the destruction process. As^{+5} is analogous to inorganic phosphate and hence offers competition for uptake through the transporter proteins intended to transport phosphate. As^{+5} is readily converted to As^{+3} by the plants using Arsenic Reductase or GSH (Finnegan and Chen 2012). The phosphate required by the plants in several biochemical reactions get replaced by As. Sometimes, iron plaque is formed due to radial transfer of oxygen from root to soil. This plaque adsorbs and co-precipitates As on the root surface, thus hindering the bioavailability of the heavy metal (Shri et al. 2019).

Triggering of the antioxidant defence mechanism to be more active is the primary pathway for As mitigation by Si NPs, ZnO NPs and Ti NPs as reported in several studies (Ahmad et al. 2020, Cui et al. 2019, Katiyar et al. 2020, Liu et al. 2014, Ma et al. 2020, Pan et al. 2019, Tripathi et al. 2016, Wang et al. 2018, Wu et al. 2020) (Fig. 5). The nanofertilizers mainly proliferate antioxidant enzymes and other non-enzymatic antioxidants like superoxide dismutase (SOD), peroxidase (POD),

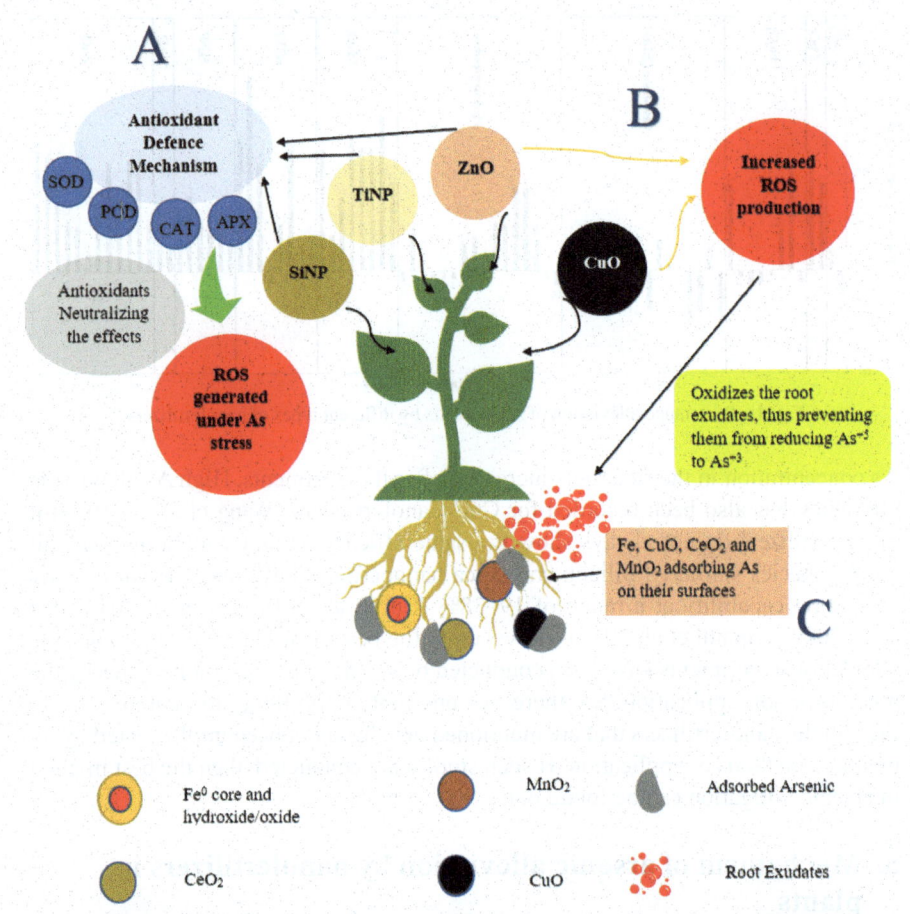

Fig. 5: Mechanisms of various nanofertilizers alleviating arsenic stress in plants. **A.** NF increasing the production of antioxidants, which is neutralizing the ROS; **B.** NF increasing the production of ROS, which hampers the reduction of As^{+5} to As^{+3}; **C.** Fe, CuO, CeO_2 and MnO_2 adsorb As on their surfaces.

ascorbate peroxidase (APX), catalase (CAT), glutathione, ascorbate, tocopherols, etc., which decreases the adverse effects of ROS such as highly reactive O_2^{*-} to H_2O_2. In some cases, ZnO and CuO NPs proliferate the production of ROS, which oxidize the root exudates, thus hampering the reduction of As^{+5} to As^{+3}. This results in a low concentration of As^{+3} than As^{+5}. These ROS also have reduced the effects of catalytic enzyme Arsenate Reductase (AR), catalyzing the conversion of As^{+5} to As^{+3} (Wang et al. 2018, 2019).

Nanofertilizers like Fe, CuO, CeO_2 and MnO_2 adsorb As on their surfaces (Gil-Diaz et al. 2014, 2016, 2017, 2019, Li et al. 2019, Wang et al. 2018, 2019) (Fig. 5). The nZVI has a core-shell structure with a hydroxide or an oxide shell. Arsenate (As^{5+}) adsorption on iron oxides in aerobic soil takes place, along with surface complexation of As with the oxide or hydroxide of nZVI that makes As^{5+} immobile. However, if pH is increased, adsorption decreases and the concentration of more mobile form of arsenic, i.e., arsenite (As^{3+}), increases. The immobilization mechanisms as documented by Vítková et al. (2018) are sorption and co-precipitation of As on added iron particles, adsorption on secondary iron oxides and hydroxides and sorption on Mn oxides and hydroxides by complex formation.

CeO_2 NPs inhibited the transportation of As^{+3} from roots to shoots by adsorption on its surface. The oxidation of As^{+3} to As^{+5} on the surface of the NPs resulted in relatively higher concentration of As^{+5} in shoots than As^{+3} (Wang et al. 2018). In CuO NPs, this adsorption may increase the hydrodynamic size of CuO NPS, thus resulting in precipitation of NP-As complex (Wang et al. 2019). As reported by Li et al. (2019), at a pH range of 4–8.2, MnO_2 nanoparticles being a strong oxidant can convert As^{+3} to As^{+5}. This leads to greater adsorption of As on iron oxyhydroxides, thus decreasing the bioavailable As. Iron and Mn plaque formations on rice roots are also responsible for decreasing the bioavailability of As. Addition of MnO_2 NPs oxidized Fe^{+2} to Fe^{+3}, which resulted in the formation of Fe^{+3} hydroxides. These Fe^{+3} hydroxides are efficient sorbents for As^{+3}.

6. Extent of commercialization

To be commercially viable, the nanofertilizers yielding positive results must undergo a proper cost-benefit analysis and environmental impact assessment. It is the duty of the industries to compare the cost-benefit analysis of conventional fertilizers and nanofertilizers. The production cost of nanofertilizers may be more compared to their conventional counterparts, but if they are required in lesser quantities then they can be beneficial to both the environment and economy. However, before commercialization of any product the public opinion and acceptance must be taken into account. The awareness of the farmers about the safe usage of nanofertilizers as well as the acceptance and willingness to pay for nanoagriculture among consumers are the decisive factors for commercialization (Younis et al. 2021). Knowledge about risks associated with nanotechnology results in development of negative opinions among people (Vandermoere et al. 2011). There is a dearth of public understanding concerning the health and environmental consequences of nanofertilizer use. Comparative field-scale cost-benefit analyses between traditional fertilizers and nanofertilizers to demonstrate the cost-effectiveness of nanofertilizer application are needed for creating awareness among the farmers.

But before all these, extensive pilot scale and field level studies must be conducted to check the effectiveness of the nanofertilizers and their remediation capacity. Then only the industries can seriously take up the products for commercial manufacture. Bardos et al. (2018) reported that nZVI has been deployed commercially in as many as 100 fields across the world after conducting pilot studies. The nanoremediation of As along with the other heavy metals in a pilot scale study was conducted in Spain (Bardos et al. 2018). However, nanofertilizers specifically for alleviating As stress or even heavy metal stress are not commercialized. The commercial nanofertilizer used was NANOFER STAR provided by NANO IRON, s.r.o. Gil-Diaz et al. (2019) conducted the field studies with commercial nZVI to investigate the decrease in leachability of As from soil after treatments. Baragano et al. (2019) also conducted field studies with commercial nZVI and Goethite nanospheres to investigate the immobilizing of As in polluted soil. Few field studies were also conducted with different nanofertilizers to obtain positive result on As remediation better plant growth (Li et al. 2019, Pan et al. 2019). These field studies, if deployed in pilot scale studies and eventually in full scale-up processes, can be fruitful in mitigating arsenic stress. The field scale studies mentioned above have been quite successful and hence need more attention. Here, we have tried to present a figure depicting the success of the field scale experiments.

In the past few years, there have been small developments in commercialization of nanofertilization in India and across the world. Dimkpa et al. (2017) listed a few names of the approved nanofertilizers which are being considered to be imported in Myanmar. In that list, two of the fertilizers are from India. Prasad et al. (2017) and Pitambara et al. (2019) also listed a few commercial nanofertilizers. Some initiatives taken in India are listed below:

1. In 2014, Indian Agriculture Research Institute (IARI) signed a Memorandum of Understanding (MoU) with Secunderabad-based Prathista Industries Limited, for commercialization of biosynthesized nanofertilizer developed by Central Arid Zone Research Institute under the Indian Agriculture Research Institute (IARI).

2. 4G nanofertilizers and nano micronutrients with protein-lacto-gluconates manufactured through biological processes were launched in August 2014 by Indian Council of Agriculture Research (ICAR) and Pratistha Industries Ltd. According to the research, the nanonutrients are required in ppm levels compared to kilograms of conventional fertilizers. The safety of human, livestock and environment was also ensured.

3. In August 2015, as a part of Make in India Initiative, a Nanocellulose Pilot Plant was inaugurated by Indian Council of Agricultural Research-Central Institute for Research on Cotton Technology (ICAR-CIRCOT).

4. In September 2019, ICAR-CIRCOT, Mumbai, and Rashtriya Chemicals & Fertilizers Limited (RCF) signed a Memorandum of Understanding (MoU) in September, 2019 for the "Development of Nano-formulations for Fertilizer Applications" with an emphasis on technology transfer and commercialization of nanofertilizers.

5. On November 2019, Indian Farmers Fertilizer Cooperative introduced nano nitrogen, nano zinc and nano copper for use in on-field trials with an objective to decrease the use of chemical fertilizers. These products were indigenously developed by the research and development team at their Kalol Unit. The nanofertilizers are expected to deliver nutrients to the plants and increase the crop production by 15–30%. Farmers were also invited at the introduction event to acclimatize them with the products.

7. Future prospects and conclusion

The future of the application of nanofertilizers depends upon balancing their pros and cons. In this assessment, the best As alleviation percentage was shown by soil application of nZVI (Fig. 4). Thus, a process may be optimized through rigorous laboratory scale experiments to device a methodology that will alleviate As consistently. The best combination of concentration, method of application and type of nanofertilizer must be chosen to optimize a process. MnO_2, ZnO and CuO have also given promising results. But to get the best results, we must perform permutation and combination between the deciding parameters. There is also an opportunity of exploring the effects to nanoparticles, which have not been applied in As mitigation. Most of the literatures consulted in this assessment have undertaken laboratory-based experiments only. The laboratory experiments which yielded optimistic results can be taken up for field studies, where again proper optimization will be necessary.

Again, to make the nanofertilizers a potent instrument for sustainable agriculture, more numbers of pilot scale processes and scale-up processes must be undertaken. Proper risk assessment, life cycle assessment and environmental impact assessment must be done before implementation of the fertilizers in large scale manufacturing of crops. Alongside these new methods, nanoparticles must also be developed by the research institutes as well as the research and development sector of the industries. There must be sharing of knowledge between the industries and the research institutes for betterment of processes. The farmers must be trained extensively in using these new fertilizers and also be discouraged in using the conventional ones. If all these are taken into consideration, then nanofertilizers have a bright future ahead. There are also new avenues of research work being opened up as the nanofertilizers are seriously considered as a tool for sustainable agriculture. Extensive research works can be done to find out the best suitable nanofertilizer, the best method of application and the perfect concentration required for individual plants. All the parameters must be assessed properly and optimized so that it can be exploited to provide food for the ever-increasing population of the world.

References

Abdel-Aziz, H.M.M., Hasaneen, M.N.A., and Omer, A.M. 2019. Impact of engineered nanomaterials either alone or loaded with NPK on growth and productivity of French bean plants: Seed priming vs foliar application. South African Journal of Botany, 125: 102–108.

Ahmad, P., Alyemeni, M.N., Al-Huqail, A.A., Alqahtani, M.A., Wijaya, L., Ashraf, M., Kaya, C., and Bajguz, A. 2020. Zinc oxide nanoparticles application alleviates arsenic (As) toxicity in soybean plants by restricting the uptake of as and modulating key biochemical attributes, antioxidant enzymes, ascorbate-glutathione cycle and glyoxalase system. Plants, 9(7): 825.

Ahmed, N.R., Sheha, A.M., and Hamd-Alla, W.A. 2019. Effect of intercropping of some legume forage crops with maize under levels of mineral NPK and nano NPK fertilizer. Journal of Plant Production, 10(10): 833–842.

Akhtar, I., Iqbal, Z., and Saddiqe, Z. 2020. Nanotechnology in pest management. pp. 69–83. *In*: Javad, S. [ed.]. Nanoagronomy. Springer, Cham. https://doi.org/10.1007/978-3-030-41275-3_5.

Ansari, M., Shahzadi, K., and Ahmed, S. 2020. Nanotechnology: A breakthrough in agronomy. pp. 1–21. *In*: Javad, S. [ed.]. Nanoagronomy. Springer, Cham. https://doi.org/10.1007/978-3-030-41275-3_1.

Babaei, K., Seyed Sharifi, R., Pirzad, A., and Khalilzadeh, R. 2017. Effects of bio fertilizer and nano Zn-Fe oxide on physiological traits, antioxidant enzymes activity and yield of wheat (*Triticum aestivum* L.) under salinity stress. Journal of Plant Interactions, 12(1): 381–389. https://doi.org/10.1080/17429145.2017.1371798.

Baragaño, D., Alonso, J., Gallego, J.R., Lobo, M.C., and Gil-Díaz, M. 2020. Zero valent iron and goethite nanoparticles as new promising remediation techniques for As-polluted soils. Chemosphere, 238: 124624.

Bardos, P., Merly, C., Kvapil, P., and Koschitzky, H.P. 2018. Status of nanoremediation and its potential for future deployment: Risk-benefit and benchmarking appraisals. Remediation Journal, 28(3): 43–56. https://doi.org/10.1002/rem.21559.

Bhumbla, D.K., and Keefer, R.F. 1994. Arsenic mobilization and bioavailability in soils. Book: Advances in Environmental Science and Technology, Ed. Nriagu J. O. John Wiley & Sons, Inc., New York (United States), 26: 51–82.

Butt, B.Z., and Naseer, I. 2020. Nanofertilizers. pp. 125–152. *In*: Javad, S. [ed.]. Nanoagronomy. Springer, Cham. https://doi.org/10.1007/978-3-030-41275-3_8.

Cai, D., Wu, Z., Jiang, J., Wu, Y., Feng, H., Brown, I.G., Chu, P., and Yu, Z. 2014. Controlling nitrogen migration through micro-nano networks. Scientific Reports, 4(1): 1–8.

Campbell, B.M., Thornton, P., Zougmoré, R., Van Asten, P., and Lipper, L. 2014. Sustainable intensification: What is its role in climate smart agriculture? Current Opinion in Environmental Sustainability, 8: 39–43.

Chhipa, H. 2017. Nanofertilizers and nanopesticides for agriculture. Environmental Chemistry Letters, 15(1): 15–22. https://doi.org/10.1007/s10311-016-0600-4.

Chatterjee, A., Das, D., Mandal, B.K., Chowdhury, T.R., Samanta, G., and Chakraborti, D. 1995. Arsenic in ground water in six districts of West Bengal, India: The biggest arsenic calamity in the world. Part I. Arsenic species in drinking water and urine of the affected people. Analyst, 120: 643–650.

Chowdhury, N.R., Das, A., Mukherjee, M., Swain, S., Joardar, M., De, A., Mridha, D., and Roychowdhury, T. 2020. Monsoonal paddy cultivation with phase-wise arsenic distribution in exposed and control sites of West Bengal, alongside its assimilation in rice grain. Journal of Hazardous Materials, 400: 123206.

Chowdhury, U.K., Biswas, B.K., Chowdhury, T.R., Samanta, G., Mandal, B.K., Basu, G.K., Chanda, C.R., Lodh, D., Saha, K.C., Mukherjee, S.K., and Roy, S. 2000. Groundwater arsenic contamination in Bangladesh and West Bengal, India. Environmental Health Perspectives, 108(5): 393–397.

Cui, H.X., Sun, C.J., Liu. Q., Jiang, J., and Gu, W. 2010. Applications of nanotechnology in agrochemical formulation, perspectives, challenges and strategies. International conference on Nanoagri, Sao pedro, Brazil, 20–25.

Cui, J., Li, Y., Jin, Q., and Li, F. 2020. Silica nanoparticles inhibit arsenic uptake into rice suspension cells via improving pectin synthesis and the mechanical force of the cell wall. Environmental Science: Nano, 7(1): 162–171.

Dimkpa, C.O., and Bindraban, P.S. 2017. Nanofertilizers: new products for the industry? Journal of Agricultural and Food Chemistry, 66(26): 6462–6473.

Duhan, J.S., Kumar, R., Kumar, N., Kaur, P., Nehra, K., and Duhan, S. 2017. Nanotechnology: The new perspective in precision agriculture. Biotechnology Reports, 15: 11–23.

El-Ghamry, A., Mosa, A.A., Alshaal, T., and El-Ramady, H. 2018. Nanofertilizers vs. biofertilizers: new insights. Environment, Biodiversity and Soil Security, 2: 51–72.

El-Henawy, A., El-Sheikh, I., Hassan, A., Madein, A., El-Sheikh, A., El-Yamany, A., Radwan, A., Mohamed, F., Khamees, M., Ramadan, M., Abdelhamid, M., Khaled, H., El-Faramawy, H., Ayoub, Y., Youssef, S., and Faizy, S. 2018. Response of cultivated broccoli and red cabbage crops to mineral, organic and nano-fertilizers. Environment, Biodiversity and Soil Security, 2: 1–25.

Finnegan, P.M., and Chen, W. 2012. Arsenic toxicity: the effects on plant metabolism. Frontiers in Physiology, 3: 182. https://doi.org/10.3389/fphys.2012.00182.

Gil-Díaz, M., Alonso, J., Rodríguez-Valdés, E., Pinilla, P., and Lobo, M.C. 2014. Reducing the mobility of arsenic in brownfield soil using stabilised zero-valent iron nanoparticles. Journal of Environmental Science and Health, Part A, 49(12): 1361–1369. https://doi.org/10.1080/10934529.2014.928248.

Gil-Díaz, M., Diez-Pascual, S., González, A., Alonso, J., Rodríguez-Valdés, E., Gallego, J.R., and Lobo, M.C. 2016. A nanoremediation strategy for the recovery of an As-polluted soil. Chemosphere, 149: 137–145.

Gil-Díaz, M., Alonso, J., Rodríguez-Valdés, E., Gallego, J.R., and Lobo, M.C. 2017. Comparing different commercial zero valent iron nanoparticles to immobilize As and Hg in brownfield soil. Science of the Total Environment, 584: 1324–1332.

Gil-Díaz, M., Rodríguez-Valdés, E., Alonso, J., Baragaño, D., Gallego, J.R., and Lobo, M.C. 2019. Nanoremediation and long-term monitoring of brownfield soil highly polluted with As and Hg. Science of the Total Environment, 675: 165–175.

Hughes, M.F. 2002. Arsenic toxicity and potential mechanisms of action. Toxicology Letters, 133(1): 1–16.

Hussain, A., Ali, S., Rizwan, M., Rehman, M.Z. ur, Qayyum, M.F., Wang, H., and Rinklebe, J. 2019. Responses of wheat (*Triticum aestivum*) plants grown in a Cd contaminated soil to the application of iron oxide nanoparticles. Ecotoxicol. Environ. Saf. 173: 156–164. https://doi.org/10.1016/J. ECOENV.2019.01.118.

Janmohammadi, M., Pornour, N., Javanmard, A., and Sabaghnia, N. 2016. Effects of bio-organic, conventional and nanofertilizers on growth, yield and quality of potato in cold steppe/Bioorganinių, tradicinių ir nanotrąšų poveikis bulvių augimui, derliui ir kokybei šaltojoje stepėje. Botanica, 22(2): 133–144.

Joardar, M., Das, A., Mridha, D., De, A., Chowdhury, N.R., and Roychowdhury, T. 2020. Evaluation of acute and chronic arsenic exposure on school children from exposed and apparently control areas of West Bengal, India. Exposure and Health. https://doi.org/10.1007/s12403-020-00360-x.

Kafkafi, U., and Kant, S. 2013. Fertigation. Encyclopaedia of Soils in the Environment, 1–9.

Kah, M., Kookana, R.S., Gogos, A., and Bucheli, T.D. 2018. A critical evaluation of nanopesticides and nanofertilizers against their conventional analogues. Nature Nanotechnology, 13(8): 677–684. https://doi.org/10.1038/s41565-018-0131-1.

Kalwani, M., Chakdar, H., Srivastava, A., Pabbi, S., and Shukla, P. 2022. Effects of nanofertilizers on soil and plant-associated microbial communities: Emerging trends and perspectives. Chemosphere, 287: 132107. https://doi.org/10.1016/J.CHEMOSPHERE.2021.132107.

Kant, S., and Kafkafi, U. 2013. Fertigation. Reference Module in Earth Systems and Environmental Sciences, pp. 1–10. DOI: 10.1016/b978-0-12-409548-9.05161-7.

Katiyar, P., Yadu, B., Korram, J., Satnami, M.L., Kumar, M., and Keshavkant, S. 2020. Titanium nanoparticles attenuates arsenic toxicity by up-regulating expressions of defensive genes in *Vigna radiata* L. Journal of Environmental Sciences, 92: 18–27.

Kaushik, S., and Djiwanti, S.R. 2019. Nanofertilizers: smart delivery of plant nutrients. pp. 59–72. *In*: Panpatte, D. G., and Jhala, Y. K. [ed.]. Nanotechnology for Agriculture: Crop Production & Protection. Springer, Singapore.

Kumari, R., and Singh, D.P. 2019. Nano-biofertilizer: An emerging eco-friendly approach for sustainable agriculture. Proceedings of the National Academy of Sciences, India Section B: Biological Sciences, 1–9. https://doi.org/10.1007/s40011-019-01133-6.

Li, B., Zhou, S., Wei, D., Long, J., Peng, L., Tie, B., Williams, P., and Lei, M. 2019. Mitigating arsenic accumulation in rice (*Oryza sativa* L.) from typical arsenic contaminated paddy soil of southern China using nanostructured α-MnO2: Pot experiment and field application. Science of the Total Environment, 650: 546–556.

Liu, C., Wei, L., Zhang, S., Xu, X., and Li, F. 2014. Effects of nanoscale silica sol foliar application on arsenic uptake, distribution and oxidative damage defense in rice (*Oryza sativa* L.) under arsenic stress. Rsc Advances, 4(100): 57227–57234.

Liu, J., Wolfe, K., Potter, P.M., and Cobb, G.P. 2019. Distribution and speciation of copper and arsenic in rice plants (Oryza sativa japonica 'Koshihikari') treated with copper oxide nanoparticles and arsenic during a life cycle. Environmental Science & Technology, 53(9): 4988–4996.

Liu, R., and Lal, R. 2014. Synthetic apatite nanoparticles as a phosphorus fertilizer for soybean (Glycine max). Scientific Reports, 4: 5686. https://doi.org/10.1038/srep05686.

Liu, R., and Lal, R. 2015. Potentials of engineered nanoparticles as fertilizers for increasing agronomic productions. Science of the Total Environment, 514: 131–139.

Loomis, R., and D. Connor. 1992. Crop Ecology: Productivity and Management in Agricultural Systems. Cambridge University Press.

Ma, X., Sharifan, H., Dou, F., and Sun, W. 2020. Simultaneous reduction of arsenic (As) and cadmium (Cd) accumulation in rice by zinc oxide nanoparticles. Chemical Engineering Journal, 384: 123802.

Ma, Z., Yue, Y., Feng, M., Li, Y., Ma, X., Zhao, X., and Wang, S. 2019. Mitigation of ammonia volatilization and nitrate leaching via loss control urea triggered H-bond forces. Scientific Reports, 9(1): 1–9. https://doi.org/10.1038/s41598-019-51566-2.

Mathur, P., and Roy, S. 2020. Nanosilica facilitates silica uptake, growth and stress tolerance in plants. Plant Physiol. Biochem. 157: 114–127. https://doi.org/10.1016/J.PLAPHY.2020.10.011.

Mir, S., Sirousmehr, A., and Shirmohammadi, E. 2015. Effect of nano and biological fertilizers on carbohydrate and chlorophyll content of forage sorghum (Speedfeed hybrid). International Journal of Biosciences, 6(4): 157–164.

Mridha, D., Paul, I., De, A., Ray, I., Das, A., Joardar, M., Chowdhury, N.R., Bhadoria, P.B.S., and Roychowdhury, T. 2021. Rice seed (IR64) priming with potassium humate for improvement of seed germination, seedling growth and antioxidant defense system under arsenic stress. Ecotoxicol. Environ. Saf. 219: 112313. https://doi.org/10.1016/j.ecoenv.2021.112313.

Najafi Disfani, M., Mikhak, A., Kassaee, M.Z., and Maghari, A. 2017. Effects of nano Fe/SiO2 fertilizers on germination and growth of barley and maize. Archives of Agronomy and Soil Science, 63(6): 817–826. https://doi.org/10.1080/03650340.2016.1239016.

Ogunkunle, C.O., Odulaja, D.A., Akande, F.O., Varun, M., Vishwakarma, V., and Fatoba, P.O. 2020. Cadmium toxicity in cowpea plant: Effect of foliar intervention of nano-TiO2 on tissue Cd bioaccumulation, stress enzymes and potential dietary health risk. J. Biotechnol. 310: 54–61. https://doi.org/10.1016/J.JBIOTEC.2020.01.009.

Pan, D., Liu, C., Yu, H., and Li, F. 2019. A paddy field study of arsenic and cadmium pollution control by using iron-modified biochar and silica sol together. Environmental Science and Pollution Research, 26(24): 24979–24987. https://doi.org/10.1007/s11356-019-05381-x.

Paramo, L.A., Feregrino-Pérez, A.A., Guevara, R., Mendoza, S., and Esquivel, K. 2020. Nanoparticles in agroindustry: applications, toxicity, challenges, and trends. Nanomater. 10: 1654 10, 1654. https://doi.org/10.3390/NANO10091654.

Pitambara, A., and Shukla, Y.M. 2019. Nanofertilizers: A Recent Approach in Crop Production. Nanotechnology for Agriculture: Crop Production & Protection, 25. https://doi.org/10.1007/978-981-32-9374-8_2.

Prasad, R., Kumar, V., and Prasad, K.S. 2014. Nanotechnology in sustainable agriculture: present concerns and future aspects. African Journal of Biotechnology, 13(6): 705–713.

Priyanka, N., Geetha, N., Manish, T., Sahi, S.V., and Venkatachalam, P. 2021. Zinc oxide nanocatalyst mediates cadmium and lead toxicity tolerance mechanism by differential regulation of photosynthetic machinery and antioxidant enzymes level in cotton seedlings. Toxicol. Reports 8: 295–302. https://doi.org/10.1016/J.TOXREP.2021.01.016.

Pullagurala, V.L.R., Adisa, I.O., Rawat, S., Kim, B., Barrios, A.C., Medina-Velo, I.A., Hernandez-Viezcas, J.A., Peralta-Videa, J.R., and Gardea Torresdey, J.L. 2018. Finding the conditions for the beneficial use of ZnO nanoparticles towards plants-A review. Environmental Pollution, 241: 1175–1181.

Ray, I., Mridha, D., and Roychowdhury, T. 2021. Waste derived amendments and their efficacy in mitigation of arsenic contamination in soil and plant systems: A review. Environ. Technol. Innov. 24: 101976. https://doi.org/10.1016/J.ETI.2021.101976.

Rizwan, M., Ali, S., Ali, B., Adrees, M., Arshad, M., Hussain, A., Zia ur Rehman, M., and Waris, A.A. 2019. Zinc and iron oxide nanoparticles improved the plant growth and reduced the oxidative stress and cadmium concentration in wheat. Chemosphere 214: 269–277. https://doi.org/10.1016/J.CHEMOSPHERE.2018.09.120.

Rossi, L., Fedenia, L.N., Sharifan, H., Ma, X., and Lombardini, L. 2019. Effects of foliar application of zinc sulfate and zinc nanoparticles in coffee (*Coffea arabica* L.) plants. Plant Physiology and Biochemistry, 135: 160–166.

Roychowdhury, T., Tokunaga, H., Uchino, T., and Ando, M. 2005. Effect of arsenic-contaminated irrigation water on agricultural land soil and plants in West Bengal, India. Chemosphere, 58(6): 799–810.

Sangeetha, J., Mundaragi, A., Thangadurai, D., Maxim, S.S., Pandhari, R.M., and Alabhai, J.M. 2019. Nanobiotechnology for agricultural productivity, food security and environmental sustainability. pp. 1–23. *In*: Panpatte, D., and Jhala, Y. [eds.]. Nanotechnology for Agriculture: Crop Production & Protection. Springer, Singapore. http://doi-org-443.webvpn.fjmu.edu.cn/10.1007/978-981-32-9374-8_1.

Sasson, Y., Levy-Ruso, G., Toledano, O., and Ishaaya, I. 2007. Nanosuspensions: emerging novel agrochemical formulations. pp. 1–39. *In*: Ishaaya, I., Nauen, R., and Horowitz, A.R. [eds.]. Insecticides Design Using Advanced Technologies. Springer, Berlin, Heidelberg. https://doi.org/10.1007/978-3-540-46907-0_1.

Sekhon, B. 2014. Nanotechnology in agri-food production: An overview. Nanotechnology, Science and Applications, 7: 31–53. 10.2147/NSA.S39406.

Sharma, D., Sharma, J., and Dhuriya, Y.K. 2019. Nanotechnology: a novel strategy against plant pathogens. pp. 245–262. *In*: Panpatte, D., and Jhala, Y. [eds.]. Nanotechnology for Agriculture: Crop Production & Protection. Springer, Singapore. https://doi.org/10.1007/978-981-32-9374-8_9.

Shri, M., Singh, P.K., Kidwai, M., Gautam, N., Dubey, S., Verma, G., and Chakrabarty, D. 2019. Recent advances in arsenic metabolism in plants: current status, challenges and highlighted biotechnological intervention to reduce grain arsenic in rice. Metallomics, 11(3): 519–532.

Siddiqui, M.H., Al-Whaibi, M.H., Firoz, M., and Al-Khaishany, M.Y. 2015. Role of nanoparticles in plants. pp. 19–35. *In*: Siddiqui, M.H. [ed.]. Nanotechnology and Plant Sciences, Springer, Cham.

Solanki, P., Bhargava, A., Chhipa, H., Jain, N., and Panwar, J. 2015. Nano-fertilizers and their smart delivery system. pp. 81–101. *In*: Rai, M., Ribeiro, C., Mattaso, L., and Duran, N. [eds.]. Nanotechnology in Food and Agriculture. Springer, Cham.

Swaminathan, M.S. 1969. Food Farming and Agriculture- April, I(10): 7–9.

Thomas, D.J., Styblo, M., and Lin, S. 2001. The cellular metabolism and systemic toxicity of arsenic. Toxicology and Applied Pharmacology, 176(2): 127–144.

Thomas, L., and Singh, I. 2019. Microbial biofertilizers: types and applications. pp. 1–19. *In*: Giri, B., Prasad, R., Wu, Q.S., Varma, A. [eds.]. Biofertilizers for Sustainable Agriculture and Environment. Springer, Cham.

Tripathi, D.K., Singh, S., Singh, V.P., Prasad, S.M., Chauhan, D.K., and Dubey, N.K. 2016. Silicon nanoparticles more efficiently alleviate arsenate toxicity than silicon in maize cultiver and hybrid differing in arsenate tolerance. Frontiers in Environmental Science, 4: 46.

Vandermoere, F., Blanchemanche, S., Bieberstein, A., Marette, S., and Roosen, J. 2011. The public understanding of nanotechnology in the food domain: The hidden role of views on science, technology, and nature. Public Underst. Sci. 20: 195–206. https://doi.org/10.1177/0963662509350139.

Vítková, M., Puschenreiter, M., and Komárek, M. 2018. Effect of nano zero-valent iron application on As, Cd, Pb, and Zn availability in the rhizosphere of metal (loid) contaminated soils. Chemosphere, 200: 217–226.

Wang, X., Sun, W., Zhang, S., Sharifan, H., and Ma, X. 2018. Elucidating the effects of cerium oxide nanoparticles and zinc oxide nanoparticles on arsenic uptake and speciation in rice (*Oryza sativa*) in a hydroponic system. Environmental Science & Technology, 52(17): 10040–10047.

Wang, X., Sun, W., and Ma, X. 2019. Differential impacts of copper oxide nanoparticles and Copper (II) ions on the uptake and accumulation of arsenic in rice (*Oryza sativa*). Environmental Pollution, 252: 967–973.

WHO. 1981. Environmental Health Criteria 18, Geneva.

Wu, F., Fang, Q., Yan, S., Pan, L., Tang, X., and Ye, W. 2020. Effects of zinc oxide nanoparticles on arsenic stress in rice (*Oryza sativa* L.): germination, early growth, and arsenic uptake. Environmental Science and Pollution Research, 27: 26974–26981.

Younis, S.A., Kim, K.H., Shaheen, S.M., Antoniadis, V., Tsang, Y.F., Rinklebe, J., Deep, A., and Brown, R.J.C. 2021. Advancements of nanotechnologies in crop promotion and soil fertility: Benefits, life cycle assessment, and legislation policies. Renew. Sustain. Energy Rev. 152: 111686. https://doi.org/10.1016/J.RSER.2021.111686.

Zhou, P., Chen, Y., Lu, Q., Qin, H., Ou, H., He, B., and Ye, J. 2018. Cellular metabolism network of *Bacillus thuringiensis* related to erythromycin stress and degradation. Ecotoxicology and Environmental Safety, 160: 328–341.

Zulfiqar, F., Navarro, M., Ashraf, M., Akram, N.A., and Munné-Bosch, S. 2019. Nanofertilizer use for sustainable agriculture: advantages and limitations. Plant Science, 289: 110270.

Zuverza-Mena, N., Martínez-Fernández, D., Du, W., Hernandez-Viezcas, J.A., Bonilla-Bird, N., López-Moreno, M.L., Komárek, M., Peralta-Videa, J.R., and Gardea-Torresdey, J.L. 2017. Exposure of engineered nanomaterials to plants: Insights into the physiological and biochemical responses-A review. Plant Physiol. Biochem. 110: 236–264. https://doi.org/10.1016/J.PLAPHY.2016.05.037.

5

Foam Glasses from Glasses of Fluorescent Lamps Waste

Isaac dos S. Nunes,[1,2,] Venina dos Santos[1] and Rosmary N. Brandalise[1]*

1. Introduction

Waste recycling is a very important resource to reuse materials and reduce extraction of raw materials from non-renewable sources (Rameshkumar et al. 2020, Ting et al. 2021). Only half of worldwide solid waste generated has potential to be recycled, which includes waste that consists of aluminum and glass (Colombo et al. 2003). Glass wastes has received increasing attention in the recent years (Chong et al. 2021).

Glasses are materials composed of non-crystalline silicates and other oxides, such as calcium oxide (CaO), sodium oxide (NaO), potassium oxide (K_2O) and aluminum oxide (Al_2O_3). The two main positive characteristics of these materials are their optical transparency (Zhang et al. 2022) and their relative ease of manufacture (Carter and Norton 2007). Silicates are materials composed mainly of silicon and oxygen, components of most soils, clays, rocks and sands (Moulson and Hebert 2003).

Vitreous waste can be 100% recycled without any loss in properties and reintroduced in production processes, reducing the use of new raw materials and saving energy in the production of new glasses (Krivtsov et al. 2004, Furlani and Tonello 2010). Millions of tons of glass wastes are produced but only marginal is fed back into recycling (Robert et al. 2021).

In spite of being recyclable, some vitreous wastes have limited applications due to their origin since they may include contaminants that could constitute a hazard

[1] Science and Technology Center, University of Caxias do Sul, C.P. 1352, 95070-560, Caxias do Sul – RS – Brazil, Emails: vsantos2@ucs.br; rnbranda@ucs.br
[2] Engineerig and Computational Science Department, Regional and Integrated University of Upper Uruguai and Missions, Av. Universidade das Missões, 464, 98802-470 – Santo Ângelo – RS - Brazil.
* Corresponding author: isaac.eq@san.uri.br

when the new products are reused. An example for residue is mercury vapor of fluorescent lamps, which restricts recycling due to the potential contamination of the glass (Furlani and Tonello 2010, Santos et al. 2010, Hu and Cheng 2012, Rey-Haap and Gallardo 2012, 2013, Pitarch et al. 2021).

In Brazil, use of fluorescent lamps soared as from 2001, pushed by government subsidies (Srinivasan 2019), this being due to the energy crisis that struck the country (Leopoldino et al. 2019). In 2014, there were in total 250 million compact fluorescent lamps marketed in this country. In this way, the search for alternatives for recycling these wastes as raw material for other articles should be considered.

Studies indicate that powdered glass from fluorescent lamps may be used in civil construction to replace concrete aggregates or prepare roofing tiles (Shao et al. 2000, Furlani and Tonello 2010, Taurino et al. 2013, Sahmenko et al. 2014, Pitarch et al. 2021, Robert et al. 2021, Sharma et al. 2021, Liu et al. 2022).

The vitrification process is a viable alternative not only for the reuse of glass but also for the disposal of hazardous waste that contains heavy metals such as mercury and lead which adhere to the amorphous structure of glass after this process (Colombo et al. 2003). Another advantage is that glass formed in this process is inactive to most chemical and biological agents, making it a viable alternative to dispose hazardous wastes in order to help reduce the volume of waste sent to waste landfills.

Another potential application of glass waste from fluorescent lamps is in the preparation of foam glasses (Aabøe and Øiseth 2004, Scarinci et al. 2005, Taurino et al. 2013, Sahmenko et al. 2014). Foam glass is a porous material with acoustic and thermal insulating characteristics besides high porosity (König et al. 2020). Regarding to the physical structure, the foam glass is a heterophasic system, consisting in a glassy matrix with gas bubbles. The solid phase is glass, that has a lot of cells (pores). These cells have thin walls, in order of micrometers, filled by gas (Fernandes et al. 2009). It presents low thermal expansion ($8.9\ 10^{-6}\ K^{-1}$), low density ($0.1–0.3\ g\ cm^{-3}$) (Scarinci et al. 2005) and is corrosion and fire resistant (Boccaccini et al. 2009). The porosity of commercial foam glasses present porosity ranging from 85 to 97% (Spiridonov e Orlova 2003, Fernandes et al. 2009) and compression resistance ranging from 0.4 to 0.6 MPa (Fernandes et al. 2009). Such properties increase the potential of its applications both indoors and outdoors (Guo et al. 2010), providing thermal comfort and even energy savings when applied as thermal insulating in regions of severe weather (Vereshchagin and Sokolova 2006).

Foam glasses find application in civil construction due to low rates of heat transfer ($0.04–0.08\ W\ m^{-1}°C^{-1}$) (Aabøe and Øiseth 2004, Scarinci et al. 2005, Sahmenko et al. 2014, Cimavilla-Román et al. 2021), allowing it to perform as a heat insulating material combined with its acoustic insulation properties (Fernandes et al. 2009, Yot and Méar 2011, Ayadi et al. 2012, Benglini et al. 2012, König et al. 2021). Compared with polymeric foams, foam glasses have the advantage of greater strength in response to high temperatures and oxidation than most industrial reagents, being applicable as photocatalysts supports (Lebullenger et al. 2010).

The first foam glasses were produced in 1930, by blowing gases into molten glass (Bernardo et al. 2007). A limiting factor for its widespread use is the high producing cost. In order to reduce it, a new technology based on the sintering of ground glass particles was developed (Vereshchagin and Sokolova 2006, Bernardo et al. 2007, König et al. 2021). Employing such methodology results in energy savings, since recycled glass can be used as raw material (powder) in the conformation process (Bernardo et al. 2007). Besides that, the sinterization temperatures applied to the system are lower than temperatures applied to the fusion of regular glasses (Colombo et al. 2003), which makes it environmentally friendly (Manevich and Subbotin 2008). According to Ayadi et al. (2012), the production of foam glasses from recycled glass leads to an economy of energy around 25%, if compared to process of blowing gases into molted glass.

Actually, foam glasses are prepared from glass powder by mixing the powdered glass with a foaming agent. The firing temperature decreases the viscosity of the vitreous phase and concomitantly decomposes the foaming agent, which releases gas into the glassy phase and results into a porous body (Fernandes et al. 2013).

The expansion of foam glasses can be explained through two different processes. The first consists in the production of the expansion gas in the sintered body. The second process consists in growing bubbles, which is a consequence of the rise of pressure in the closed pores (Steiner 2006, Owoeye et al. 2020).

This study aimed at investigating the potential applications of glass from fluorescent lamps to obtain foam glasses while considering the following variables of the production process: the mass percentage of $CaCO_3$, forming pressure, firing temperature, heating rate and particle size of powdered glass. The best process conditions were identified based on the volumetric expansion of foams.

2. Materials and methods

2.1 Materials

Glass from fluorescent lamps was obtained from Apliquim Brasil Recicle, Paulínia - São Paulo, Brazil. An equiproportional mixture of glass from compact and tubular fluorescent lamps (the mercury contaminants being removed via thermal decontamination) were milled in a porcelain ball mill for 2 h and subsequently granulometrically separated with sieves. The average particle size of the fractions of powdered glass powder (PG) was determined by laser diffraction (CILAS model 1180) (FR), resulting in 11, 29 and 91.0 μm. Calcium carbonate (QuimVale, RJ, Brazil), with minimum purity of 97% and particle size of 3 μm (determined by laser diffraction), was used as foaming agent.

2.2 Methods

2.2.1 Characterization of the powdered glass and calcium carbonate

The chemical composition of the PG and calcium carbonate mixture was determined by X-ray fluorescence (XRF Shimadzu-1800) (JP)). The chemical composition was analyzed using the Lakatos software (Department of Materials Engineering, Modena University, Italy) to determine the glass viscosity, the glass transition (T_g) and softening temperatures as well as the glass theoretical density (Lakatos et al. 1976). This method of analysis is based on a theoretical calculation that considers a linear dependence of the glass components, expressed in molar percentages, on the viscosity of the material. This method showed a standard deviation of approximately

3.01°C (Fernandes et al. 2013). Thermogravimetric analysis (TGA) and differential scanning calorimetry (DSC) were performed on PG and $CaCO_3$ from room temperature to 900°C using a gravimetric equipment (Netzsch STA 449 Jupiter) and heating rate of 5°C min^{-1}.

2.2.2 Preparation of foam glasses

Initially, foam glasses were obtained by preparing a mixture of glass powder with $CaCO_3$ (2, 3, 4 and 8 wt %) in a planetary mixer (Pavitest 3010 model C) (Brazil), followed by dry mixing for 5 minutes. Subsequently, 15 mL of poly(vinyl alcohol) (5 wt% solution) was added for each 100 g of glass powder, and the resulting mixture was mixed again for 5 min. The samples were prepared by compacting (Pokorny et al. 2011, Atilla et al. 2013, Fernandes et al. 2013) 65 g of the mixture in a press (Bovenau 10 t) (Brazil) at pressures of 20, 30 and 40 MPa using a cylindrical mould (ø = 54 mm). Drying was carried out for 24 h at room temperature (Pokorny et al. 2011), followed by 6 h and 18 h in an electrical oven at 60°C and 105°C, respectively. The obtained samples (foam glasses) were fired in the electrical oven at 700, 750, 800 and 850°C (Pokorny et al. 2011, Fernandes et al. 2013) at a heating rate of 5°C min^{-1} (Chen et al. 2011, Fernandes et al. 2013) for 30 min (Chen et al. 2011, Pokorny et al. 2011). The coding adopted for the study was GX.CY.PZ, where X stands for the content (98, 97, 96 and 92wt %) of glass (G), Y stands for the content (2, 3, 4, 8wt %) of $CaCO_3$ (C) and Z stands for the pressure (P) level (20, 30 and 40 MPa).

2.2.3 Characterization of the foam glasses

Thermomechanical analysis (TMA) (Shimadzu TMA-60) (JP), from room temperature to 900°C, using 5 g of load was conducted to determine the softening temperature of the glass and foam glasses made with different percentages of $CaCO_3$. The foam glasses were characterized by the volumetric expansion determined by the volume of samples immersed in a sand-filled cube of precalculated density. The determination of volume was obtained through the weight difference between the sand-filled cube and the foam glasses-containing cube. A similar methodology was reported in the literature (Pokorny et al. 2011); the volumetric expansion was determined according to Equation (1), where % E is the percentage of volumetric expansion, V_2 the volume after firing and V_1 the volume of green samples.

$$\% E = \frac{V_2 - V_1}{V_1} \cdot 100 \tag{1}$$

In order to assess density, the foam glasses were cut into prismatic shape. Measurements for volume calculation were performed with the aid of a digital pachymeter and the mass was ascertained by means of an analytical scale. Density was calculated by the ratio between foam mass and volume.

The morphological analysis of the foam glasses was performed with the aid of scanning electronic microscopy (SEM) in a Shimadzu SSX-550 (JP) instrument at 15 kV. In order to facilitate the analysis operation, samples were cut, polished and recovered with a fine gold layer (plasma deposited). Mapping of chemical elements was performed with the aid of a Dispersive Energy Spectroscopy (DES) instrument coupled to the microscope. Micrographs obtained from SEM were applied to determine the average pore size, using the software Image J.

Thermal conductivity of the foam glasses was assessed with the aid of the heat flow technique based on the ASTM E 1225/09 Method. In order to perform thermal conductivity tests, foam glasses were cut into 70×70 mm^2 test specimens. The height of the test specimens varied as a function of the maximum height reached by each composition; however, this figure is to be taken into consideration for the calculation of the conductivity of the material. The test was performed from room temperature and up to 250°C, with levels of 50°C and level period of 2 hours and heating rate of 2.5°C min^{-1}. Temperatures were recorded every 5 seconds.

The foam glasses G96.C4.P40, fired at 850°C, and G97.C3.P40, fired at 800°C, were analyzed taking as reference the methodology from NBR 10.006-04 (ABNT 2004). Milled foams were analyzed by ICP-OES, which allowed monitoring the presence of metals (mercury, lead and cadmium, in this case). In order to perform a comparison, the same analysis was performed to the PG.

3. Results and discussion

3.1 Characterization of materials

The chemical composition of the glass and calcium carbonate mixture was determined by X-ray fluorescence as shown in Table 1.

Figure 1 shows the thermograms of glass and $CaCO_3$ obtained using TGA and DSC. The results (Fig. 1a) indicated that the glass lost 0.5 weight % over the entire temperature range of the analysis. According to Fig. 1b, $CaCO_3$ showed weight loss of 2.1% up to 500°C based on TGA. Thermal decomposition of $CaCO_3$ starts near 700°C (Bernardo and Albertini 2006). As shown in Fig. 1b, the sample showed a severe loss weight at this temperature according to the TGA data, and stabilization occurred soon after 800°C; the mass of the sample remaining constant until the final analysis temperature (1,100°C) and maximum mass loss 55.7% were reached. The observed weight loss originated from the endothermic decomposition of calcium carbonate (Fig. 1b), which released CO_2 and formed CaO as confirmed by DSC analysis (Bernardo and Albertini 2006).

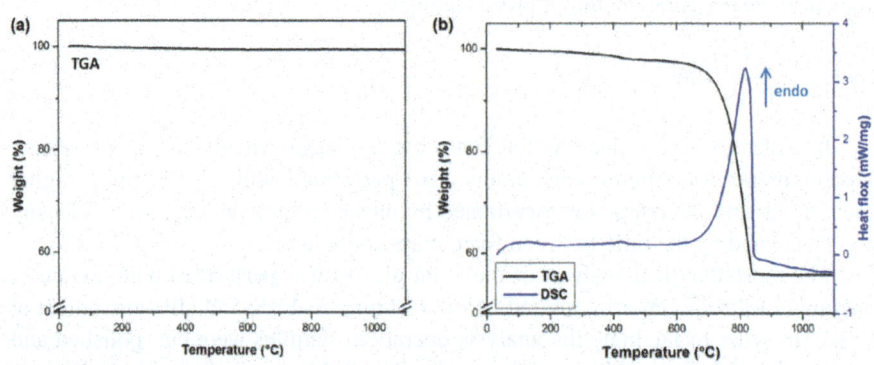

Fig. 1: Thermal analysis of (a) mixture of fluorescent lamps glass (PG) (TGA) and (b) $CaCO_3$ (DSC and TGA).

Table 1: Chemical composition of glass and calcium carbonate determined by FRX.

Oxide	% wt.	
	Glass	CaCO$_3$
SiO$_2$	74.93	0.43
CaO	6.13	56.81
BaO	4.07	-
K$_2$O	3.82	0.02
PbO	3.78	-
Na$_2$O	2.38	-
Al$_2$O$_3$	2.35	0.26
MgO	0.74	-
SrO	0.45	-
Fe$_2$O$_3$	0.29	0.07
SO$_3$	0.18	0.06
P$_2$O$_5$	0.09	0.11
Rb$_2$O	0.06	-
Ra	0.03	-
CuO	0.02	-
ZnO	0.02	-
ZrO$_2$	-	0.11
CO$_2$	0.66	42.13
	100.00	100.00

3.2 Evaluation of foam glasses

Figure 2 shows the results of the volumetric expansion obtained for the foam glasses prepared with 2, 4 and 8 wt % CaCO$_3$ at 20, 30 and 40 MPa pressure and fired at 800 to 850°C.

Figure 2b shows that volumetric expansion increased as the CaCO$_3$ content increased from 2 to 4%. The G98.C2.P40 foams showed higher volumetric expansion (257.2%) when fired at 800°C than when fired at 850°C (100.6%). Increasing the firing temperature above the softening point may decrease the viscosity of the glass phase, which might reduce expansion (Atilla et al. 2013), and probably resulted in the expansion observed for this foam at 850°C. Volumetric expansion was maximized for the foams prepared with 4% CaCO$_3$. At the temperature of 800°C, the G96.C4.P40 foams showed volumetric expansion average of 289.0%. Increasing the temperature to 850°C resulted in a slight variation in the percentage of expansion to 243.9%. The G96.C4.P30 foams fired at 850°C showed volumetric expansion of 273.1%.

The increase of the expansion of foam prepared with 4 wt% de CaCO$_3$, when compared to those prepared with 2 wt% (Fig. 2), can be explained by the greater amount of CO$_2$ to be released in the glass structure, since more CaCO$_3$ was added to the mix. Furthermore, according to Fernandes et al. (2009), the viscosity in glassy phase increases as the amount of CaCO$_3$ increases. Taking it into account, it is possible

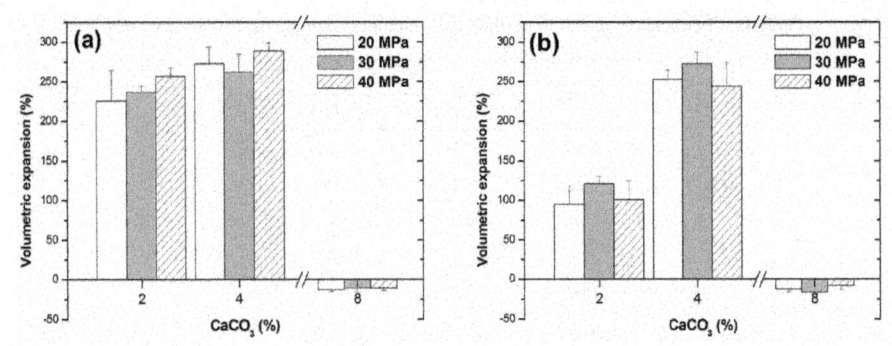

Fig. 2: Volumetric expansion of foam glasses prepared with 2, 4 and 8 wt% $CaCO_3$ compacted at 20, 30 and 40 MPa and fired at (a) 800°C and (b) 850°C.

to explain that increasing the temperature to 850°C resulted in greater expansion. As it can be seen in Fig. 1(b), this increase ensures the complete decomposition of $CaCO_3$ and the release of greater amount of CO_2. Besides that, changing the viscosity makes the glassy structure to withstand higher temperatures.

It seems like the presence of 4 wt% of $CaCO_3$ resulted in a more appropriated viscosity to the expansion of foams. Despite the increase in temperature, this composition retained more gas in the interior of the glassy phase, as was observed in the range of temperature analyzed (800 and 850°C).

The use of 8 wt% $CaCO_3$ resulted in the contraction of the samples for both analyzed temperatures, with an average contraction of –11.6%. The increase in the percentage of CaO, due to decomposition of $CaCO_3$, stimulates devitrification and increases the glassy phase viscosity. According to the literature (Fernandes et al. 2013), the addition of high amounts of $CaCO_3$ in the preparation of foam glasses contributes to glass crystallization and increases the glassy phase viscosity, thus reducing expansion.

The evaluation of T_g can inform about the changes in the glassy phase viscosity. Figure 3 shows the TMA results for glass and G98.C2.P40, G96.C4.P40 and G92. C8.40 foams.

The evaluation of the glass composition (Table 1) was used to determine thermal properties through theoretical analysis with the Lakatos software. The T_g (inflection point) of glass was 561.6°C (Fig. 3a), this result being in agreement with the result obtained from analysis with the Lakatos software. The T_g of G98.C2.P40 foam glasses (Fig. 3b) was 619.6°C. Increasing the percentage of $CaCO_3$ in the preparation of foam glasses to 4wt% (Fig. 3c) and 8wt% $CaCO_3$ (Fig. 3d) increased the glass transition temperature to 632.0°C and 637.9°C, respectively. Therefore, the percentage of $CaCO_3$ positively correlated with the glassy phase viscosity, which allowed the point of softening to be reached at higher temperatures. This result is in agreement with the literature (Fernandes et al. 2009).

The temperature during the foam glasses expansion process is important for two phenomena: the decreased viscosity of the glass and the thermal decomposition of the foaming agent (Atilla et al. 2013). The increase in $CaCO_3$ percentage raised T_g, (Fig. 3(b), (c) and (d)) contributing to the increase in glassy phase viscosity of foams, requiring heat treatment at higher temperatures to reach the same viscosity level.

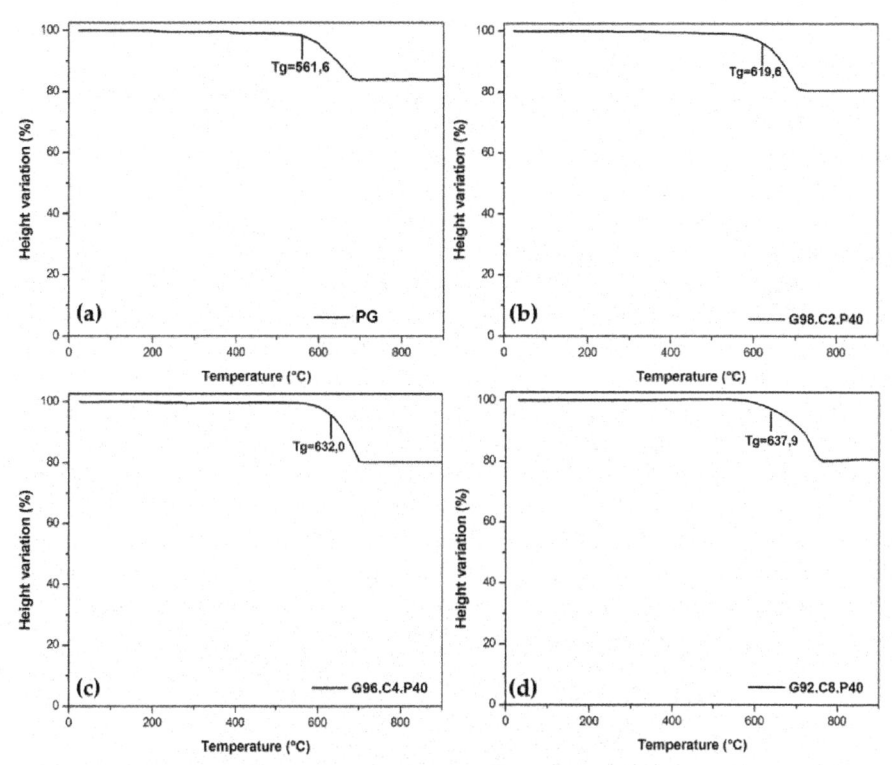

Fig. 3: Thermomechanical analysis of (a) neat PVD and foam glasses prepared with (b) 2 wt%,(c) 4 wt% and (d) 8 wt% $CaCO_3$ compacted at 40 MPa.

This observation corroborates the discussion regarding the effect of temperature on foams expansions at $CaCO_3$ percentages of 2, 4 and 8 wt%. As noted by other authors (Fernandes et al. 2009), beyond viscosity change, high $CaCO_3$ percentages promote glass crystallization. The results obtained from the addition of 8 wt% $CaCO_3$ suggest that the glassy phase viscosity is likely not sufficiently low to allow expansion at the studied temperatures (up to 850°C). At temperatures above 850°C, $CaCO_3$ will be fully decomposed (Chen et al. 2012) (Fig. 1(b)); thus, CO_2 will no longer be released (T_{onset} = 700°C and T_{endset} = 830°C) and increases in temperature will not result in further expansion of this sample due to the total absence of a gas source. The rise in $CaCO_3$ percentage changed the glassy phase viscosity; the mixture containing 4 wt% $CaCO_3$ resulted in a more suitable viscosity for foams expansion over the studied temperature range. The mixture retained more gas in the glassy phase even as the firing temperature was increased. The observed rise in viscosity as the $CaCO_3$% was increased may explain the result obtained for the mixture containing 8 wt% $CaCO_3$, which did not show expansion.

Figure 4 shows the 8 wt% $CaCO_3$ samples during the firing at different temperatures. In the temperatures of 750, 800 and 850°C, cracks were observed on the surface of the samples and the expansion did not occur. The results showed that for this composition, the increase in temperature does not lead to expansion of the samples. Other authors had similar results when they studied the preparation of foam glasses using $CaCO_3$.

Glass of cathode-ray-tube panel (CRT) was mixed with $CaCO_3$ as foaming agent in percentages of 1.5, 2, 4 and 10 wt% (König et al. 2014). The results showed that by using 10% $CaCO_3$ the density of the glass body is 1.6 g cm^{-3} while the best composition (with 4 wt% of $CaCO_3$) showed 0.4 g cm^{-3}. According to the authors, these results can be explained by the difficult sinterization of glassy particles, rising the volumes of gases out of the glass body by the open porosity.

Arcaro et al. (2016a) tested the use of 10 wt% of $CaCO_3$. Results showed that for this percentage of foaming agent did not occur expansion. Authors justified this by the increase of internal pressure in pores, that results in the rupture of the walls, evolving gas from the structure. This effect can also be observed in Fig. 4.

The samples prepared with 8 wt% $CaCO_3$ did not show significant variations even when the temperature was increased. This composition showed cracks on its top surface when fired at 800°C, as can be seen in Fig. 5. When the firing temperature was increased to 850°C (Fig. 6), the ceramic body did not show cracks on the top surface. The presence of cracks at 800°C can be attributed to the need of CO_2 release from the inner portion. As the temperature increased, the glassy mass likely shrank more and particles began to coalesce, which eliminated cracking on the top surface.

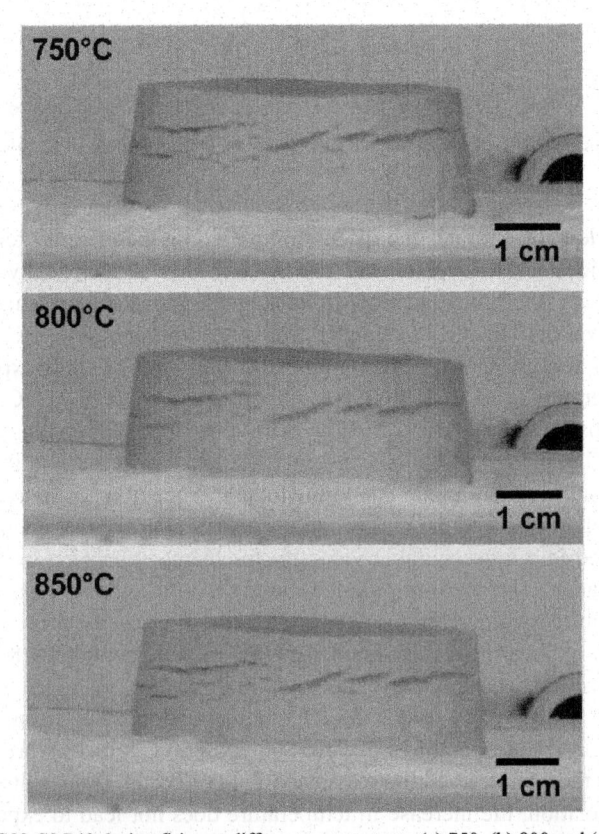

Fig. 4: G92.C8.P40 during firing at different temperatures (a) 750, (b) 800 and (c) 850°C.

Similar results were also reported in the literature (Liao and Huang 2012), which were justified by the need of CO_2 expulsion out of the sample cracks due to the high viscosity of glass. As related to the cracks observed at the sides of the test specimens

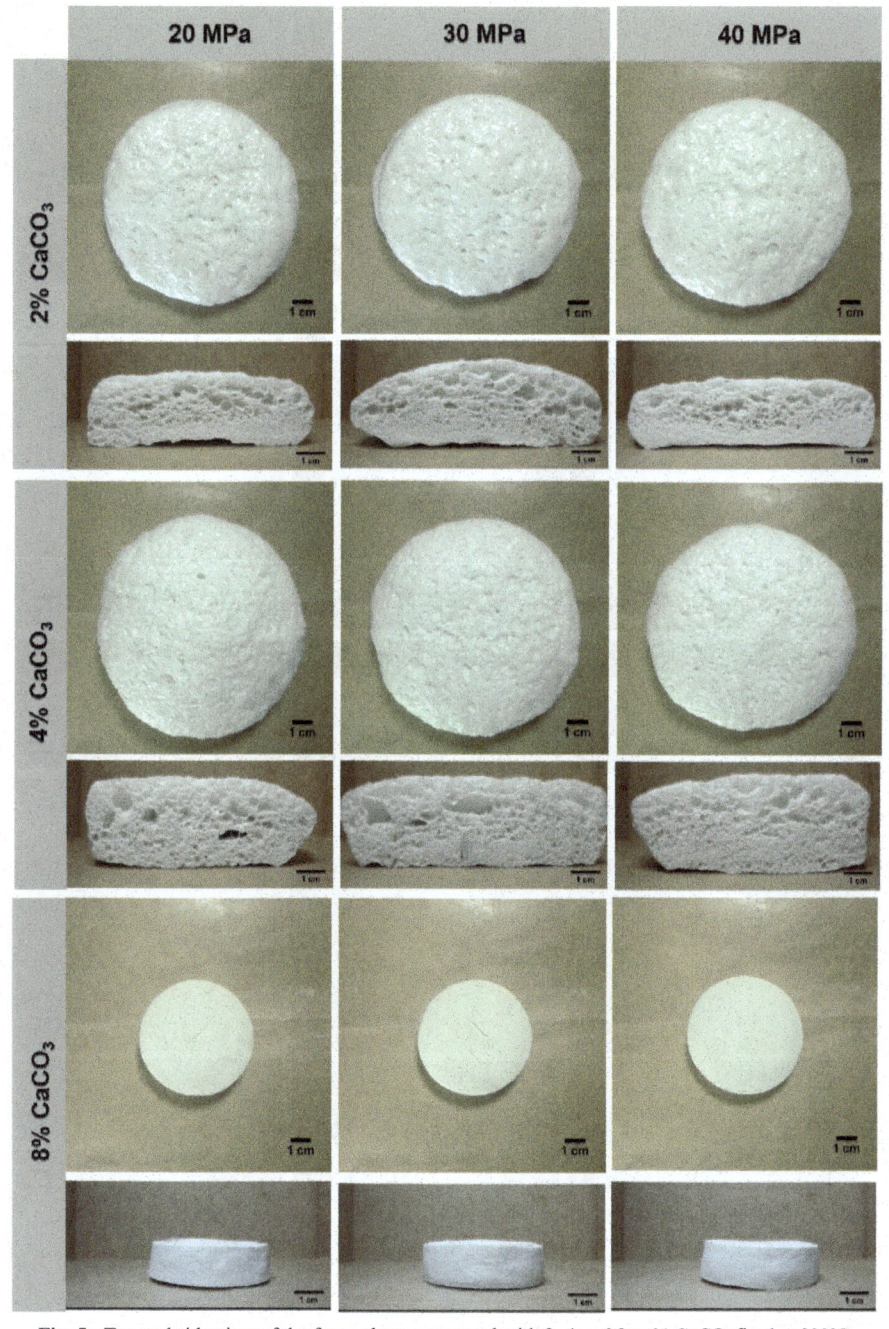

Fig. 5: Top and side view of the foam glasses prepared with 2, 4 and 8 wt% $CaCO_3$ fired at 800°C.

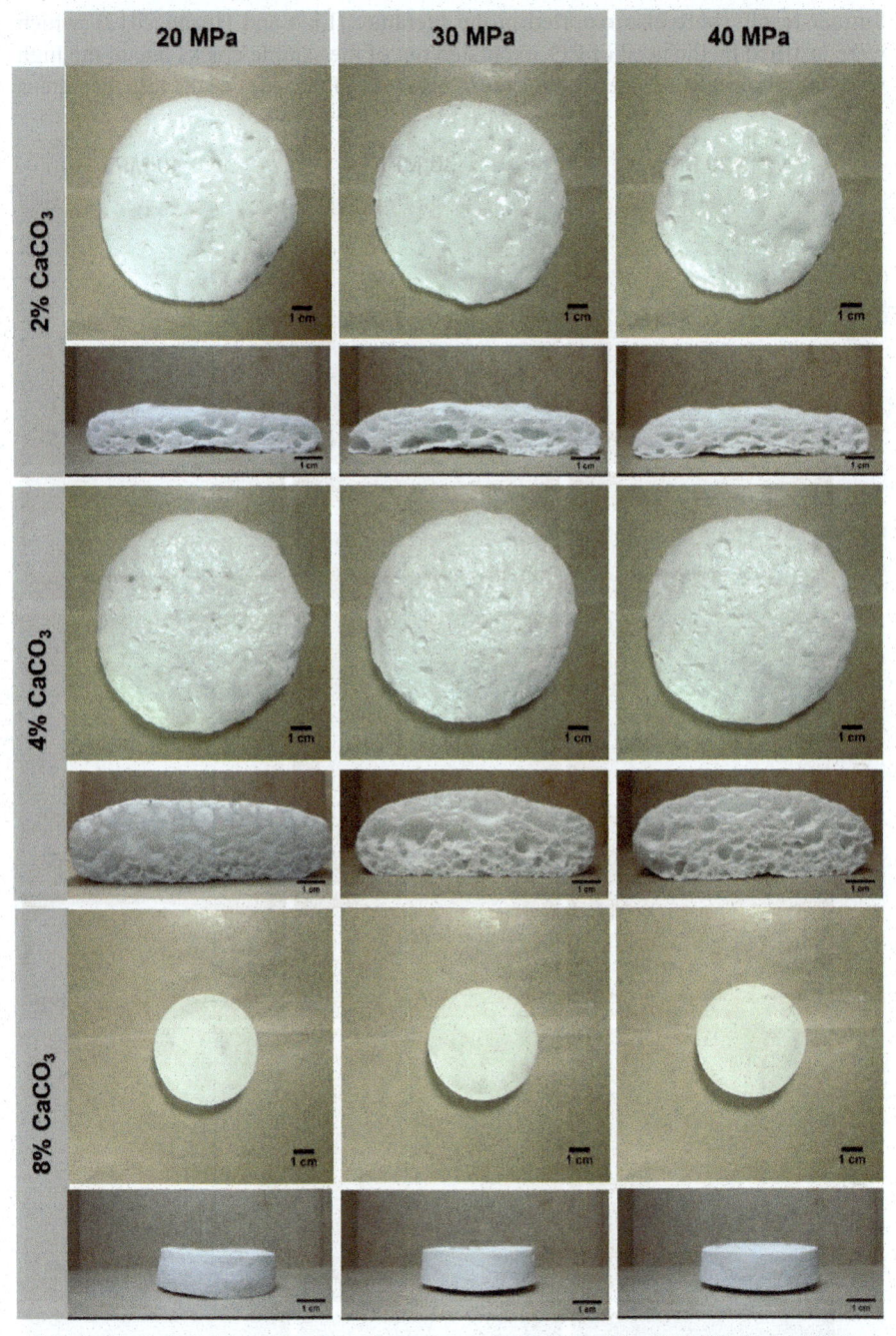

Fig. 6: Top and side view of the foam glasses prepared with 2, 4 and 8 wt% CaCO$_3$ fired at 850°C.

(Fig. 4), it is suggested that by increasing the temperature the cracks from which gas is expelled tend to diminish.

Compositions with 8 wt% $CaCO_3$ (Figs. 5 and 6) did not show any modifications in their shape after firing at the temperatures of 800 and 850°C, both at the top view and the side view. In Figs. 5 and 6, the effect of the temperature rise on the 2 and 4 wt% $CaCO_3$ compositions can be observed. These compositions had open pores on the surface, which indicated gas emission out of the glass body. Similar results were reported in the literature (Pokorny et al. 2008).

Upon temperature increase, significant reduction in foam height occurred for 2 wt% $CaCO_3$ and side views enabled one to infer that the rise in temperature led to increased pore size resulting from the coalescence of small pores as reported in the literature (Méar et al. 2006a).

For 4 wt% $CaCO_3$ foams temperature rise also led to increased pore size as reported in the literature (Fernandes et al. 2013). In this case, as the viscosity of the glassy phase of the 4 wt% $CaCO_3$ composition was higher (Fig. 3b) as compared with that at 2 wt% (Fig. 3c), the rise in temperature did not reduce sufficiently the temperature so as to allow most of the CO_2 to leave the structure under expansion, resulting in more highly expanded structures.

As related to the structure exhibited by the foam glasses, as can be seen in the side views of Figs. 5 and 6, mainly in cases where expansion occurred, pore size heterogeneity can be noticed. The glass particle size employed is 91 μm, while the foaming agent average particle size is 3μm. According to the literature (Scarinci et al. 2005) homogeneity in pore size distribution was compromised by the different size of the raw materials.

Figure 7 shows the SEM micrographs obtained for G92.C8.P30 (Fig. 7a) and G96.C4.P30 (Fig. 7b) foams, both fired at 800°C. Figure 7a shows small spherical pores having inside whitish particles. As it can be seen in Fig. 7a, the average pore size is 16.1 μm. According to the DES mapping performed in the micrograph, the whitish particles have high calcium concentration, which could mean CaO from the foaming agent decomposition. Abdollahi and Yekta (2020) also observed the presence of reduced metals in the pore walls, resulting from the decomposition of

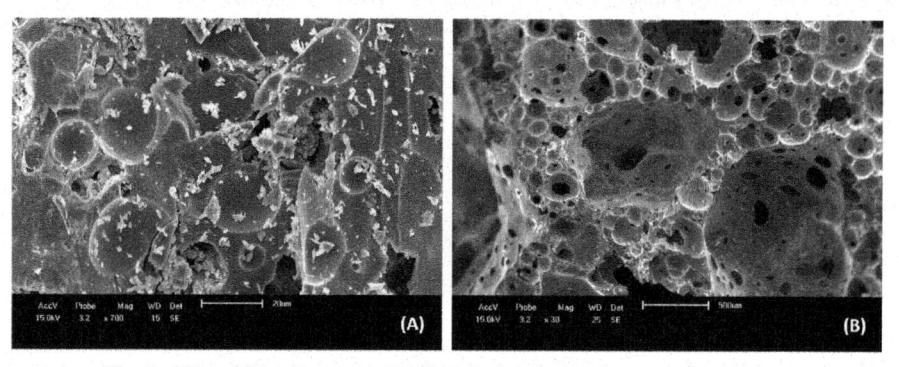

Fig. 7: SEM of foam glasses (a) G92.C8.P40 and (b) G96.C4.P40, fired at 800°C.

foaming agents. The distribution of particles originating from the decomposition of foaming agents was also observed by (Pokorny et al. 2008).

The occurrence of spherical pores is characteristic of the volumetric expansion initial phase. Chen et al. (2012) demonstrated that the expansion process was incipient, further expansion having been hindered by the high viscosity of the glassy phase.

Figure 7b exhibited larger pores as compared with those of Figure 7a. Large pores, with an average size of 1000.1 μm, presenting small pores in their walls, with an average size of 160.1 μm, have been noticed. Since the glassy phase viscosity was lower in view of the lower $CaCO_3$ percentage, this made it possible for the pores to grow and interconnect. Such observations were combined with pore shape change, starting from spherical and turning to a polyhedral formation. The heterogeneous distribution of pore size points out the coalescence phenomenon. It occurs when large cells arise from the junction of small pores (Owoeye et al. 2020). It is possible to say that bigger pores resulted from the coalescence of the small ones. The remaining small pores are those that were not adhered to by the bigger ones. This scenario characterizes a wide distribution of pore sizes. These results are in agreement with literature reports (Chen et al. 2012, Mugoni et al. 2015, Owoeye et al. 2020, Cimavilla-Román et al. 2021). Besides pore shape change, small pores on the walls of larger pores could be observed, these being formed by the coalescence of the spherical small pores. Similar results are described in the literature (Méar et al. 2007, König et al. 2020, Cimavilla-Román et al. 2021).

Table 2 shows the results for statistical analysis (ANOVA), evaluated with confidence of 95%. Referring to the compaction pressure employed in the compaction of green samples, the evaluation showed that it is not a significant parameter for the expansion process of foam glass.

The results in Table 2 also show that the firing temperature of the foams, the percentage of $CaCO_3$ and the interaction of these two variables were significant for the expansion process. However, the highest value of the quadratic average was

Table 2: Statistical analysis of the influence of process variables on the percentage of expansion of foam glass with 2, 4 and 8 wt% CaCO3, compacted at 20, 30 and 40 MPa and fired at 800 to 850°C.

	SQ	GDL	MQ	Teste F		P	S
				Calc.	Tab.		
Temperature	22443.68	1	22443.68	29.37	0.00398	0.00002	Yes
%CaCO$_3$	1422157.13	2	711078.56	930.68	0.05135	0.000000	Yes
Compaction Pressure	2140.29	2	1070.15	1.40	0.05135	0.256977	No
Temperature*%CaCO$_3$	37741.48	1	37741.48	49.40	0.00398	0.000000	Yes
Temperature*Compaction	1438.18	1	1438.18	1.88	0.00398	0.176871	No
%CaCO$_3$*Pressure	1102.16	1	1102.16	1.44	0.00398	0.236014	No
Error	34381.95	45	764.04				
Total	1521404.87	53					

MQ = mean square; GDL = degrees of freedom; Calc = F calc; Tab = F tab; P = p value; S = SQ = sum of squares; S = significance.

obtained for the percentage of the $CaCO_3$ variable, and showed that this was the most significant variable for the process, almost separately determining the result. Considering that 2 and 4 wt% $CaCO_3$ showed expansion for the foam glasses and that increasing the amount of $CaCO_3$ leads to a negative expansion, a new composition was developed using 3 wt% $CaCO_3$ at compaction pressure of 40 MPa.

Figure 8(a) shows the results obtained for the volumetric expansion of the G97.C3.P40 foam glasses fired at 800 and 850°C compared with G98.C2.P40 and G96.C4.P40 foams. The foams prepared with 3 wt% $CaCO_3$ fired at 800°C showed volumetric expansion of 370.9%, which was much higher than the expansion of foams prepared with 2 wt% (257.2%) and 4 wt% $CaCO_3$ (289.0%), fired at the same temperature. As the firing temperature was increased to 850°C, volumetric expansion decreased for all compositions, reaching a maximum of 243.9% for 4 wt% $CaCO_3$ and 243.2% for 3 wt%, while the foams prepared with 2 wt% showed volumetric expansion of 100.7%. The decrease in the volumetric expansion response to the increased temperature can be attributed to the expulsion of the gaseous phase (beginning during the collapse of the structure) out of the porous body (König et al. 2021). The increase in firing temperature results in increased gas pressure in the pores, which interconnects the pores (Méar et al. 2006b) and subsequently results in the collapse of the porous body as the temperature increases (Chen et al. 2012) and the gas is expelled. In relation to the amount of $CaCO_3$, 3 wt% revealed itself to be the best composition, even better than 4 wt% that was initially considered better than 2 and 8 wt%.

The evaluation of the firing temperature in the G97.C3.P40 composition is shown in Fig. 8(b). The largest expansion observed was 370.8% at a temperature of 800°C. Firing temperatures of 750 and 850°C produced average volumetric expansions of 269.2% and 243.2%, respectively. A temperature of 700°C did not result in significant volumetric expansion, reaching an average value of 8.0%. Same results are reported in literature (Mugoni et al. 2015). According to Fig. 1b, calcium carbonate initiated the process of thermal decomposition at this temperature and released a small quantity of gas, which did not allow for significant expansion of the glassy body. A temperature of 750°C provided for less volumetric expansion than that observed for foams fired at 800°C because the vitreous structure cannot retain

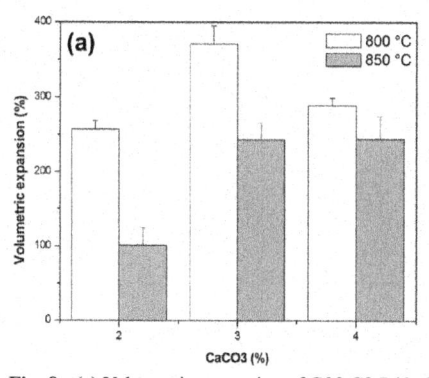 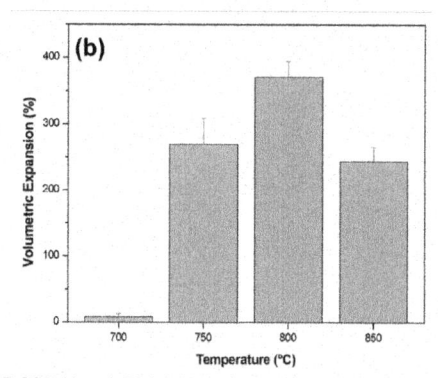

Fig. 8: (a) Volumetric expansion of G98.C2.P40, G97.C3.P40 and G96.C4.P40 foam glasses fired at 800 and 850°C, (b) Effect of firing temperature on the G97.C3.P40 foam glass.

the gas and/or the gas phase is less abundant than when the material is fired at 800°C. The vitreous structure may not have been able to retain the gas phase because the calcium carbonate was not completely decomposed and glass particles may not have been sintered (according to the Lakatos software analysis, the softening temperature of the glass is 790°C).

The results suggest that the amount of 3 wt% of $CaCO_3$ leads to greater expansion of the foam glasses at the firing temperature of 800°C. Therefore, these results were used to evaluate other variables of the process, such as heating rate and particle size of PG, keeping the firing temperature at 800°C.

Figure 9 shows the results for foam glasses G97.C3.P40, fired at 800°C, with different particle sizes for glass at different heating rates. Regarding the heating rate, results have shown that as the heating temperatures were increased from 2.5°C min⁻¹ to 5.0 and 7.5°C min⁻¹, the expansion of foam glasses reduced. Considering the variation of particle size, results have shown that the biggest expansions were observed in bigger particles (91.0 μm). Reducing the particle size to 29.0 μm resulted in substantial reduction of expansion. However, when the average particle size was reduced to 11.0 μm, a slight increase of expansion in comparison to the foams prepared with particle size of 29.0 μm was observed.

The best results were observed when the particle size of PG and heating rate were adjusted to 91.0 μm and 2.5°C min⁻¹, respectively. In this combination, the volumetric expansion reached the maximum value of 593.6%.

The increase of the heating rate resulted in reducing the volumetric expansion. These results are not in accordance with the ones suggested in literature (Pokorny et al. 2011, Bento et al. 2013). According to Pokorny et al. 2011, when reduced heating rates are applied, bodies remain in higher temperatures for longer time. This results in coalescence of pores and the structure collapses. Other variable that contributes

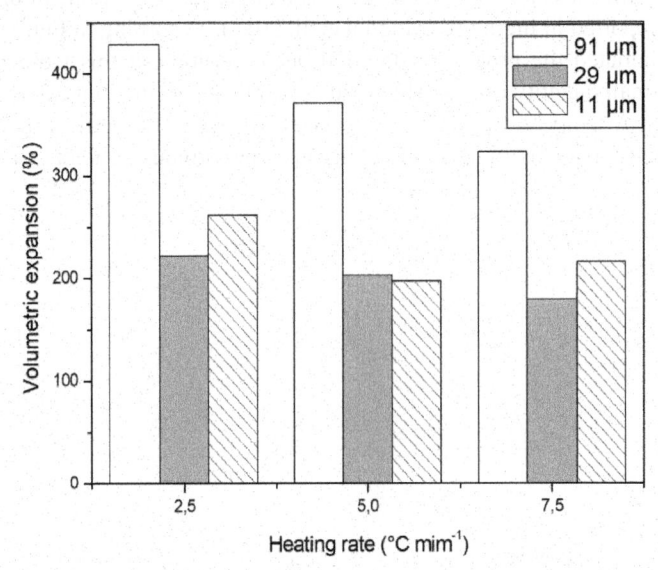

Fig. 9: Volumetric expansion for foam glasses G97.C3.P40 fired at 800°C, using heating rates of 2.5, 5.0 and 7.5°C min⁻¹ and powdered glass with medium particle sizes of 91.0, 29.0 and 11.0 μm.

to the collapse of the structure is the increase of time of isotherm, as suggested by Bernardo and Albertini (2006). Results may be justified by the time of isotherm employed, that is, 30 min. It was probably not long enough to allow the gas out of the foams, reducing the expansion of the bodies. Kurtulus et al. (2021) observed that reducing the heating rate, employing an isotherm time of 30 min, resulted in foam glasses with better physical, thermal and mechanical properties.

Reducing the particle size of PG also reduced the expansion of foam glasses. This result agreed with results presented in literature (Pokorny et al. 2008, König et al. 2014). Bernardo et al. (2007) related the reduction of the volumetric expansion of foam glass with small particles (< 37 µm) to the glass crystallization. In accordance with König et al. (2014), the particle size of glass employed presented marginal influence in the expansion process. Furthermore, the main variable of the process is the decomposition kinetics of $CaCO_3$, which were evaluated varying the particle size of PG.

As already mentioned, combining a particle size of 29.0 µm with heating rate of 5.0°C min^{-1} resulted in the most reduced volumetric expansion, when comparing all the combinations tested. On the other hand, the combination of particle size of 91.0 µm and heating rate of 2.5°C min^{-1} resulted in the greatest volumetric expansion, 428.4%, and, consequently, the body with greatest variation of diameter after fire. For this particle size, less energy is needed for the processing of the raw materials.

The assessment of density, thermal conductivity and mechanical strength to compaction was performed aiming at the characterization and comparison of properties of the G97.C3.P40 foam fired at 800°C as compared with glassy foams reported in the literature. The density of the G97.C3.P40 foam was 0.207 ($\pm$ 0.09) g·cm^{-3} with total porosity of 92.0 ($\pm$ 0.36)%. This density is less than the best result reported by Kurtulus et al. (2021). The porosity was higher than shown by Arcaro et al. (2016b), Teixeira et al. (2017a) and Kurtulus et al. (2021), but was similar to Souza et al. (2017) and König et al. (2020). According to literature (Scarinci et al. 2005), densities of commercial foam glasses vary from 0.1 to 0.3 g cm^{-3}. However, foam glasses of higher density values can be encountered in the literature (Méar et al. 2006b, Ayadi et al. 2012, Fernandes et al. 2013, Konig et al. 2014). The G97.C3.P40 foam glass was assessed in terms of mechanical strength to compression which was 0.24 ($\pm$ 0.02) MPa. According to the literature (Méar et al. 2006b, 2007), high porosity values reduce mechanical strength to compression in view of the higher gas volume within the structure.

Figure 10 illustrates the assessment of the thermal conductivity for G97.C3.P40 foam glass fired at 800°C as a function of temperature. Data show that the thermal conductivity of foam glasses fall upon increased test temperatures.

According to the literature (Pokorny et al. 2011), the thermal conductivity of foam glasses falls with rising temperatures caused by the intensification of the glass atomic vibrations. The rising behavior from a certain temperature threshold is a feature of ceramic materials. At low temperatures, conductivity was 0.33 W m^{-1} C^{-1}. This value is similar to that reported by (König et al. 2016). The lowest value found was 0.06 W m^{-1} °C^{-1} at 125°C, which is less than that presented in the foam glasses prepared by Teixeira et al. (2017b) and König et al. (2021). Souza et al. (2017) produced foam glasses with thermal conductivities between 0.177 and

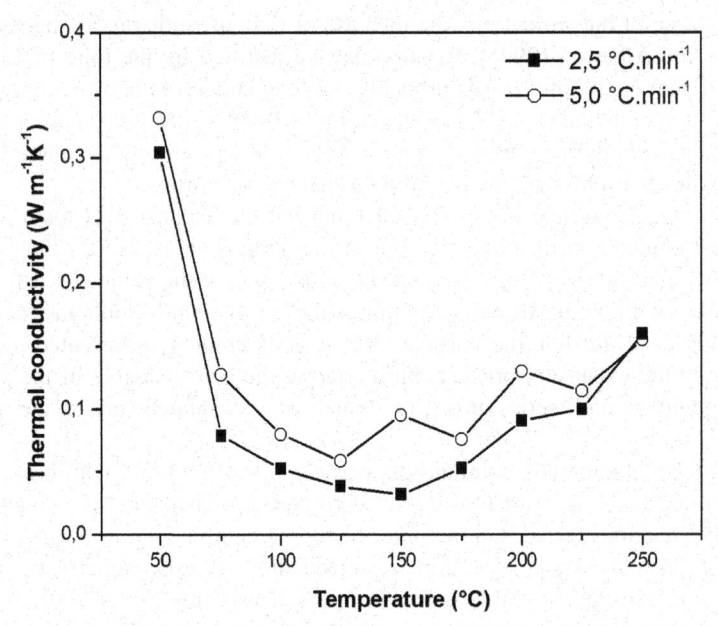

Fig. 10: Assessment of the thermal conductivity of the G97.C3.P40 foam glass fired at 800°C as a function of temperature.

0.055 W m^{-1} °C^{-1}. The foam glasses obtained by Arcaro et al. (2016a) presented conductivity ranging from 0.5 to 1.25 W m^{-1} °C^{-1}. Authors suggested its application as insulating material in the construction industry.

According to several studies (Bernardo and Albertini 2006, Méar et al. 2006b, Pokorny et al. 2011, Atilla et al. 2013, Konig et al. 2014), foam glasses thermal conductivity varies from 0.04 to 0.55 W m^{-1} °C^{-1}. Foam glasses having thermal conductivities below 0.25 W m^{-1} °C^{-1} are classed as thermal insulators (Méar et al. 2006b) so that the G97.C3.P40 foam glasses obtained in this study can be considered as a thermal insulator.

3.3 Metal analysis

Table 3 shows the results of evaluation for metal analysis in glasses of compact and tubular lamps used in this study. The evaluation was carried out by ICP-EOS.

Results have shown that neither cadmium nor mercury were detected in the samples employed in the preparation of foams. High levels of lead were detected in compact lamps. According to Santos et al. (2010) and Xie et al. (2012), lead is present in the composition of glass of these lamps.

The NBR 10.004 (ABNT 2004a) defines the fluorescent lamps as hazardous waste after use. Not only in Brazil, but other countries' regulations and standards classify this waste as hazardous, such as the Commission Decision 2001/118/EC, from European Union, and in United States, regulated by Code of Federal Regulations, Title 40 (US EPA 2012). These wastes undergo a thermal decontamination, aiming at changing its classification to non-hazardous. To get this classification, the metal solubilization analysis is required, according to NBR 10.006 (ABNT 2004b). ASTM D3987-85 is employed to address the same goal (ASTM 2020).

Table 4 shows the results for metal analysis of powdered glass and foam glasses G97.C3.P40, fired at 800°C and G96.C4.P40, fired at 850°C, after solubilization, quantified by ICP-EOS.

Results have shown that both PG and foam glasses did not release cadmium during the solubilization test. An amount of 0.910 mg L^{-1} of mercury from PG was released, whereas even traces of this metal were not detected in the solubilization test for foam glass. The highest concentration of lead was released from PG (121 mg L^{-1}). For the foam glass G97.C3.P40 lead concentration was reduced in 76.4%, reaching an amount of 28,6 mg L^{-1}, while the foam glass G96.C4.P40 reached an amount of 1.26 mg L^{-1}, representing a reduction of 98.7%.

In accordance with Colombo et al. (2003), the vitrification allows incorporating heavy metals adhered into the amorphous structure of glasses throughout chemical bonds. The literature suggests that after the expansion of foam prepared with lead contaminated glasses, the release of lead is reduced due to vitrification (Yot and Méar 2011). These observations justify the reductions of 79.4% and 98.7% of lead concentration and the absence of lead in the solubilization test for the foam glasses, when comparing it to the analysis performed to PG.

The variation of concentration of lead released by the foams prepared with 3 and 4 wt% of $CaCO_3$ is probably related to the fire temperature applied in each of the cases. Considering that foam glasses with 4 wt% present 1% less of glass in its composition, the reduction of metal released was expected. However, the release ratio between the two compositions is greater than 22, which points out that even if it is not the same composition, the temperature probably influences the process.

On regard of parameters presented in NBR 10.004 (ABNT 2004a), the possible wastes of foam glasses may be classified as non-hazardous − non-inert. This classification is consequence of the release of mercury being greater than the amount allowed in the solubilization tests.

Table 3: Metal analysis of glasses employed in preparation of foam glasses.

Lamp type	Parameter (ppm)					
	Cadmium	LOQ	Lead	LOQ	Mercury	LOQ
Compact	ND	0.098	32.2	0.393	ND	0.094
Tubular	ND	0.094	6.42	0.378	ND	0.098

Table 4: Metal analysis of powdered glass and foams G97.C3.P40 and G96.C4.P40.

Sample	Parameter (mg L^{-1})					
	Cadmium	LOQ	Lead	LOQ	Mercury	LOQ
Powdered glass	ND	0.178	121	0.714	0.910	0.178
G97.C3.P40	ND	0.199	28.6	0.795	ND	0.199
G96.C4.P40	ND	0.194	1.26	0.776	ND	0.194

4. Conclusion

The possibility of preparing foam glasses using mercury-decontaminated glass from fluorescent lamps residues provides for the development of a product with potential application as thermal insulator in view of its high porosity allied to low density and furthermore minimizes extraction of raw materials from non-renewable sources. The results showed that the percentage of foaming agent ($CaCO_3$) was crucial to the process, and a value of 3 wt% resulted in higher expansion. The addition of 8 wt% $CaCO_3$ increased the T_g of the mixture, which did not result in volumetric expansion. The heating rate and particle size of powdered glass are important characteristics to the expansion of foam glasses. The reduction of heating rate to 2.5°C min^{-1} and the increase of particle size of PG to 91.0 μm resulted in greater expansion. The largest expansion (428.4%) was obtained for the G97.C3.P40 foam (97% glass, 3% $CaCO_3$ and 40 MPa) fired at 800°C, using heating rate of 2.5°C min^{-1}. Values obtained for density and thermal conductivity and density for the foam glass G97.C3.P40 enable it to be compared with commercial foam glasses. The fire process of foams reduced the concentration released in the metal analysis up to 98.7%. Higher firing temperatures (850°C) resulted in reduced release of lead, which allowed to classify the foam glasses as non-hazardous – non-inert materials. Metal analysis has also shown that was not detected cadmium and mercury contamination at foam glasses.

Acknowledgments

Authors would like to thank the University of Caxias do Sul, the Federal University of Rio Grande do Sul, Apliquim Brasil-Recicle and Fernanda Andreola, of Universitá degli Studi di Modena e Reggio Emilia (IT).

References

Aabøe, R., and Øiseth, E. 2004. Foamed glass: an alternative lightweight and insulating material. pp. 167–176. *In*: Limbachyia, Mukesh C., and Roberts, John J. [eds.]. Sustainable Waste Management and Recycling: Glass Waste, London. Proceedings of the International Conference Organized by the Concrete and Masonry Research, London: Thomas Telford Publishing; London.

Abdollahi, S., and Yekta, B.E. 2020. Predicton of foaming temperature of glass in the presence of various oxidizers via thermodynamics route. Ceram. Int., 46: 25626–25632.

ABNT - ASSOCIAÇÃO BRASILEIRA DE NORMAS TÉCNICAS. NBR 10.004: Resíduos sólidos - classificação. Rio de Janeiro, 2004a, 71 p.

ABNT - ASSOCIAÇÃO BRASILEIRA DE NORMAS TÉCNICAS. NBR 10.006: Procedimento para obtenção de extrato solubilizado de resíduos sólidos. Rio de Janeiro, 2004b, 7p.

Arcaro, S., Albertini, A., Cesconeto, F.R., Oliveira Maia, B.G., Siligardi, C., and Novaes de Oliveira, A.P. 2016a. Thermal insulators produced from industrial solid wastes. Cerâmica, 62: 32–37.

Arcaro, S., Oliveira Maia, B.G., Souza, M.T., Cesconeto, F.R., Granados, L., and Novaes de Oliveira, A.P. 2016b. Thermal insulating foams produced from glass from glass waste and banana leaves. Materials Research, 19(5): 1064–1069.

ASTM D 3987-12. 2020. Standard Test Method for Shake Extraction of Solid Waste with Water, ASTM International, West Conshohocken, PA, 2020.

Attila, Y., Günden, M., and Tasdesmirci, A. 2013. Foam glass processing using a polishing glass powder residue. Ceram. Int., 39: 5869–5877.

Ayadi, A., Stiti, N., Boumchedda, K., Rennai, H., and Lerari, Y. 2011. Elaboration and characterization of porous granules based on waste glass. Powder Technol., 208: 423–426.

Benglini, G.A., Busto, M., Fantoni, M., and Fino, D. 2012. Eco-efficient waste glass recycling:integrated waste management and green product development through LCA. Waste Manage, 32: 1000–1008.

Bento, A.C., Kubaski, E., Sequinel, T., Pianaro, S., Varela, J.A., and Tebcherani, S.M. 2013. Glass foam of macroporosity using glass waste and sodium hydroxide as the foaming agent. Ceram. Int., 39: 2423–2430.

Bernardo, E., and Albertini, F. 2006. Glass foams from dismantled cathode ray tubes. Ceram. Int., 32: 603–608.

Bernardo, E., Cedro, R., Florean, M., and Hreglich, S. 2007. Reutilization and stabilization of wastes by the production of glass foams. Ceram. Int., 33: 963–968.

Boccaccini, A.R., Rossetti, M., Roether, J.A., Zhein, S.H.S., and Ferraris, M. 2009. Development of titania coatings on glass foams. Constr. Buil. Mat., 23: 2554–2558.

Carter, C.B., and Norton, M.G. 2007. Ceramic Materials: Science and Engineering, Springer.

Chen, B., Luo, Z., and Lu, A. 2011. Preparation of sintered foam glass with high fly ash content. Mater. Lett., 65: 3555–3558.

Chen, B., Wang, K., Chen, X., and Lu, A. 2012. Study of foam glass with high content of fly ash using calcium carbonate as foaming agent. Mater. Lett., 79: 263–265.

Chong Fu, Jianwei Liang, Gao Yang, Abd alwahed Dagestani, Wei Liu, Xudong Luo, Baobao Zeng, Haidong Wu, Meipeng Huang, Lifu Lin, and Xin Deng. 2021. Recycling of waste glass as raw materials for the preparation of self-cleaning, light-weight and high-strength porous ceramics. J. Clean. Prod., 317: 128395.

Cimavilla-Román, P., Villafañe-Cavo, J., López-Gil, A., König, J., and Rodríguez-Perez, M.A. 2021. Modelling of the mechanisms of heat transfer in recycled glass foams. Constr. Build. Mat., 274: 122000.

Colombo, P., Brusatin, G., Bernardo, E., and Scarinci, G. 2003. Inertization and reuse of waste materials by vitrification and fabrication of glass-bassed products. Curr. Opin. Solid St. M., 7: 225–239.

Commission Decision 2001/118/EC of 16 January 2001 amending Decision 2000/532/EC as regards the list of wastes (notified under document number C(2001) 108). Official Journal of the European Union L47, 2001, pp. 1–31.

Fernandes, H.R., Tulyaganov, D.U., and Ferreira, J.M.F. 2009. Preparation and characterization of foams from sheet glass and fly ash using carbonates as foaming agents. Ceram. Int., 229–235.

Fernandes, H.R., Andreola, F., Barbieri, L., Lancelotti, I., Pascual, M.J., and Ferreira, J.M.F. 2013. The use of egg shells to produce Cathode Ray Tube (CRT) glass foams. Ceram. Int., 39: 229–235.

Furlani, E., and Tonello, G. 2010. Recycling of steel slag and glass cullet from energy saving lamps by fast firing production of ceramics. Waste Manage, 30: 1714–1719.

Guo, H.W., Wang, X.F., Gong, Y.X., Liu, X.N., and Gao, D.N. 2010. Improved mechanical property of foam glass composites toughened by glass fiber. Mater. Lett., 64: 2725–2727.

Hu, Y., and Cheng, H. 2012. Mercury risk from fluorescent lamps in China: current state and future perspective. Environ. Int., 44: 141–150.

König, J., Petersen, R.R., and Yue, Y. 2014. Influence of the glass-calcium carbonate mixture's characteristics on the foaming process and properties of the foam glass. J. Eur. Ceram. Soc. 34: 1591–1598.

König, J., Petersen, R.R., and Yue, Y. 2016. Influence of the particle size on the foaming process and physical characteristics of foam glasses. J. Non-Cryst. Solids, 447: 190–197.

König, J., Lopez-Gil, A., Cimavilla-Roman, P., Rodriguez-Perez, M.A., Petersen, R.R., Østergaard, M.B., Iversen, N., Yue, Y., and Spreitzer, M. 2020. Synthesis and properties of open- and closed-porous foamed glass with low densisty. Constr. Build. Mat., 247: 118574.

König, J., Petersen, R.R., Iversen, N., and Yue, Y. 2021. Application of foaming agent-oxidizing agent couples to foamed-glass formation. J. Non-Cryst. Sollids, 553: 120469.

Krivtsov, V., Wäger, P.A., Dacombe, P., Gilgen, P.W., Heaven, S., Hilty, L.M., and Banks, C.J. 2004. Analysis of energy footprint associated with recycling of glass and plastic-case studies for industrial ecology. Ecol. Model., 174: 175–189.

Kurtulu, C., Kurtulus, R., and Kavas, T. 2021. Foam glass derived from ferrochrome slag and waste container glass: Synthesis and extensive characterizations. Ceram. Int. (in press).

Lakatos, T., Johansson, L.G., and Simmingskold, B. 1976. Viscosity-temperature relations in the glass system SiO_2-Al_2O_3-Na_2O-K_2O-CaO-MaO in the composition range of technical glasses. Glass Technol., 13: 88–95.

Lebullenger, R., Chenu, S., Rocherullé, J., Merdrignac-Conacec, O., Cheviré, F., Tessier, F., Bouzaza, A., and Brosillon, S. 2010. Glass foams for environmental applications. J. Non-Cryst. Solids, 356: 2562–2568.

Leopoldino, C.C.L., Mendonça, F.M. de, Siqueira, P.H. de L., and Borba, E.L. 2019. The disposal of fluorescent lamps of industries of the metropolitan region of Belo Horizonte - MG, J. Clean. Prod., 233: 1486–1493.

Liao, Y., and Huang, C. 2012. Glass foam from the mixture of reservoir sediment and Na_2CO_3. Ceram. Int., 38: 4415–4420.

Liu, J., Li, S., Gunasekara, C., Fox, K., and Tran, P. 2022. 3D-printed concrete with recycled glass: Effect of glass gradation on flexural strength and microstructure. Constr. Build. Mater. 314, Part B: 125561.

Manevich, V.E., and Subbotin, K.Y. 2008. Mechanism of foam-glass formation. Glass and Ceramics, 65: 5–6.

Méar, F., Yot, P., Cambon, M., Caplain, R., and Ribes, M. 2006a. Characterisation of porous glasses prepared from Cathode Ray Tube (CRT). Powder Technol., 162: 59–63.

Méar, F., Yot, P., and Ribes, M. 2006b. Effects of temperature, reaction, time and reducing agent content on the synthesis of macroporous foam glasses from waste funnel glasses. Mater. Lett., 60: 929–934.

Méar, F., Yot, P., Viennois, R., and Ribes, M. 2007. Mechanical behavior and thermal and electrical properties of foam glass. Ceram. Int., 33: 543–550.

Moulson, A.J., and Hebert, J.M. 2003. Eletroceramics: Materials, Properties, Aplications. WYLEY.

Mugoni, C., Montorsi, M., Siligardi, C., Andreola, F., Lancellotti, I., Bernardo, E., and Barbieri, L. 2015. Design of glass foams with low environmental impact. Ceram. Int., 41: 3400–3408.

Pitarch, A.M., Reig, L., Gallardo, A., Soriano, L. Borrachero, M.V., and Rochina, S. 2021. Reutilisation of hazardous spent fluorescent lamps glass waste as supplementary cementitious material. Constr. Build. Mater., 292: 123424.

Pokorny, A., Vicenzi, J., and Bergmann, C.P. 2008. Effects of the alumina addition in the microstructure of foam glass. Cerâmica., 54: 97–102.

Pokorny, A., Vicenzi, J., and Bergmann, C.P. 2011. Influence of heating rate on the microestructure of glass foams. Waste Manage. Res., 29: 172–179.

Rameshkumar, S., Pednekar, M., Sarat Chandra T., James J. Doyle, and Ramesh Babu, P. 2020. Advanced separation process for recovery of critical raw materials from renewable and waste resources. Encyclopedia of Renewable and Sustainable Materials, 5: 1–9.

Rey-Raap, N., and Gallardo, A. 2012. Determination of mercury distribution inside spent compact fluorescent lamps by atomic absorption spectrometry. Waste Manage, 32: 944–948.

Rey-Raap, N., and Gallardo, A. 2013. Removal of mercury bonded in residual glass from spent fluorescent lamps. J. Environm. Manag., 115: 175–178.

Robert, D., Baez, Ed., and Setunge, S. 2021. A new technology of transforming recycled glass waste to construction components. Constr. Build. Mater., 313: 125539.

Sahmenko, G., Toropovs, N., Sutinis, M., and Justus, J. 2014. Properties of high performance concrete containing waste glass micro-filler. Key Eng. Mat., 604: 151–164.

Santos, E.J., Hermann, A.B., Vieira, F., Sato, C.S., Corrêa, Q.B., Maranhão, T.A., Tormen, L., and Curtius, A.J. 2010. Determination of Hg and Pb in compact fluorescent lamp by slurry sampling inductively coupled plasma optical emission spectrosmetry. Microchem. J., 96: 27–31.

Scarinci, G., Brusatin, G., and Bernardo, E. 2005. Glass foams. pp. 158–176. *In*: Scheffler, M., and Colombo, P. (eds.). Cellular Ceramics: Structure, Manufacturing, Properties And Applications, Weinheim: WILEY-VCH Verlag GmbH & Co. KGaA, Weinheim, Germany.

Shao, Y., Lefort, L., Moras, S., and Rodriguez, D. 2000. Studies on concrete containing ground waste glass. Cement Concrete Res., 30: 91–100.

Sharma, L., Taak, N., and Bhandari, M. 2021. Influence of ultra-lightweight foamed glass aggregate on the strength aspects of lightweight concrete. Materials Today: Proceedings, 45: 3240–3246.

Souza, M.T., Maia, B.G.O., Teixeira, L.B., Oliveira, K.G., Teixeira, A.H.B., and Oliveira, A.P.N. 2017. Glass foams produced from glass from glass bottles and eggshell wastes. Process Saf. Environ., 111: 60–64.

Spiridonov, Y.A., and Orlova, L.A. 2003. Problens of foam glass production. Glass and Ceramics, 60: 9–10.

Srinivasan, S. 2019. The light at the end of the tunnel: Impact of policy on the global diffusion of fluorescent lamps. Energy Policy, 128: 907–918.

Steiner, A.C. 2006. Foam glass production from vitrified municipal waste fly ashes, 2006, 223f. Tesis – Technische Universiteit Eindhoven, Dusseldorf.

Taurino, R., Pozzi, P., Lucchetti, G., Paterlini, L., Zanasi, T., Ponzoni, C., Schivo, F., and Barbieri, L. 2013. New composite materials based on glass waste. Composites: Part B 45: 497–503.

Teixeira, L.B., Fernandes, V.K., Maia, B.G.O., Arcaro, S., and Novaes de Oliveira, A.P. 2017a. Vitrocrystalline foams produced from glass and oyster shell wastes. Ceramics International, 43: 6730–6737.

Teixeira, L.B., B.G. de O. Maia, Arcaro, S., Sellin, N., and de Oliveira, A.P.N. 2017b. Production and characterization of vitrocrystalline foams from solid wastes. Matéria, v.22, n.4.

Ting, G.H.A., Tay, Y.W.D., and Tan, M.J. 2021. Experimental measurement on the effects of recycled glass cullets as aggregates for construction 3D printing. J. Clean. Prod., 300: 126919.

US EPA Title 40 - Protection of Environment, Chapter I - Environmental Protection Agency Solid Wastes, Hazardous Waste Management System. 2012.

Vereshchagin, V.I., and Sokolova, S.N. 2006. Formation of a porous structure in granulated glass ceramic material from zeolite-bearing rock with alkali additives. Glass and Ceramics, 63: 227–229.

Xie, F., Liu, L., and Li, J. 2012. Recycling of leaded glass: scrap cathod ray glass and fluorescent lamp glass. Procedia Environmental Sciences, 16: 585–589.

Yot, P.G., and Méar, F.O. 2011. Characterization of lead, barium and strontium leachability from foam glasses elaborated using waste cathode ray-tube glasses. J. Hazard. Mater., 185: 236–241.

Zhang, Q., Wang, R., Shen, Y., Zhan, L., and Xu, Z. 2022. An ignored potential microplastic contamination of a typical waste glass recycling base. J. Hazard. Mater., 422: 126854.

6

Recovery and Disposal of Tannery Waste Containing Toxic Metals

Caroline Agustini,* Taysnara Simioni, Nadini Pinheiro, Éverton Hansen, Victória Kopp and *Mariliz Gutterres*

1. Leather processing and waste generation

Tanneries are one of the oldest types of industries and are important in both developed and developing countries. Global leather trade reaches 150 billion dollars annually (Kanagaraj et al. 2020). Besides its great importance in the economy, this industry also processes, as its raw material, a by-product of the slaughtering industry: animal hide. However, the economic importance of tanneries has as a counterpoint the environmental issues related to the tanning process and the generated waste, both in the form of liquid effluents and in the form of solid waste (Agustini and Gutterres 2017a, Piccin et al. 2016).

The process performed in tanneries consists of transforming the animal hide into leather. The processing is divided into three main phases: beamhouse, tanning, and finishing. Beamhouse operations aim to clean and prepare the hide or skin for tanning. Tanning is the unitary operation in which tanning agents react with the hide matrix, stabilizing the collagen, so that the hide becomes resistant to chemical and physical changes and to biological degradation. The tanning processes are classified according to the type of tanning agent used to bind the collagen fibers (Gutterres and Mella 2015). Due to the wide variety of leathers, there are many possible types of tanning, chrome tanning being the most used: it uses basic chromium sulfate

Federal University of Rio Grande do Sul, Chemical Engineering Department, Laboratory for Leather and Environmental Studies – LACOURO, Porto Alegre, Brazil.
Corresponding author: agustini@enq.ufrgs.br

as a tanning agent, which is the chemical more discussed of the tanning industry regarding environmental impact. The finishing step operations give the leather its final characteristics, such as firmness, color, and softness, depending on the purpose for which the leather will be destined (Dettmer et al. 2010a, Winter et al. 2017).

Since the transformation of raw hides into finished leather takes place in drums where the hide is submerged in water (average tannery water consumption is between 25–80 m³ per tons of processed raw material) as well as the many washing operations carried out during the leather manufacture or by means of finishing products in aqueous dispersions applied to the leather surface, most of the chemicals added during tanneries process remain in the wastewater, claiming properly treatments before discard (Agustini and Gutterres 2017a). The wastewaters are treated by association of mechanical, chemical and biological methods, in wastewater treatment plants (WWTPs) (Vazifehkhoran et al. 2018). There are advanced alternative treatments that can increase the wastewater's chromium removal efficiency, reuse of chromium and improve the water quality in the many aspects and requirements.

The main solid waste with chromium consists of shavings generated in the thickness adjustment stage of leather and the residual sludge produced in the distinct operations in WWTPs. Sludge WWTPs are commonly generated from the physical sedimentation of solids in wastewater pre-treatment operation, after the physical-chemical coagulation/flocculation and after biological treatment processes, used to reduce the concentrations of sedimentable solids, organic loads and chemical compounds. The leather industry faces (Sathish et al. 2019) serious problems in the incomplete use of solid waste generated. The current method of disposing of the chromium containing wastes is through its disposal in landfills for hazardous waste, which implies labor costs and pollution of the place (Aarthi et al. 2019). Beyond the demand of land areas, soil and groundwater are subjected to be contaminated with chromium that can accumulate in the food chain, in addition to gradual or increased risk of uncontrolled generation of biogas, which releases CH_4 directly into the atmosphere. A diagram of the process carried out in tanneries with details of the steps performed and indications of the chromium added and remaining in the residues generated in each step is shown in Fig. 1. Solid tannery residues can be used to produce new products with benefit—such as adsorbents—and to generate energy through sustainable technologies—such as biogas.

2. Chromium in wastewater

The tanning operation is carried in the pickling bath, with 50 to 100% of bath volume over the hide weight, at room temperature and bath pH between 2.8 and 3.0. Basic chromium sulfate (generally 33% basic) is used in an amount of about 2.0 % Cr_2O_3 on the hide weight, that is, a concentration between 40 and 47 g/L of Cr_2O_3. The drum rotates continuously for about 5 hours.

The chemical processing carried out in tanneries from beamhouse to wet finishing occurs with the hide in an aqueous submerged bath, which ends up generating a large amount of wastewater. The added chromium is not fully absorbed and crosslinked to the hide, remaining partially in the wastewater, about 6–8 g/L, that is, 13–20% of

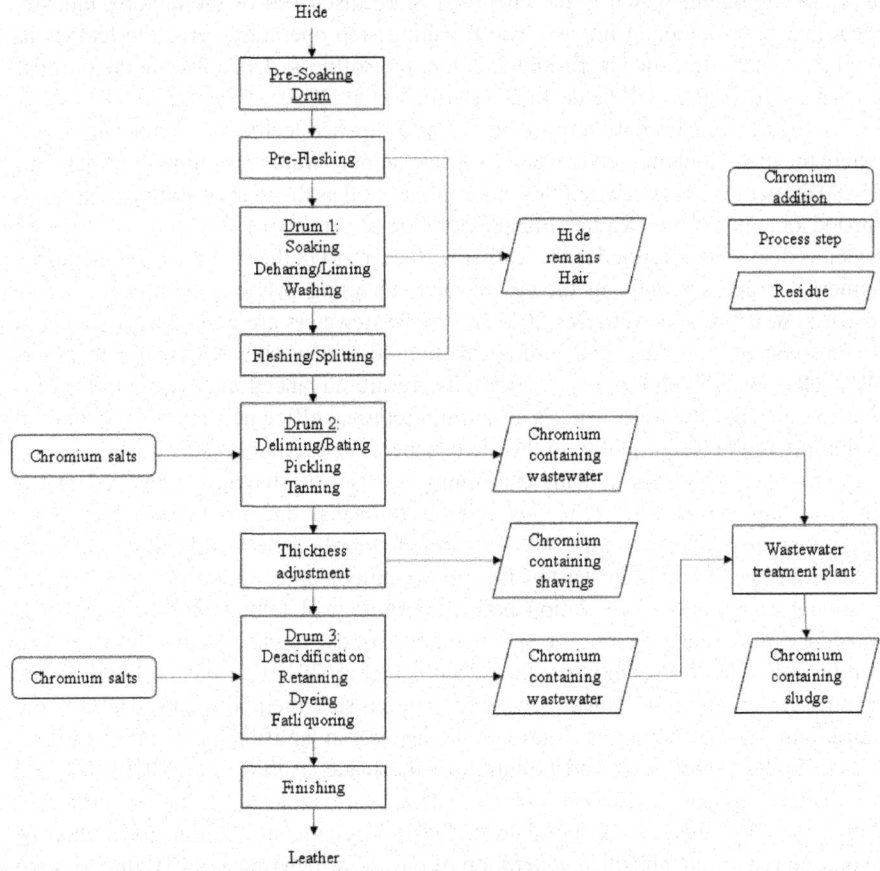

Fig. 1: Flowchart of the production of leather with an indication of chromium addition and presence in wastes (Agustini and Gutterres 2017a).

Cr_2O_3 remains in the residual bath. Thus, alternatives are studied, and applied for the removal, use or recovery of chromium.

2.1 *Tanning bath recycling*

Wastewater reuse is a rational practice applied in industrial processes to reduce water consumption and the discharge of liquid effluents. In tanneries, tanning bath recycling can provide additional benefits, since remaining chromium in the tanning bath can be reused, minimizing the consumption of chemicals and the pollution load of the raw effluent generated by this industry. However, recycling of residual tanning baths must be carefully investigated, as it should not interfere with the quality of the final product obtained.

Pilot and industrial-scale studies have shown the feasibility of recycling chrome-containing baths. A study of direct reuse of tanning baths containing chromium (Aquim et al. 2019) obtained 42% reduction in water consumption, in addition to reducing the cost of the tanning formulation by 10%, considering reductions achieved for basic chromium sulfate and sodium chloride supply. The volume of water used in the formulation was replaced by the reuse of tanning baths, with tests adding from 4.0 to 5.1% of basic chromium sulfate, while the original formulation used 5.5% of basic chromium sulfate. Tanning baths after reuse came out cleaner when compared to its composition before reuse, and each residual float was reused only once, because after that the concentration of Cr_2O_3 in some floats was too low to be reused again. The chromium content in leather met the required specification of 3.5% (ISO 5398-1, 2007), i.e., reuse did not decrease the chromium absorption. Shrinkage tests also showed satisfactory results for the leather obtained. Regarding the sodium chloride concentration in the baths (added to control hide swelling), the amount of salt could be reduced from 5.5% to 4.0%, since reuse tanning baths already contained sodium chloride.

The increase of the chromium exhaustion in the tanning bath was tested (Zhang et al. 2016) using a salt-free process, 5.0% chrome offer and pH 5.0 bath, and obtained 98% exhaustion of chromium. The spent chrome tanning liquors from the process were completely collected and reused for ten cycles. Reduction of 82.4% chromium concentration in the spent float was observed. The tanning process and the leather quality were not negatively affected by the recycling of the spent float from the tanning process (Zhang et al. 2017a).

The residual chrome tanning bath can also be treated before reuse, especially when the concentration of organic components is not suitable for direct reuse, which would give leather poor appearance and physical properties that do not meet the established requirements. The application of peroxy-electrocoagulation process was investigated to remove chemical oxygen demand and protein from chrome-containing baths, achieving 53 and 100% removal efficiency, respectively (Benhadji et al. 2018). Tanning tests with the treated chromium bath showed leather of good hydrothermal stability measured by the shrinkage temperature.

Tanneries willing to apply the recycling of chrome-containing baths should perform quality tests of the leather on a pilot scale before its application on an industrial scale, since any change in the supply of chemicals or operational conditions may affect the quality of the leather. In addition to reducing the amount of water used in the process, recycling chrome tanning baths promotes the reuse of chromium from wastewater, which would otherwise be discarded after treatment to meet the standards for the disposal of effluents.

2.2 Chromium precipitation for recycling

Chemical precipitation is an effective method and it is widely used in industrial processes, as it is simple and of low operational cost. In the precipitation process, the dosed product reacts with the dissolved metal ions and these are converted to an insoluble precipitate. The precipitate can be separated from the water through sedimentation or filtration and the treated wastewater can be reused or discarded (Mella et al. 2015).

Chemical precipitation of chromium is often used in tanneries. Among the various products that can be used, the main ones are magnesium oxide, for carrying out the process in a short time, and sodium hydroxide, for its low cost. Chemical precipitation with sodium hydroxide is one of the techniques most used in tanneries to remove chromium from wastewaters, due to the ease and simplicity of operation, the low cost and availability on the market of chemicals, and the lack of sophisticated equipment for carrying out the process. Despite these advantages, chemical precipitation requires a high amount of chemicals to reduce the concentration of metals for disposal. Another disadvantage is the excessive generation of sludge that requires additional treatment or adequate final disposal (Mella et al 2016).

Mella et al. (2015) studied the removal of chromium from tanning wastewater through chemical precipitation technique and its reuse in the tanning process. Efficient chromium removal from wastewater samples was achieved, with up to 99.74% removal efficiencies. The recovered chromium was reused as a tanning agent in leather processing, where good permeation of chromium was reached in leathers tanned with the liquors prepared from the sludge. The hides tanned with the chromium-containing liquor recovered by chemical precipitation showed good hydrothermal stability.

It is difficult to completely remove Chromium (III) from tannery effluent by alkaline precipitation due to the abundance of strong organic ligands. Thereby, the speciation of the residual Chromium (III) after alkaline precipitation is of crucial significance to guide the selection and design of further treatment processes. Wang et al. (2016) revealed the speciation of the residual Chromium (III) with the aid of comprehensive analytical techniques and they found the structures of Chromium (III) complexes.

2.3 Chromium (III) adsorption

Adsorption is a mass transfer process between a solid, which is referred to as adsorbent, and a fluid phase that carries the adsorbate, which is the molecule to be removed (Wang and Guo 2020). It is a unit operation that represents a surface phenomenon where the adsorbate is attracted to the adsorbent's surface and accumulates on it (Foust et al. 1982). The interactions between adsorbate and adsorbent will specify the type of adsorption that occurs. If there is no chemical bonding and desorption process is easily possible, adsorption is classified as physisorption in which the interactions are represented by weaker forces such as Van der Walls. However, if electrons are transferred and chemical bonding is established, chemisorption occurs, representing a higher interaction by ionic or covalent bonds and the desorption process is more difficult (Piccin et al. 2017). Nevertheless, adsorption efficiency can be influenced by process conditions such as pH of the medium, adsorbent dosage, temperature and initial adsorbate concentration (Soliman and Moustafa 2020).

Adsorption is a very promising technique for tannery wastewater treatment containing organic or inorganic substances, such as dyes, aromatic compounds and heavy metals (Crini 2005). Although chromium is used in its trivalent form as tanning agent, living beings' exposure to trivalent or hexavalent forms can lead to measurable DNA–protein crosslinks formation, indicating exposure of the cells to reactive forms of this metal. Therefore, chromium needs appropriate removal from wastewater (Sebestyén et al. 2019). Chromium (III) adsorption has been proposed using different types of adsorbents. Clay has attracted attention of researchers as an adsorbent due to its large availability, surface area and low production cost. In aqueous solutions, it is possible to achieve 95.21% of chromium removal using 1 g.L^{-1} of bentonite clay and even 97.37% using 1.5 g.L^{-1} of modified bentonite clay with MnFe$_2$O$_4$ particles. Effect of pH has a large impact on the process, and increasing pH is prejudice to Chromium (III) adsorption, so the pH value of 6 was reported as optimum using those adsorbents. Contact time also plays an important rule, and length of times beyond 60 and 90 min could cause saturation of active sites. In addition, temperatures higher than 25°C decreased the removal, which could be caused by changed surface area of active sites on the adsorbents (Ahmadi et al. 2020).

Trivalent chromium can also be adsorbed by multiwall carbon nanotubes modified with anionic surfactant sodium lauryl sulfate, showing removal efficiency around 92%. Electrostatic interactions between Chromium (III) particles and adsorbents of negatively charged surface make adsorption process very easy and thermodynamics studies revealed a physisorption process, but optimum values were obtained fixing pH at 5 to avoid chromium precipitation. Also, a dosage of 2.0 g.L^{-1}, contact time of 60 min and 25°C are needed. However, increasing Chromium (III) initial concentration to values higher than 150 ppm decreases adsorption capacity (Dokmaji et al. 2020). Other materials are effectively used for Chromium (III) adsorption in aqueous solutions such as gallic acid-functionalized graphene hydrogel (Liu et al. 2019), xanthate pine bark (Arin et al. 2018) and sodium polyacrylate (Ouass et al. 2018), for instance.

Chromium removal from real tannery wastewater has been proposed using low-cost adsorbents such as phosphate mine waste. Removal of 99.21% and adsorption capacity of 97.23 mg.g^{-1} could be achieved at 50°C and at the equilibrium dose of adsorbent of 40 g.L^{-1} (2 g), with initial chromium concentration of 3.92 g.L^{-1}, besides a removal of more than 66% of chemical oxygen demand (COD). Adsorbent surface area of phosphate mine waste showed an average of 15.12 m^2.g^{-1}. Besides that, adding phosphate mine to the effluent increased pH around 6. In addition, mechanism is attributed to adsorption and ion exchange on the solid surface, precipitation as Cr(OH)$_3$ and co-precipitation of aqueous species of the adsorbent (such as phosphate, silicate) combined with Chromium (III) (Oumani et al. 2019). Water hyacinth biochar had also been successfully used for chromium adsorption from real tannery wastewater (Hasem et al. 2020).

Synthetic effluent has also been studied for chromium removal and using activated carbon as adsorbent. Researches highlight low-cost materials as activated carbon precursors, such as *Parthenium hysterophorus* weed, an Ethiopian plant, with the adsorbent prepared through pyrolysis at 500°C. Synthetic effluent was formulated with cooper, lead, zinc calcium, potassium, sodium and chromium. Conditions of the optimum point (pH =2, contact time = 90 min, adsorbent dose = 90 g.L^{-1}) resulted in a chromium removal of 90% at initial chromium concentration of 100 mg.L^{-1}. Authors also investigated chromium removal from real tannery wastewater at optimum point conditions, resulting in 89% of removal, which was very close to synthetic effluent, a fact that was attributed to similar composition between the effluents (Bedada et al. 2020).

3. Use and treatment of solid waste containing chromium

The main chromium-containing solid residues generated in the tanning process are the leather shavings generated in the thickness regulation stage and the sludge generated in the WWTPs. The common management of this waste is its disposal in industrial landfills. However, due to its high organic load and physic-chemical properties, there are several alternatives for the recovery and reuse of these residues.

3.1 Landfill disposal—leaching of chromium in the soil

The leather industry generates large amounts of chromium-containing solid waste. Historically, the most common way to manage solid tannery wastes is by disposing them of at landfill sites. The increase in environmental restrictions for the disposal of this waste in landfills and the increase in operation/maintenance costs have motivated the scientific community in the search for alternative treatments for these materials. Even so, landfill remains a common method of management and, in such cases, it is necessary to monitor and treat liquid effluents (leachate) and the generated gases (Agustini and Gutterres 2017a, Cabrera-codony et al. 2020, Polizzi et al. 2018, Zupančič and Jemec 2010).

After the waste is disposed of in landfills, organic compounds begin to be degraded through biochemical reactions by microorganisms present in the medium. In the presence of atmospheric air, which occurs soon after the sealing of the landfill cells, organic compounds are oxidized via aerobic route, through reactions similar to combustion, since the products generated are carbon dioxide and water vapor. However, the main biotransformation occurring in these landfills is anaerobic digestion (AD) that occurs after the complete exhaustion of the oxygen supply (Priebe et al. 2011).

The confinement of tannery wastes in industrial hazardous waste landfill, where the wastes are slowly degraded under anaerobic conditions, ultimately generates high volumes of slurry and gas (biogas) which must be adequately collected and treated. Priebe et al. (2011) have characterized the gases generated in cells of an industrial hazardous waste landfill located in the South of Brazil, containing only tannery waste (30% of sludge from the effluent treatment plants of the tanneries and 70% of wet-blue shavings). The concentrations of methane (CH_4), carbon dioxide (CO_2), nitrogen (N_2), oxygen (O_2) and hydrogen sulfide (H_2S) were analyzed in cells that had been closed between 5 to 120 months before sample collection. For cells sealed less than two years before sample collection, the authors observed a drop in the concentration of oxygen (O_2) and an increase in the concentration of carbon dioxide (CO_2), indicating the occurrence of the aerobic decomposition process. However, they also observed an increase in methane concentration for the same time period, suggesting the simultaneous occurrence of anaerobic methanogenic processes, which increased considerably for cells with a sealing time greater than two years, reaching, in these cases, the proportion of 60% of methane. Natural and uncontrolled AD of waste into landfill releases methane (CH_4) and carbon dioxide (CO_2) into the atmosphere, which is undesirable due to the effects of global warming, resulting from the release of greenhouse gases (Agustini and Gutterres 2017a).

Leachate migration from uncontrolled municipal and industrial solid waste disposal sites leads to significant subsoil and groundwater pollution. In order to combat this geo-environmental problem, landfill construction must be done in such a way that the solute transport along with leaching is minimal and significantly attenuated (Ghosh et al. 2012). Landfill leachates present variable and complex composition, usually exhibiting strong dark brown color, high amounts of biodegradable and recalcitrant organic compounds, ammonium, suspended solids and heavy metals. Typically, landfill leachates are retained in lagoons located onsite and then transferred to centralized WWTPs, practices that have been increasingly questioned by water authorities. The onsite landfill leachate treatment usually comprises the implementation of multistage treatment strategies based on biological/ physical/chemical techniques due to the complexity of landfill leachates and the need for cost effective treatments. The definition of the best multistage treatment strategy is a highly challenging task that depends on the leachate characteristics, legal disposal standards to be met, treatment flexibility, treatment costs, and environmental and social sustainability. An integrated treatment embracing the following stages proved to be efficient for the remediation of the leather tannery landfill leachate: (i) initial biological process for removal of biodegradable organics, ammonium and alkalinity, (ii) coagulation/flocculation process for total elimination of chromium and partial removal of recalcitrant organic compounds and suspended solids, (iii) advanced oxidation process (AOP) or electrochemical AOP (EAOP) for degradation of recalcitrant organic compounds and biodegradability enhancement, and (iv) final biological polishing step. In the referenced study, the final leachate fulfilled the discharge requirements into water bodies (Webler et al. 2019).

Landfilling of tannery waste could represent a sustainable solution only if its long-term residual emissions reach an acceptable level in one-generation time. To reach sustainable landfilling, mobile substances should be degraded, removed or stabilized, landfill uncontrolled emissions should be minimized and the deposited materials should reach a final storage quality in equilibrium with the environment. An important role can therefore be played by mechanical and biological pre-treatments of tannery sludge on obtaining a condition of geological deposit of inorganic substances (i.e., heavy metals) and non or slowly biodegradable organic compounds, during final sludge landfilling. The results obtained by Alibardi and Cossu (2016) indicated that the combination of an aerobic stabilization process followed by a compaction and drying process applied to tannery sludge allows to (i) extend the landfill life time by reducing volume and mass of tannery sludge to be disposed of, (ii) minimize long-term environmental impacts of tannery sludge landfilling due to an increased biological stability of treated sludge and a reduced mobility of organic and inorganic compounds, (iii) maximize the role of tannery sludge landfilling as carbon and metal sink, and also (iv) the overall process allowed reducing leachability of metals due to a reduced content of dissolved organic carbon (Alibardi and Cossu 2016).

3.2 *Anaerobic digestion and biogas production—chromium inhibition in the process*

In response to strict regulations and environmental policies that encourage alternative treatments and considering its high organic load, anaerobic digestion (AD) has become an attractive solution from the perspective of sustainable and integrated management of solid waste (Agustini and Gutterres 2017a, Lazaroiu et al. 2018, Polizzi et al. 2018). AD is a biochemical process of decomposition of organic matter, carried out by a consortium of microorganisms that live symbiotically in the absence of oxygen (Seadi et al. 2008, Xu et al. 2019). The process involves four stages: hydrolysis, acidogenesis, acetogenesis and methanogenesis, which occur sequentially and in parallel, because the different microbial communities involved in each stage work in sequence, with the products from one stage serving as a substrate for the next one (Hagos et al. 2017, Khalid et al. 2011, Parsaee et al. 2019). First, fermentative bacteria hydrolyze the complex organic macromolecules turning it into simpler and more soluble compounds. Afterwards, these simpler molecules are degraded by acidogenic (fermentative) bacteria into short-chain fatty acids (propionate, acetate, butyrate and lactate), alcohols, and gaseous by-products (NH_3, H_2, CO_2 and H_2S). In the next stage, acetogenic bacteria convert the high molecular weight organic acids and alcohols produced in the previous stage, into acetic acid, CO_2 and H_2. Finally, in the final phase of AD, methane is produced by two groups of methanogenic bacteria: acetoclastic bacteria, which are responsible for the decomposition of acetate into methane and carbon dioxide, and hydrogenotrophic methanogens, which produce methane using hydrogen as an electron donor and carbon dioxide as a receptor (Appels et al. 2008, Li et al. 2019, Mirmohamadsadeghi et al. 2019, Neshat et al. 2017). The biological process of AD is complex and depends on several factors, including substrate characteristics (chemical composition, nutrient balance, particle size) and process parameters (pH, temperature, organic loading rate, retention time) (Siddique and Wahid 2018, Zhang et al. 2019).

AD is a promising alternative for the management of organic materials from the technological point of view, since it is capable of converting practically all sources of biomass, including different types of waste, into a highly energetic biogas (Agustini and Gutterres 2017b). Biogas produced in AD processes is composed mainly of CH_4 (50–70%) and CO_2 (30–50%). The relative content of CH_4 and CO_2 in biogas is dependent on the nature of the substrate and the parameters used in the process (Angelidaki et al. 2018, Gao et al. 2018, Khan et al. 2017, Mihaescu et al. 2018, Wu et al. 2015). This biogas can be used for the production of fuel, electricity, chemicals and heat (Appels et al. 2008). AD also originates a by-product that, because of its elemental composition, which typically contains large amounts of ammonia, is increasingly being used as a bio fertilizer in the agricultural sector (Agustini and Gutterres 2017b, Zupančič and Jemec 2010).

A report of an attempt to produce energy through the AD of tannery residues was in 1982 by Cenni et al. (1982). Since then, several authors have investigated the feasibility of anaerobic processes for waste and tannery sludge, especially in those countries where the leather industry plays a prominent role in the national economy, such as in Brazil (Agustini et al. 2018c, b, Bonoli et al. 2014, Ganesh et al. 2006,

Gao et al. 2015, Kameswari et al. 2014, Priebe et al. 2016, Zupančič and Jemec 2010). As a general conclusion, studies on AD of solid leather wastes agree on the viability of the process, warning against possible operational problems related to the imbalance of the C/N ratio and the inhibitory conditions of ammonia, fatty acids and sulfides (Polizzi et al. 2018).

The influence of chromium in tannery sludge and shavings in AD processes of these residues was determined by Agustini et al. (2018a). The low concentration of chromium in sludge proved to be more suitable for AD than other widely used tanning agents, such as vegetable tannins which, because they are of vegetable origin, give the idea of being more bioavailable. The difference in the supply of chromium (3%) and tannin (30%) during the tanning process overall greatly influences the remaining amount of the respective tanning agent in the residue and, consequently, its biodegradability. Even with all the problems expressed concerning the wide use of chromium, a small amount of chromium remains in sludge (7.35 mg/l) and shavings (7.79%). The presence of chromium in the shavings did not influence AD at all, due to their small proportion in the assays—same proportion that these residues produced in tanneries. In another study, Agustini et al. (2018b) found that the low concentrations of chromium in sludge and in shavings are even more suitable for biogas and methane production than in tannery residues with its absence. Tests with chromium produced about 3 times more biogas, with 55% of CH_4 on average, than tests with tannery residues without it.

3.3 Leather shavings as adsorbent material

Leather industries produce an average of 450–730 kg of solid waste for each ton of raw hides (Pei et al. 2019). During thickness, adjustment leather shavings are generated (Basegio et al. 2006). This solid leather waste has a majority collagen composition and its other components will depend on the chemicals used in the process but especially on tanning agent used on tanning step (Shakilanishi et al. 2017). The most produced shavings are chrome tanned ones, since more than 90% of raw hides are tanned with chromium salts (Sasia et al. 2019). It is estimated that 2.6 million tons of chrome shavings are generated worldwide annually (Arcibaz-Orozco et al. 2019). Commonly, shavings are disposed in landfills or incinerated, but those methods are not environmentally suitable. In addition, chrome shavings are classified as a hazardous waste because of chromium toxicity and low biodegradability. Adsorption is a simple, low-cost and efficient way to handle leather shavings, and its use has been reported in several studies in recent years.

Chrome shavings have been used for dye adsorption, especially with anionic leather dyes such as Acid Red 357, since chrome shavings are positively charged. One of the main factors for leather shavings adsorption is pH, due to the influence on charge surface of the shavings, so low pH values contribute to enhance anionic dye removal. In aqueous solutions, it is possible to reach up to 58% of removal for this dye at pH = 3 and adsorbent dosage of 6.0 g L^{-1}, which shows an adsorption capacity of 68.2 mg g^{-1} at equilibrium (Piccin et al. 2016). Acid Brown 414, another anionic leather dye, has shown better affinity with chrome shavings and 95.4% of removal using 50 mg of adsorbent mass for a contact time of 60 minutes has been reported (Carvalho Pinheiro et al. 2020).

Chrome shavings can be used for dye removal from synthetic effluent in a laboratory-scale adsorption process. Formulation containing soaking, deacidification, washing, dyeing and fatliquoring and retanning steps can be treated and variables of adsorbent and dye concentration, pH and rotation speed have significant effect on the adsorption. Even though the amount of chemicals in the formulation can interfere on the adsorption and cause competition for active sites, anionic dye removal can reach up to 94.36% under the conditions of pH 3, adsorbent dosage of 16.8 gL^{-1} and 35 rpm, for an initial dye concentration of 104.1 mgL^{-1}. By optimization, it is possible to obtain low adsorbent dosages (around 8.0 gL^{-1}), but dye removal will also decrease (87.37%) and conditions will not be always possible to reproduce in laboratory, such as lowest pH of 2.3, since synthetic effluents are more difficult to handle (Gomes et al. 2016). However, toxicity of synthetic effluents can be reduced by up to 90% using chrome shavings (Piccin et al. 2016).

In addition, some researchers have investigated the transformation of leather shavings into activated carbon through pyrolysis since the solid waste in its original form does not have large surface area for adsorption. It is important to note that pyrolysis of chrome shavings can generate toxic gaseous byproducts such as hydrogen cyanid and isocyanic acid, in addition to yields lower than 40% in some cases (Fang et al. 2018, Poletto et al. 2016). However, surface area can be increased and even achieve microporous region values, depending on pyrolysis, activation and mineralization conditions, going from surface areas lower than 5 m^{2}g^{-1} (shavings form) to 927.4 m^{2}g^{-1} and so, being suitable for Methylene Blue adsorption, for instance (Manera et al. 2016). Other compounds can also be adsorbed by chrome shavings or their activated carbon, such as polyphenols (Marsal et al. 2012) and tannic acid (Chaudhary and Pati 2017).

Vegetable tanned shavings, pickled hide shavings and wet-white shavings have been proposed as adsorbents for dye removal from aqueous solutions. It is possible to achieve percentages of dye removal higher than 95% depending on dyes properties, whereas cationic or anionic dye characteristic, and without any leather shavings surface modification before adsorption. The presence or absence of tanning agent has an important influence on shavings adsorption process, since it is related to their functional groups available for bonding with dyes and determines if electrostatic interactions between adsorbent and adsorbate will be favorable for adsorption. Vegetable tannins have a phenolic structure that gives them a strong affinity with cationic dyes, while pickled hide and wet-white shavings have a cationic nature, being more suitable for anionic dyes removal (Carvalho Pinheiro et al. 2020).

3.4 *Obtaining activated carbon from sludge containing chromium*

Commercial activated carbon is the most used adsorbent material by their large specific surface area, high porosity and surface containing functional groups that increases the adsorption capacity (Jiang et al. 2013). Usually, the application of activated carbon increases the cost of the wastewater treatment process. For this reason, it is economically interesting to use cheaper raw materials to produce activated carbon sorbents, such as sludge from tannery wastewater (Rengarag et al. 2002).

In literature, there are few studies related to production of activated carbon from tannery solid wastes. Puchana-Rosero et al. (2016) produced activated carbon from the sludge of tannery WWTP through microwave-assisted pyrolysis. Mella et al. (2019) produced activated carbon using tannery solid waste mixed with cattle hair waste from unhairing steps, as a new alternative sorbent to remove leather dyes from wastewaters.

The tannery sludge seems to serve as a good precursor for production of activated carbon. The activated carbon is efficient and presents excellent adsorptive capacity for leather dyes, up to 70%. The adsorbent produced possesses mesoporous pore size distribution that makes tannery sludge an excellent alternative eco-friendly precursor for the production of activated carbon (Puchana-Rosero et al. 2018).

Kantarli and Yanik (2010) investigated the utilization of tannery shavings as raw material for activated carbon production. Activated carbons were produced from chromium and vegetable tanned leather shaving wastes by physical and chemical activation methods. The activated carbon produced from vegetable tanned leather shaving waste produced showed higher surface area and micropore volume than the activated carbon produced from chromium tanned leather shaving waste. The potential application of activated carbons obtained from vegetable tanned shavings as adsorbent for removal of water pollutants have been checked for phenol, methylene blue, and Chromium (VI). Adsorption capacities of activated carbons were found to be comparable to that of activated carbons derived from biomass.

Han et al. (2020) investigated the fabrication of nitrogen-doped activated carbons from leather solid wastes by a different activation method with KOH. This method increased the ratio of surface nitrogen species. The high surface area, abundant micropores, and plentiful surface pyridinic N guaranteed a superior nitrogen-doped activated carbon that could serve as an excellent adsorbent to remove phenols from wastewater as well as an outstanding electrode material with a high and stable charge/discharge capacity.

3.5 *Dechroming of chromium-containing wastes*

The disposal of chromium-tanned leather wastes, such as shavings and finished leather cuts, represents high cost for industries since large quantities are daily-generated (Ferreira et al. 2010). Dechroming is a technique that recovers the chromium content from those wastes to make them less hazardous for the environment. Also, by the dissolution of chromium (III) from the complex collagen–chromium, it is possible to reuse the collagen extracted at several applications such as coagulant and poultry feed production, depending on molecular mass, purity (about 80–95%) and chromium content of the collagen recovered, which commonly needs a purification step (Chaudhary and Pati 2016, Villena-Mozzo et al. 2018). However, since chromium is strongly bonded to collagen, dechroming is not very easy to be made (Wang et al. 2019). Hydrolysis is the most applied method for dechroming and consists in breaking collagen polypeptide chains through chemical or biological treatments, since using only water the reactions would need high temperatures and pressures (Amaral 2008).

Chemical hydrolysis is very usual for dechroming and can be performed using alkaline or acid agents such as sodium hydroxide and sulfuric acid, respectively (Chaudhary and Pati 2016). Chrome shavings have high amount of collagen and alkaline hydrolysis with NaOH solution can be used for recovering it and their chromium salts. At first, chrome shavings are submerged with alkaline solution, pH and temperature need to be controlled and the reaction time begins. After that, centrifugation is used to separate the collagen hydrolysate from the chromium cake and each of these resulting products need the appropriate treatment. On a pilot scale, it is possible to recover 87.16% of the collagen content, but residual chrome is also present (1.17%). Despite the alkaline, hydrolysis is the most common commercial method and presents some advantages as lower reaction times and easier pH control, the chromium from the filtration cake is mostly in hydroxide form, being necessary posterior treatment to obtain basic chromium sulfate to reintroduce it to the tanning process (Barra Hinojosa and Marrufo 2020). Scopel et al. (2016) studied an innovative approach to changing chromed leather waste into a raw material for polymeric film production. Collagen hydrolysate was extracted through alkaline hydrolysis. The films produced with the addition of glycerol and collagen hydrolysate presented mechanical properties similar to the ones of commercial biodegradable films applied as mulches in agriculture. Dettmer et al. (2010a) handled the scrap-leather waste through thermal treatment (gasification and combustion). The chromium-rich ashes were utilized as a source of chromium for the synthesis of sodium chromate (Na_2CrO_4). The authors used sodium nitrate to oxidize the trivalent chromium to the hexavalent form. The resulting conversion of chromium (III) to chromium (VI) achieved was over 94%, and the sodium chromate obtained from the ashes showed physical properties similar to the commercial product. Dettmer et al. (2010b) studied obtaining sodium chromate, followed by the production of basic chromium sulfate to application in leather tanning. The typical characteristics of tanned hides were achieved, that is, the hydrothermal stability of the tanned samples and their structure modification (distance between fibers) verified using electron microscopy.

Acid hydrolysis also can be performed and the obtained acid chromium extracts can be used in tanning as addition to chromium sulfate, but dechroming degrees are around 30–60% using H_2SO_4 (Ferreira et al. 2010). Using a combination of inorganic and organic acids, H_2SO_4 and $H_2C_2O_4$, it is possible to dechrome chrome shavings. Initially, dechroming occurs by leaching the shavings with the mixed acids and then purification by salting out with NaCl solution, for instance. It is possible to vary the acid mass ratios and times of reaction and so obtain different kinds of collagen with the advantage of some of them with higher molecular mass compared to alkaline hydrolysis. However, optimum chromium degrees around 95% present the total amount of acids spent and the highest process cost, which reduces the viability of the process (Tian et al. 2020).

Secondary pollution is visible in chemical hydrolysis methods either by incomplete dechroming that generates new chrome-containing waste or by chromium solutions produced. Biological hydrolysis is known as a cleaner method for dechroming using microorganisms or enzymes that could grow on collagen-base wastes and hydrolysis products can be separated through filtration. However, leather manufacture has the purpose to prevent biodegradation, which makes this hydrolysis more difficult to proceed. *Penicillium* is a microorganism that can grow on chrome-tanned leather and destroy C-N and especially N-H collagen bonds. The mechanism is supported on the initial growth of *Penicillium* nourished by uncross-linked collagen, damaging collagen structure and releasing chromium to the medium, leading to microbe growth, the microorganisms will proceed the dechroming. Also, the chrome content is important for a biodegradation degree. Using *Penicillium* sp. in liquid medium, an increase in chrome content from 1.75% to 2.74% decreases the biodeterioration from 10.3% to 4.8%. After 96 h, a decline in microbe growth is observed due to secondary fungal metabolites produced. In addition, the pH of the medium becomes very acid until it reaches up to 3.36, which also inhibits microbe growth (Zhang et al. 2017b). In order to obtain higher dechroming degrees and protein recovery, biological agents can be used in combination with chemical alkalis-based hydrolysis in a two-step method, using the alkaline agents first and then enzymes. However, enzymes' costs and their process conditions are not always economically attractive. The use of conventional bating enzyme has been proposed as a low-cost alternative, showing a protein recovery of 79.45%, collagen hydrolysate and filter cake with an average of 1.28 ppm and 16.84% of chromium, respectively, being at chromium hydroxide form at the filter cake (Sasia et al. 2019).

3.6 Use of chromium waste as a fertilizer

Leather solid waste has been tested to produce organic fertilizer and as a soil conditioner due to its organic carbon content and suitability as a nitrogen source, since it contains, on average, 140 g/kg of N (dry weight) (Nogueira et al. 2011, 2010). The use of these residues as a source of nutrients for cultivated plants would provide less use of conventional fertilizers and less accumulation of residues in landfills. Thus, studies on the characterization of residues from the leather industry as well as on the feasibility of using these residues as an alternative source of N for cultivated plants are of great interest to agriculture and leather industry sectors (De Oliveira et al. 2008).

Many attempts have been developed to use the leather waste in the production of fertilizers without any treatment. Konrad and Castilhos (2002) investigated the soil chemical changes and growth of corn, after application of tannery waste (primary sludge with chromium at the rates of 13.8 and 27.6 mg ha^{-1}, and the sludge from liming, at the rates of 10.25 and 20.50 mg ha^{-1}). Corn yield with both treatments was similar to that obtained with commercial NPK fertilizer. The Chromium (III) present in the tannery sludges applied to the soil showed low mobility and presented no reactions of oxidation. Castilhos et al. (2002) evaluated the yields of wheat, lettuce, and radish, as well as the chemical changes in the soil due to the addition of tannery wastes (primary tannery sludge, chromium-tanned leather shavings and finished leather shreds). Yields of the three crops in microplots treated with the wastes were similar to that obtained in microplots treated with lime plus NPK fertilizer. Chromium concentrations in the soil and crops cultivated with tannery sludge varied from 40.7 to 71.2 and from 0.08 to 2.71 mg kg^{-1}, respectively. Additions of chromium-tanned leather shavings and finished leather shreds did not decrease the crop yields or change chromium concentrations in soil or plants. Daudt et al. (2007) investigated the possibility of using wet-blue leather sawdust as a component for growing media of seedlings of a garden plant (*Tagetes patula* L.). The addition of wet-blue leather sawdust reduced bulk density while increasing the total porosity and the amount of water in micropores. The seedlings showed good tolerance to the presence of wet-blue leather sawdust up to a fraction of 50% of the mixture with carbonized rice husks and fine vermiculite.

Although these research studies demonstrated the efficiency of leather solid wastes on plant growth, Chromium (III) is not essential for plants and its use can cause serious environmental problems (Lima et al. 2010). In general, the accumulation of chromium in plants alters enzymatic functions and structure of membranes and chloroplasts, thereby damaging cells, impairing root growth, causing chlorosis in young leaves, and decreasing pigmentation. A high chromium content in soil can affect biochemical processes in which symbiotic microorganisms are involved and subsequently the growth of plants. Chromium is a persistent heavy metal in soil, and its bioavailability depends on soil properties and metal retention time (Bavaresco et al. 2019).

As the solid residues of chrome-tanned leather are a complex of collagen and chromium and, knowing that the presence of chromium limits the application of these residues in the fertilizer industry, a potential solution could be the removal of chromium with the recovery of a solid collagen material, which contains high levels of nitrogen and could be used safely as a nitrogen source in agriculture (Majee et al. 2019, Nogueira et al. 2010, Pati and Chaudhary 2015).

De Oliveira et al. (2008) studied the use of leather residues after chromium (collagen) extraction as a nitrogen source for elephant grass. Collagen proved to be a good alternative source of nitrogen for the growth of elephant grass, in doses up to 16 t ha^{-1}, providing the need for nitrogen in a similar way to fertilization with mineral nitrogen. Pati and Chaudhary (2015) used purified protein hydrolysate derived from chrome-tanned leather shavings in a fertilizer formulation. The formulated fertilizer (1–3 t ha^{-1}) was employed as a nitrogen source in the production of soybean and it provided an encouraging result for productivity (plant height, number of leaves, number of pods and weight of seeds) of the soybean plant, similar to that of commercial fertilizer. In order to obtain a better use for the collagen and to aggregate its value, some works (Majee et al. 2019, Nogueira et al. 2011, 2010) aimed at evaluating the efficiency of collagen (wet-blue leather waste after chromium extraction) enriched with mineral phosphorus and potassium on the growth of plants. The application of PK enriched-collagen formulations showed promising agronomic results, equivalent or superior to those obtained with urea and commercial NPK formulations.

4. Other metals in leather processing

4.1 Tanning

In leather processing, the hide is turned into leather by the tanning step. The tanning agents react with the collagen matrix and are fixed into the hide, leading to the stabilization (Fuck et al. 2011). Chrome tanning is the most used method because the product is a light, less expensive leather of high hydrothermal and bacterial resistance (Agustini et al. 2018b). However, chromium has become a concern in the leather industry, due to the presence of hexavalent chromium in leather, owing to the oxidation of trivalent chromium and environmental pressure (Fuck et al. 2011, Gao et al. 2020). Due to the main concern of the environmental pollution problems of traditional chrome tanning method (Gao et al. 2020), there are other mineral tanning agents alternatives for chromium, such as aluminum, zirconium and titanium tanning agents.

Aluminium as a tanning agent was used a long time ago by the Romans (Musa et al. 2013). Its operation is similar to chrome tanning (Liu et al. 2020). The tanned leather is pure white, fine and tight, soft and stretchable (Gao et al. 2020, Haroun et al. 2009). The disadvantages of the leather are flat, thin, slightly underfilled, poorly tacky, and also not resistant to water washing (Gao et al. 2020). Hydrothermal stability of aluminium tanned leather is usually lower than other tanning agents (Onem et al. 2017), it is limited to 75°C (China et al. 2019). This happens because the aluminum (III) salts do not form strong covalent bonds (Onem et al. 2017) such as chromium, and the reaction is electrostatic. So, there is a weak nature of links with carboxyl groups of collagen molecules (China et al. 2019). Hence, aluminum salt is used in association with other tanning agents, and not as the only one (Gao et al. 2020).

Zirconium (IV) salts are a potential chrome-free tanning agent because of its good dyeing property, excellent filling and wear resistance properties (Li et al. 2021). It is also used in tanning especially in wet white production (Onem et al. 2017). However, there are still insufficiency and drawback in zirconium-tanned leather tear strength, permeability, usage amount and so on, compared to chrome-tanned leather (Li et al. 2021). The growth of zirconium complex in the fiber is based on its hydrolysis in the presence of water. They create more stable coordination compounds with collagen. However, the hydrophilic nature and the cost of these salts limit their applications (Onem et al. 2017). The tanning with zirconium is equivalent to vegetable tanning, because of the presence of so many hydrogen bonding on the complex (Heidemann 1993). Meanwhile, large amount of zirconium tanning agent (usually 12%) is employed in the tanning process in order to achieve ideal tanning effect (Li et al. 2021).

Titanium tanning agent has been designed as one of the most promising substitutes for chromium tanning in leather industry (Seggiani et al. 2014). Titanium (IV) salts have a polymeric nature that leads to a softer leather (Heidemann 1993). The leather is characterized by a positive charge so that most of the current anionic retanning agents, fatliquors, and dyestuff can be used to further enhance the technical performances. The upper leathers produced showed physical properties comparable with those of the conventionally chrome-tanned leathers used as control but a lower hydrothermal stability with shrinkage temperatures of 75–80°C due to the weak chemical interactions with collagen. The interaction with collagen carboxyls is similarly electrovalent, rather than covalent (Heidemann 1993). About the industrial applications of titanium tanning for high-quality bovine upper leather, it is a fact that they are very limited because the chrome-tanned leather still remains unmatchable with respect to its hydrothermal stability and excellent technical properties (Seggiani et al. 2014).

Distinct characteristics of physical properties and thermal stability happen due to different tanning agents and their molecular structure and interactions they make with collagen (Onem et al. 2017). The advantage of using these agents is that the leather shavings can be disposed of in landfills, due to the low toxicity. Thus, they can be used in pre-tanning or retanning to achieve a better performance of leather.

Although less toxic, the efficiency of the tanning with alternative mineral tanning agents is not as great as with chromium. The excellent technical properties and hydrothermal stability associated with the use of Chromium (III) tanning salts for the production of high-quality bovine leather make the chromium-based tanning in a dominant position with respect to the alternative chromium-free mineral-tanning processes that have been developed in the past decades (Seggiani et al. 2014).

4.2 Finishing

In leather finishing formulation chemicals containing heavy metals, like chromium, lead, cobalt, cadmium sulfate, copper and nickel may also be used in the pigment coat and alkyl tin compounds (Sreeram et al. 2009). The dye Acid Red 357, used nowadays in the leather industry, contains chromium in the structure (Piccin et al. 2012). Lead chromate-based pigments—chrome yellow—is used in preparation of transition colors and color matching during finishing. Cobalt complex is used within a metal complex dye. Also, cadmium sulfate is a metal salt used to prepare certain pigment colors. Copper is used to increase some properties such as color depth and light fastness (Basaran et al. 2002). These inorganic pigments are used due to their solidity and brilliant color but these are toxic heavy metals. However, they can be replaced by organic pigments or pigments from rare earth colorants, which are free of carcinogenic aryl amines (Dixit et al. 2015). Cadmium, cobalt and copper were found at quite low levels in finished leather (Basaran et al. 2002).

In leather finishing, several metals are used. Silver and titanium nanoparticles are applied for an antimicrobial leather for footwear industry (Carvalho et al. 2018). TiO_2 doped with N and Fe made a leather with self-cleaning properties (Petica et al. 2015). Copper nanoparticles improved wet and rub fastness, color fastness to water and adhesion strength (Kothandam et al. 2016).

5. Conclusion

The transformation of animal hide into leather usually uses basic chromium sulfate. It generates liquid and solid waste containing metal. This chapter presented different ways of recovery and disposal of the tannery waste. When the metal is present in wastewater, it is possible to reuse the effluent containing chromium, minimizing the consumption of chemicals and generation of pollution load. Chromium precipitation is a treatment that allows the reuse of the chromium and it is often used in tanneries. Adsorption removes the molecule with the adsorbent, such as clay and carbon nanotubes. Most of the tannery solid wastes are disposed of in landfill sites. This common solution does not contribute to the environment. Thus, one sustainable solution is anaerobic digestion, which produces biogas that can be used for the production of fuel and electricity. Chrome shavings can be used for dye adsorption and leather shavings can be transformed into activated carbon through pyrolysis.

Dechroming of chromium-containing wastes make the waste less hazardous for the environment. Chromium waste can produce fertilizer, but the removal of chromium seems to be best for the plants. Beyond chromium, it is possible to use other metals in leather processing. The tanning step can use aluminum, zirconium or titanium. However, the efficiency does not compare with chromium. In finishing step, some metals are presented in the dyes but they can be replaced. To avoid environmental pollution, it is necessary to find the right treatment of the wastes from tanneries. Different technologies are available for liquid and solid wastes and depends on the end goal.

References

Aarthi, A., Mahalingam, Umadevi, Parimaladevi, R., and Vasanth, G. Sathe. 2019. Polyvinyl thiol assisted Ag NPs as an efficient SERS analyzer and visible light photocatalyst for tannery waste landfill leachate. Vacuum, v. 161, n. December 2018, p. 125–129.

Agustini, C., and Gutterres, M. 2017a. Biogas production from solid tannery wastes. *In*: Artemio, A.V., and N. (eds.). Biogas: Production, Applications and Global Developments. 1. ed. No: Nova Science Publishers, p. 79.

Agustini, C., and Gutterres, M. 2017b. The range of organic solid wastes from anaerobically digested tanneries. *In*: Vic, A., and Artemio, N. (eds.). Biogas: Production, Applications and Global Developments. Nova Science Publishers, Nova York, p. 241.

Agustini, C., Michael, Meyer, Marisa da Costa, and Mariliz, Gutterres. 2018a. Biogas from anaerobic co-digestion of chrome and vegetable tannery solid waste mixture : Influence of the tanning agent and thermal pretreatment. Process Saf. Environ. Prot., 118: 24–31.

Agustini, C., Franciela Spier, Marisa da Costa, and Mariliz, Gutterres. 2018b. Biogas production for anaerobic co-digestion of tannery solid wastes under presence and absence of the tanning agent. Resour. Conserv. Recycl., 130: 51–59.

Agustini, C., Marisa da Costa, and Mariliz, Gutterres. 2018c. Biogas production from tannery solid wastes – Scale-up and cost saving analysis. J. Clean. Prod., 187: 158–164.

Ahmadi, A., Rauf, Foroutan, Hossein, Esmaeili, and Sajad, Tamjidi. 2020. The role of bentonite clay and bentonite clay@MnFe2O4 composite and their physico-chemical properties on the removal of Cr(III) and Cr(VI) from aqueous media. Environ. Sci. Pollut. Res., 27: 14044–14057.

Alibardi, L., and Cossu, R. 2016. Pre-treatment of tannery sludge for sustainable landfilling. Waste Manag., 52: 202–211.

Amaral, L. 2008. Alternatives for the treatment of chrome-tanned leather wastes - enzymatic hydrolysis and bacterial action. MsC. Dissertation, Federal University of Rio Grande do Sul, Porto Alegre, RS.

Angelidaki, I., Laura, Treu, Panagiotis, Tsapekos, Gang, Luo, Stefano, Campanaro, Henrik, Wenzel, and Panagiotis, G. Kougias. 2018. Biogas upgrading and utilization: Current status and perspectives. Biotechnol. Adv., 36: 452–466.

Appels, L., Jan, Baeyens, Jan, Degrève, and Raf, Dewil. 2008. Principles and potential of the anaerobic digestion of waste-activated sludge. Prog. Energy Combust. Sci., 34: 755–781.

Aquim, P., Patrice, M. de Aquim, Éverton, Hansen, and Mariliz, Gutterres. 2019. Water reuse: An alternative to minimize the environmental impact on the leather industry. J. Environ. Manage., 230: 456–463.

Arcibaz-Orozco, J., Barajas-Elias, B.S., Caballero-Briones, F., Nielsen, L., and Rangel-Mendez, J.R.. 2019. Hybrid carbon nanochromium composites prepared from chrome-tanned leather shavings for dye adsorption. Water Air Soil Pollut., 230: 142–159.

Arim, A., Margarida, J. Quina, and Licínio, M. Gando-Ferreira. 2018. Uptake of trivalent chromium from aqueous solutions by xanthate pine bark: characterization batch and column studies. Process Saf. Environ. (in press).

Barra Hinojosa, J., and Marrufo, L. 2020. Optimization of alkaline hydrolysis of chrome shavings to recover collagen hydrolysate and chromium hydroxide. Leather Footwear J., 20: 15–28.

Basaran, B., Iscan, M., Behzat, Oral Bitlisli, and Ahmet, Aslan. 2002. Heavy metal contents of various finished leathers. Journal of the Society of Leather Technologies and Chemists, 90.

Basegio, T., Haas, C., Pokorny, A., Bernardes, A.M., and Bergmann, C.P. 2006. Production of materials with alumina and ashes from incineration of chromium tanned leather shavings: Environmental and technical aspects. J. Hazard. Mater. B, 137: 1159–1164.

Bavaresco, J., Jessé, R. Fink, Moacir, T. Moraes, Antonio, R. Sánchez-Rodríguez, and Ibanor, Anghinoni. 2019. Chromium from hydrolyzed leather affects soybean growth and nodulation. Pedosphere, 29: 95–101.

Bedada, D., Kenatu, Angassa, Amare, Tiruneh, Helmut, Kloos, and Jemal, Fito. 2020. Chromium removal from tannery wastewater through activated carbon produced from *Parthenium hysterophorus* weed. Energ. Ecol. Environ., 5: 184–195.

Benhadji, A, Mourad, Taleb Ahmed, Hayet, Djelal, and Rachida, Maachi. 2018. Electrochemical treatment of spent tan bath solution for reuse. J. Water Reuse Desalination 8, 123–134.

Bonoli, M., Salomoni, C., Caputo, A., and Francioso, O. 2014. Anaerobic Digestion of High-Nitrogen Tannery By-products in a Multiphase Process for Biogas Production. Chem. Eng. Trans., 37: 271–276.

Cabrera-Codony, A., Ruiz, B., Gil, R.R., Lucia Alexandra Popartan, Eric Santos Clotas, Maria Martin, and Fuente, E. 2020. From biocollagenic waste to efficient biogas purification : Applying circular economy in the leather industry. Environ. Technol. Innov.

Carvalho, I., Ferdov, S., Mansilla, C., Marques, S.M., Cerqueira, M.A., Pastrana, L.M., Henriques, M., Gaidau, C., Ferreira, P., and Carvalho, S. 2018. Development of antimicrobial leather modified with Ag–TiO2 nanoparticles for footwear industry. Science and Technology of Materials, 30: 60–68.

Carvalho Pinheiro, N., Oscar W. Perez-Lopez, and Mariliz Gutterres. 2020. Solid leather wastes as adsorbents for cationic and anionic dye removal. Environ Technol., 27: 1–9.

Castilhos, D., Tedesco, M.J., and Vidor, C. 2002. Rendimentos de culturas e alterações químicas do solo tratado com resíduos de curtume e crômo hexavalente. Rev. Bras. Ciência do Soloncia do Solo, 26: 1083–1092.

Cenni, F., Dondo, G., and Tombetti, F. 1982. Anaerobic digestion of tannery wastes. Agricultural Wastes, May 1982, 4(3): 241–243.

Chaudhary, R., and Pati, A. 2016. Purification of protein hydrolyzate recovered from chrome tanned leather shaving waste. J. Am. Leather Chem. Assoc., 111: 10–16.

Chaudhary, R., and Pati, A. 2017. Adsorption isotherm and kinetics of tannic acid on to carbonized chrome tanned leather solid waste. J. Am. Leather Chem. Assoc., 112: 198–206.

China, C., Askwar Hilonga, Mihayo M. Maguta, Stephen S. Nyandoro, Swarna V. Kanth, Gladstone C. Jayakumar, and Karoli N. Njau. 2019. Preparation of aluminium sulphate from kaolin and its performance in combination tanning. SN Applied Sciences, 1(8): 1–8.

Crini, G. 2005. Recent developments in polysaccharide-based materials used as adsorvents in wastewater treatment. Prog. Polym. Sci., 30: 38–70.

Daudt, R., Cirilo Gruszynski, and Atelene Normann Kämpf. 2007. Uso de resíduos de couro wet-blue como componente de substrato para plantas. Cienc. Rural, 37: 91–96.

De Oliveira, D., Diana Q. Lima, Kele Tatiane Gomes Carvalho, Ana Rosa Ribeiro Bastos, Luiz Carlos Alves de Oliveira, João José Marques, and Robervone Nascimento. 2008. Use of leather industry residues as nitrogen sources for elephantgrass. Rev. Bras. Ciência do Solo, 32: 417–424.

Dettmer, A., Keila Nunes, Mariliz Gutterres, and Nilson Romeu Marcilio. 2010a. Production of basic chromium sulfate by using recovered chromium from ashes of thermally treated leather. Journal of Hazardous Materials, 176(1–3): 710–14.

Dettmer, A., Nunes, K., Gutterres, M., and Marcilio, N. 2010b. Tanning using basic chrome sulfate obtained from ash produced in the thermal treatment of leather wastes. The Journal of the American Leather Chemists Association, v. 105: 280–288.

Dixit, S., Ashish Yadav, Premendra Dwivedi, and Mukul Das. 2015. Toxic hazards of leather industry and technologies to combat threat: a review. Journal of Cleaner Production, 87: 39–49.

Dokmaji, T., Taleb Ibrahim, Mustafa Khamis, Mohamed Abouleish, and Isra Alam. 2020. Chemically modified nanoparticles usage for removal of chromium from sewer water. Environ. Nanotechnol. Monit. Manag., 14: 100319.

Fang, C., Xuguang Jiang, Guojun Lv, Jianhua Yan, and Xiaobing Deng. 2018. Nitrogen-containing gaseous products of chrome-tanned leather shavings during pyrolysis and combustion. Waste Manage, 78: 553–558.

Ferreira, M., Manuel F. Almeida, Sílvia C. Pinho, and Isabel C. Santos. 2010. Finished leather waste chromium acid extraction and anaerobic biodegradation of the products. Waste Manage., 30: 1091–1100.

Fuck, W., Gutterres, M., Marcílio, N.R., and Bordingnon, S. 2011. The influence of chromium supplied by tanning and wet finishing processes on the formation of Cr(VI) in leather. Brazilian Journal of Chemical Engineering, 28(2): 221–28.

Ganesh, R., Balaji, G., and Ramanujam, R.A. 2006. Biodegradation of tannery wastewater using sequencing batch reactor—Respirometric assessment. Bioresour. Technol., 97: 1815–1821.

Gao, D., Yiming Cheng, Pingping Wang, Fan Li, Yingke Wu, Bin Lv, Jianzhong Ma, and Jianbin Qin. 2020. An eco-friendly approach for leather manufacture based on P(POSS-MAA)-aluminum tanning agent combination tannage. Journal of Cleaner Production, 257: 120546.

Gao, S., Mingxing Zhao, Yang Chen, Meijuan Yu, and Wenquan Ruan. 2015. Tolerance response to in situ ammonia stress in a pilot-scale anaerobic digestion reactor for alleviating ammonia inhibition. Bioresour. Technol., 198: 372–379.

Gao, Y., Jiang Jianguo, Yuan Meng, Feng Yan, and Aikelaimu Aihemaiti. 2018. A review of recent developments in hydrogen production via biogas dry reforming. Energy Convers. Manag., 171: 133–155.

Ghosh, S., Sunil Kumar, Somnath Mukherjee, Dipankar Dey Tarafder, and Joseph P.A. Hettiaratchi. 2012. Adsorptive chromium removal by some clayey soil for abatement of tannery waste pollution. J. Hazardous, Toxic, Radioact. Waste, 16: 243–249.

Gomes, C.S., Piccin, J.S., and Gutterres, M. 2016. Optimizing adsorption parameters in tannery-dye-containing effluent treatment with leather shaving waste. Process Saf. Environ., 99: 98–106.

Gutterres, M., Mella, B. 2015. Chromiumin in tannery wastewater. In Heavy Metals in Water: Presence, Removal and Safety, 315–44.

Hagos, K., Jianpeng Zong, Dongxue Li, Chang Liu, and Xiaohua Lu. 2017. Anaerobic co-digestion process for biogas production: Progress, challenges and perspectives. Renew. Sustain. Energy Rev., 76: 1485–1496.

Han W., Hongliang Wang, Kedong Xia, Shanshuai Chen, Puxiang Yan, Tiansheng Deng, and Wanbin Zhu. 2020. Superior nitrogen-doped activated carbon materials for water cleaning and energy storing prepared from renewable leather wastes. Environment International, Volume 142, September, 105846.

Haroun, M., Palmina Khristova, Gurshi, A., and Anthony Dale Covington. 2009. Potential of Vegetable Tanning Materials and Basic Aluminum Sulphate in Sudanese Leather Industry. (March).

Heidemann, E. 1993. Fundamentals of Leather Manufacturing. Darmstadt: Eduard Roether KG Druckerei und Verlag.

ISO 5398-1:2007, 2007. Leather. Chemical Determination of Chromic Oxide.

Jiang Z., Garg, V.K., and Kadirvelu, K. 2013. Chromium (VI) removal from aqueous system using *Helianthus annuus* (sunflower) stem waste. Langmuir, 19: 731.

Kameswari, K. Sri Bala, Chitra Kalyanaraman, Umamaheswari, B., and Thanasekaran, K. 2014. Enhancement of biogas generation during co-digestion of tannery solid wastes through optimization of mix proportions of substrates. Clean Technol. Environ. Policy, 16: 1067–1080.

Kanagaraj, J., Rames C. Panda, and Vinodh M. Kumar. 2020. Trends and advancements in sustainable leather processing: Future directions and challenges—A review. Journal of Environmental Chemical Engineering, 8(5) October: 104379.

Kantarli, I., and Yanik, J. 2010. Activated carbon from leather shaving wastes and its application in removal of toxic materials. Journal of Hazardous Materials, 179(1–3): 348–356, 15 July 2010.

Khalid, A., Muhammad Arshad, Muzammil Anjum, Tariq Mahmood, and Lorna Dawson. 2011. The anaerobic digestion of solid organic waste. Waste Manag., 31: 1737–1744.

Khan, Imran Ullah, Mohd Hafiz Dzarfan Othman, Haslenda Hashim, Takeshi Matsuura, Ismail, A.F., Rezaei-DashtArzhandi, M., and Wan I. Azelee. 2017. Biogas as a renewable energy fuel—A review of biogas upgrading, utilisation and storage. Energy Convers. Manag., 150: 277–294.

Konrad, E., and Castilhos, D. 2002. Soil chemical changes and corn growth as affected by the addition of tannery sludges. Rev. Bras. Ciência do Solo, 26: 257–265.

Kothandam, R., Pandurangan, M., Jayavel, R., and Gupta, S. 2016. A novel nano-finish formulations for enhancing performance properties in leather finishing applications. Journal of Cluster Science, 27(4): 1263–72.

Lazaroiu, G., Lucian Mihaescu, Gabriel Negreanu, Constantin Pana, Ionel Pisa, Alexandru Cernat, and Dana-Alexandra Ciupageanu. 2018. Experimental investigations of innovative biomass energy harnessing solutions. Energies, 11.

Li, Wenbo, Jianzhong Ma, Yongxiang Zhou, Xiaodan Sun, and Dang-Ge Gao. 2021. The Application of Sulfonated Tetraphenyl Calix[4] Resorcinarene as a Novel, multi-functional and eco-friendly ligand in zirconium tanning system. Journal of Cleaner Production, 280: 124337.

Li, Y., Chen, Yinguang, and Wu, Jiang. 2019. Enhancement of methane production in anaerobic digestion process: A review. Appl. Energy, 240: 120–137.

Lima, D., Oliveira, L.C.A., Bastos, A.R.R., Carvalho, G.S., Marques, J.G.S.M., Carvalho, J.G., and de Souza, G.A. 2010. Leather industry solid waste as nitrogen source for growth of common bean plants. Appl. Environ. Soil Sci., 1–7.

Liu, G., Ruiquan Yu, Tianxiang Lan, Zheng Liu, Peng Zhang, and Ruifeng Liang. 2019. Gallic acid-functionalized graphene hydrogel as adsorbent for removal of chromium (III) and organic dye pollutants from tannery wastewater. RSC Adv., 9: 27060–27068.

Liu, Y., Bin Song, Jinwei Zhang, Carmen Gaidau, and Haibin Gu. 2020. Aluminum tanning of hide powder and skin pieces under microwave irradiation. Journal of Leather Science and Engineering, 2(1).

Majee, S., Halder, G., and Mandal, T. 2019. Formulating nitrogen-phosphorous-potassium enriched organic manure from solid waste: A novel approach of waste valorization. Process Saf. Environ. Prot., 132: 160–168.

Manera, C., Poli, J., Poletto, P., Ferreira, S., Dettmer, A., Wander, P., and Godinho, M. 2016. Activated carbon from leather shaving waste, part ii. effect of char demineralization and activation time on surface area and pore size distribution. J. Am. Leather Chem. Assoc., 111: 413–421.

Marsal, A., Maldonado, Fernando, Cuadros, Sara, and Elena Bautista, M. 2012. Adsorption isotherm, thermodynamic and kinetics studies of polyphenols onto tannery shavings. Chem. Eng. J., 183: 21–29.

Mella, B., Ana Cláudia Glanert, and Mariliz Gutterres. 2015. Removal of chromium from tanning wastewater and its reuse. Process Safety and Environmental Protection, 95: 195–201.

Mella, B., Ana Cláudia Glanert, and Mariliz Gutterres. 2016. Removal of chromium from tanning wastewater by chemical precipitation and electrocoagulation. Journal of the Society of Leather Technologists and Chemists, 100: 55–61.

Mella, B., Jaqueline Benvenuti, Renata F. Oliveira, and Mariliz Gutterres. 2019. Preparation and characterization of activated carbon produced from tannery solid waste applied for tannery wastewater treatment. Environmental Science and Pollution Research, 26: 6811–6817.

Mihaescu, L., Lazaroiu Gheorghe, Gabriel Negreanu, and Ionel Pîşă. 2018. Influence of the characteristics of biogas generated in the leather industry on combustion quality. Therm. Sci., 22: S1349–S1357.

Mirmohamadsadeghi, S., Keikhosro Karimi, Meisam Tabatabaei, and Mortaza Aghbashlo. 2019. Biogas production from food wastes: A review on recent developments and future perspectives. Bioresour. Technol. Reports 100202.

Musa, A.E., and Gasmelseed, G.A. 2013. Eco-friendly vegetable combination tanning system for production of hair-on shoe upper leather. Industrial Sciences, 2(1): 5–12.

Neshat, S.A., Mohammadi, Maedeh, Najafpour, Ghasem D., and Lahijani, Pooya. 2017. Anaerobic co-digestion of animal manures and lignocellulosic residues as a potent approach for sustainable biogas production. Renew. Sustain. Energy Rev., 79: 308–322.

Nogueira, Francisco G.E., Isabela A. Castro, Ana R.R. Bastos, Guilherme A. Souza, Janice G. de Carvalho, and Luiz C.A. Oliveira. 2011. Recycling of solid waste rich in organic nitrogen from leather industry: Mineral nutrition of rice plants. J. Hazard. Mater., 186: 1064–1069.

Nogueira, Francisco, G.E., Nayara T. do Prado, Luiz C.A. Oliveira, Ana R.R. Bastos, João H. Lopes, and Janice G. de Carvalho. 2010. Incorporation of mineral phosphorus and potassium on leather waste (collagen): A new NcollagenPK-fertilizer with slow liberation. J. Hazard. Mater., 176: 374–380.

Onem, E., Yorancioğlu, A., Karavana, H.A., and Yılmaz, O. 2017. Comparison of different tanning agents on the stabilization of collagen via differential scanning calorimetry. Journal of Thermal Analysis and Calorimetry, 129(1): 615–22.

Ouass, A., Lamya Kadiri, Youness Essaadaoui, Rida Allah Belakhmima, Mohammed. Cherkaoui, Ahmed Lebkiri, and El Housseine Rifi. 2018. Removal of trivalent chromium ions from aqueous solutions by Sodium polyacrylate beads. Mediterr. J. Chem., 7: 125–134.

Oumani, A., Mandi, L., Berrekhis, F., and Ouazzani, N. 2019. Removal of Cr3+ from tanning effluents by adsorption onto phosphate mine waste: Key parameters and mechanisms. J. Hazard. Mater., 378: 120718.

Parsaee, M., Kiani, M.K.D., and Karimi, K. 2019. A review of biogas production from sugarcane vinasse. Biomass and Bioenergy, 122: 117–125.

Pati, A., and Chaudhary, R. 2015. Soybean plant growth study conducted using purified protein hydrolysate-based fertilizer made from chrome-tanned leather waste. Environ. Sci. Pollut. Res., 22: 20316–20321.

Pei, Y., Shan Chu, Yiran Zheng, Jiahui Zhang, Hui Liu, Xuejing Zheng, and Keyong Tang. 2019. Dissolution of collagen fibers from tannery solid wastes in 1-allyl-3-methylimidazolium chloride and modulation of regenerative morphology. ACS Sustainable Chem. Eng., 7: 2530–2537.

Petica, A., Gaidău, C., Ignat, M., Şendrea, C., and Anicai, L. 2015. Doped TiO2 nanophotocatalysts for leather surface finishing with self-cleaning properties. Journal of Coatings Technology and Research, 12(6): 1153–63.

Piccin, J.S., Gomes, C.S., Feris, L.A., and Gutterres, M. 2012. Kinetics and isotherms of leather dye adsorption by tannery solid waste. Chemical Engineering Journal, 183: 30–38.

Piccin, J.S., Carolina S.Gomes, Bianca Mella, and Mariliz Gutterres. 2016. Color removal from real leather dyeing effluent using tannery waste as an adsorbent. Journal of Environmental Chemical Engineering, 4: 1061–1067.

Piccin, Jeferson Steffanello, Tito Roberto Sant'Anna Cadaval Jr., Luiz Antonio Almeida de Pinto, and Guilherme Luiz Dotto. 2017. Adsorption isotherms in liquid phase: experimental, modeling, and interpretations. pp. 19–52. *In*: Bonilla-Petriciolet, A., Mendoza-Castillo, D.I., and Reynel-Ávila, H.E. [eds.]. Adsorption Processes for Water Treatment and Purification. Springer International Publishing, Cham, Switzerland.

Poletto, P., Aline Dettmer, Vinicius Marcondes Bacca, Gabriela Collazzo, Edson Luiz Foletto, and Marcelo Godinho. 2016. Activated carbon from leather shaving waste. Part I. pyrolysis and physical activation. J. Am. Leather Chem. Assoc., 111: 325–333.

Polizzi, C., Felipe Alatriste-Mondragón, and Giulio Munz. 2018. The role of organic load and ammonia inhibition in anaerobic digestion of tannery fleshing. Water Resour. Ind., 19: 25–34.

Priebe, G.P.S., Marcílio, N.R., and Gutterres, M. 2011. Landfill Gas Generation of Final Leather Wastes Disposal, in: XXXI International Union of Leather Technologists and Chemists Societies.

Priebe, G.P.S., Kipper, E., Gusmão, A.l., Marcilio, N.r., and Gutterres, M. 2016. Anaerobic digestion of chrome-tanned leather waste for biogas production. J. Clean. Prod., 129: 410–416.

Puchana-Rosero, M.J., Adebayo, Matthew A., Lima, Eder C., Fernando M. Machado, Thue, P.S., and Umpierres, Cibele S. 2016. Microwave-assisted activated carbon obtained from the sludge of tannery-treatment effluent plant for removal of leather dyes. Colloids and Surfaces. A, Physicochemical and Engineering Aspects (Print), 504: 105–115.

Puchana-Rosero, M.J., Eder C. Lima, Bianca Mella, Dimitrius da Costa, Eduardo Poll, and Mariliz Gutterres. 2018. A coagulation-flocculation process combined with adsorption using activated carbon obtained from sludge for dye removal from tannery wastewater. Journal of the Chilean Chemical Society, 63: 3867.

Rengarag, S., Seung Hyeon Moon, R. Sivabalan, B. Arabindoo, and V. Murugesan. 2002. Agricultural solid waste for the removal of organics: adsorption of phenol from water and wastewater by palm seed coat activated carbon. Waste Manag. 22: 543–548.

Sasia, A.A., Paul Sang, and Arthur Onyuka. 2019. Recovery of collagen hydrolysate from chrome leather shaving tannery waste through two-step hydrolysis using magnesium oxide and bating enzyme. J. Soc. Leath. Tech Ch., 103: 80–84.

Sathish, M., Balaraman Madhan, and Jonnalagadda Raghava Rao. 2019. Leather solid waste : An eco-benign raw material for leather chemical preparation– A circular economy example. Waste Management, 87: 357–367.

Scopel, B., Lamers, D., Matos, E., Baldasso, C. and Dettmer, A. 2016. Collagen hydrolysate extraction from chromed leather waste for polymeric film production. The Journal of the American Leather Chemists Association, 111: 30–40.

Seadi, Teodorita Al, Dominik Rutz, Heinz Prassl, Michael Köttner, Tobias Finsterwalder, Silke Volk, and Rainer Janssen 2008. Biogas Handbook. University of Southern Denmark Esbjerg, Esbjerg.

Sebestyén, Z., Jakab, E., Badea, E., Barta-Rajnai, E., Sendrea, C., and Czégény, Z. 2019. Thermal degradation study of vegetable tannins and vegetable tanned leathers. J. Anal. Appl. Pyrolysis, 138: 178–187.

Seggiani, M., Monica Puccini, Sandra Vitolo, Cinzia Chiappe, Christian Silvio Pomelli, and Castiello, D. 2014. Eco-friendly titanium tanning for the manufacture of bovine upper leathers: pilot-scale studies. Clean Techn Environ Policy, 16: 1795–1803.

Shakilanishi, S., Narasimhan Kannan Chandra Babu, and Chittibabu Shanthi. 2017. Exploration of chrome shaving hydrolysate as substrate for production of dehairing protease by *Bacillus cereus* VITSN04 for use in cleaner leather production. J. Clean. Prod., 149: 797–804.

Siddique, M., and Wahid, Z. 2018. Achievements and perspectives of anaerobic co-digestion: A review. J. Clean. Prod., 194: 359–371.

Soliman, N., and Moustafa, A. 2020. Industrial solid waste for heavy metals adsorption features and challenges; a review. J. Mater. Res. Technol., 9: 10235–10253.

Sreeram, K.J., Jonnalagadda Raghava Rao, and Balachandran Unni Nair. 2009. Leather Pigments : Towards a Cleaner Greener Approach. In AAQTIC China.

Tian, Z., Ying Wang, Hao Wang, and Kang Zhang. 2020. Regeneration of native collagen from hazardous waste: chrome-tanned leather shavings by acid method. Environ. Sci. Pollut. Res., 27: 31300–31310.

Vazifehkhoran, Ali Heidarzadeh, Seung Gu Shin, and Jin M. Triolo. 2018. Use of tannery wastewater as an alternative substrate and a pre-treatment medium for biogas production. Bioresource Technology, 258(February): 64–69.

Villena-Mozzo, M., Néstor Caracciolo, and Susana Boeykens. 2018. Reuse of leather waste: collagen hydrolyzate for the treatment of tanneries effluents. J. Urban Environ. Eng., 12: 287–292.

Wang, D., Shiya He, Chao Shan, Yuxuan Ye, Hongrui Ma, Xiaolin Zhang, Weiming Zhang, and Bingcai Pan. 2016. Chromium speciation in tannery effluent after alkaline precipitation: Isolation and characterization. Journal of Hazardous Materials, 316: 169–177, 5 October.

Wang, J., and Guo, X. 2020. Adsorption kinetic models: Physical meanings, applications, and solving methods. J. Hazard. Mater., 390: 122–156.

Webler, A.D., Shiya He, Chao Shan, Yuxuan Ye, Hongrui Ma, Xiaolin Zhang, Weiming Zhang, and Bingcai Pan. 2019. Development of an integrated treatment strategy for a leather tannery landfill leachate. Waste Manag., 89: 114–128.

Winter, C., Caroline Agustini, M. Schultz, and Mariliz Gutterres. 2017. Influence of pigment addition on the properties of polymer films for leather finishing. Journal of the Society of Leather Technologists and Chemists, 101: 78–85.

Wu, B., Xiangping Zhang, Yajing Xu, Di Bao, and Suojiang Zhang. 2015. Assessment of the energy consumption of the biogas upgrading process with pressure swing adsorption using novel adsorbents. J. Clean. Prod., 101: 251–261.

Xu, F., Yangyang Li, Mary Wicks, Yebo Li, and Harold Keener. 2019. Anaerobic digestion of food waste for bioenergy production. Encycl. Food Secur. Sustain., 2: 530–537.

Zhang, C., Jiang Lin, Xinju Ji, and Biyu Peng. 2016. A salt-free and chromium discharge minimizing tanning technology: the novel cleaner integrated chrome tanning process. J. Clean. Prod., 112: 1055–1063.

Zhang, C., Fuming Xia, Jiajun Long, and Biyu Peng. 2017a. An integrated technology to minimize the pollution of chromium in wet-end process of leather manufacture. J. Clean. Prod., 154: 276–283.

Zhang, J., Zhangwei Han, Bo Teng, and Wuyong Chen. 2017b. Biodeterioration process of chromium tanned leather with *Penicillium* sp. Int. Biodeterior. Biodegradation, 116: 104–111.

Zhang, L., Kai-Chee Loh, and Jingxin Zhang. 2019. Enhanced biogas production from anaerobic digestion of solid organic wastes: Current status and prospects. Bioresour. Technol. Reports, 5: 280–296.

Zupančič, G., and Jemec, A. 2010. Anaerobic digestion of tannery waste: Semi-continuous and anaerobic sequencing batch reactor processes. Bioresour. Technol., 101: 26–33.

7

Treatment of Water Contaminated by Heavy Metal using Membrane Separation Processes

Wendel Paulo Silvestre* and *Camila Baldasso*

1. Introduction

The exact meaning of the term 'heavy metal' varies among the several fields of study that deals with these substances. However, it is widely agreed that heavy metals are naturally occurring elements that have a high atomic mass and a high density relative to water. Besides this, these elements are also considered to be toxic, especially to vertebrates (Appenroth 2009, Tchounwou et al. 2012). Arsenic, cadmium, mercury, and lead are examples of 'classic' heavy metals, especially due to the severe acute toxicity. Nevertheless, strongly radioactive elements, as polonium, radium, and uranium may also be considered as heavy metals, in this case, due to the toxicity that arises from their strong radioactivity (Castellote et al. 2002, Bai et al. 2017, Pourret and Hoursthouse 2019).

Besides the toxicity concerns, heavy metals also have important environmental impact, contaminating water, soil, water tables, and, in several times, biomagnifying in food chains and disrupting ecological niches (Baird and Cann 2012). Figure 1 presents the elements commonly classified as 'heavy metals' in Environmental Sciences.

Postgraduate Program in Process Engineering and Technologies (PGEPROTEC), University of Caxias do Sul, RS, Brazil. Street Francisco Getúlio Vargas, 1130, Petrópolis, Caxias do Sul, RS, Brazil.
Email: cbaldasso@ucs.br.
* Corresponding author: wpsilvestre@ucs.br

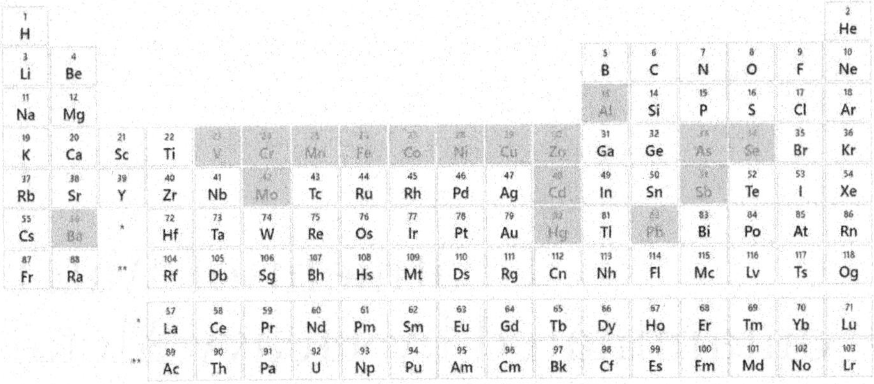

Fig. 1: Elements that are commonly classified as 'heavy metals' in Environmental and Biological Sciences.

Due to the deleterious nature of the heavy metals when released to the environment, the use of efficient and cost saving methods to remove these substances from waste streams and waters is paramount, especially when considering the hydric resources, such as drinking water and water for agricultural uses, such as irrigation water (Ahmed et al. 2019, Chaoua et al. 2019, Silva et al. 2018).

Historically, the removal of heavy metals from water used a chemical reaction that induced a physical-chemical change in the metal ions, rendering them insoluble, chelated, or associated with other chemical species; posteriorly, a separation process is used to obtain both the clean water and the heavy metal-associated material (Maraschin et al. 2020, Tunçsiper 2020). Common separation processes used in heavy metal removal from water after chemical treatment are coagulation followed by precipitation and/or flocculation; at very small scales, centrifugation may also be carried out. In several parts of the world, the treatment of water to render it potable is carried out by these methods (Cainglet et al. 2020, Pivokonský et al. 2020, Wu et al. 2021). Table 1 presents some methods used to remove heavy metals from water.

The use of separation processes that employ membranes has become widespread in the last decades of the 20th century. Membrane separation processes are acknowledged as very efficient and reliable, with lower energy and maintenance cost; however, the deployment costs are generally quite expressive (Ahmed et al. 2020, Ezugbe and Rathial 2020). Figure 2 presents the main advantages and drawbacks of membrane design and separation processes using membranes.

Nowadays, membrane separation processes, such as reverse osmosis and electrodialysis, are employed in the treatment of water, mainly seawater desalination. Regions with small water reserves, such as the Middle East, rely on membrane process to obtain drinking and irrigation water from the sea. This kind of process is acknowledged to achieve high efficiency and low maintenance cost; however, the main drawback is the implantation cost, which may be quite expressive (Ahmed et al. 2020, Elsaid et al. 2020, Suwaileh et al. 2020).

Membrane separation processes are also widely used in industry for a range of applications that ranges from pharmaceuticals and food industries to the treatment of wastes and recovery of interest materials (Buonomenna et al. 2012, Dhineshkumar

Table 1: Methods of treatment of water/wastewater in the removal of heavy metals (Abdullah et al. 2019).

Method	Description	Advantages	Disadvantages
Precipitation	Chemical agents convert metal ions into insoluble precipitates of either hydroxide, sulfide, carbonate and phosphate. The solid precipitate is later separated by filtration process.	- Simple method and high degree of selectivity; - Precipitants are relatively inexpensive.	- Ineffective to treat water containing high concentration of heavy metals; - Requires large amount of precipitate agents; - Production of large quantity of toxic sludge; - Chemical stabilization and proper precipitant disposal are needed; - Slow metal precipitation and sedimentation.
Coagulation-flocculation	A positively charged coagulant is introduced to reduce surface negative charge of particles and allow them to aggregate. Anionic flocculant is then added to react with the positively charged aggregates, binding them to form larger group that can be separated by filtration process.	- Relatively economic as alum is an inexpensive coagulant; - Simple operation.	- Incomplete heavy metals removal; - Often needs to be coupled with precipitation method to ensure effective removal; - Production of sludge.
Adsorption	Materials with adsorptive properties (i.e., highly porous, large surface area, active functional groups) entrap metal ions through physical or chemical interactions. The adsorbents are later separated from solution by filtration process and undergo regeneration process.	- Wide selection of adsorbents; - Relatively inexpensive; - Simple operation.	- Tedious post treatment process; - Nanosized adsorbents are unable to give promising results due to very strong Van der Waal's forces; - Certain adsorbents need to be hybridized for maximum binding capacity.
Ion-exchange	Solid ion exchange resin having strong sulfonic acid group ($-SO_3H$) or carboxylic acid group ($-COOH$) is mostly used in this process. Reversible exchange of ions between solid and liquid phase could take place where H^+ is released from the functional groups to allow complexation of metal with the free functional group.	- Fast kinetic; - Convenient process; - Economic process as it uses low-cost materials and resin can be re-generated.	- Fouling of metal ions on ion exchange media; - Only suitable for low concentration of metals; - Highly sensitive to pH; - Presence of free acids may result in low binding affinity.

and Ramasamy 2017). Nowadays, treatment of wastewater and contaminated water is generally composed by a stage that employs one or more membrane-based process to achieve higher process efficiency, also aiming to save costs (Ezugbe and Rathial 2020).

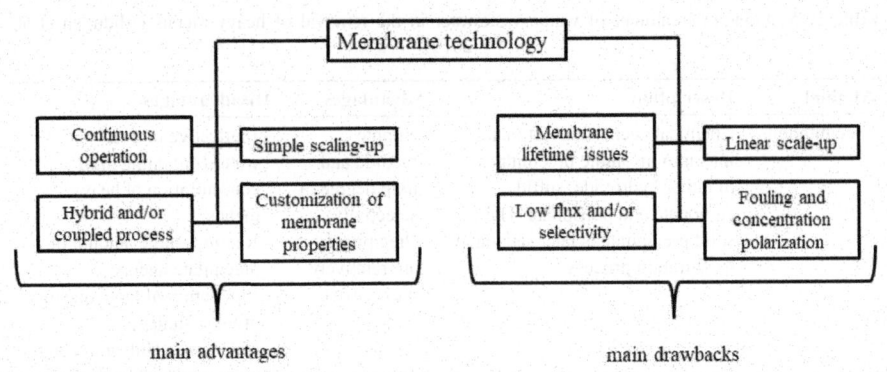

main advantages   main drawbacks

Fig. 2: Main advantages and drawbacks relative to membrane technology.

Among the several membrane separation process, micro and ultrafiltration, reverse osmosis, and electrodialysis are the most used ones in the purification of wastewater and the removal of heavy metals. These methods rely on membrane porosity or/and the formation of electric charges, fostering or hindering ion diffusion, and allowing for the separation to occur (Ezugbe and Rathial 2020, Gurreri et al. 2020).

2. Microfiltration

Microfiltration is one of the first separation processes based on membranes that were employed in wastewater treatment. Using porous membranes with pore size in the range of micrometers (10^{-6} m), microfiltration is suitable to be employed in the removal of heavy metals after chemical treatment, especially coagulation (Ibrahim et al. 2019, Mulder 1996, Musale et al. 2008, Nunes and Peinemann 2001).

Relative to separation process using porous membranes, microfiltration, ultrafiltration, and nanofiltration rely on the same physical-chemical principle of size exclusion, in which species with sizes larger than the pore of the membrane are retained. Although being different processes, the operational setup may be considered quite similar for the three processes; the main differences are the membrane constitution (i.e., membrane microstructure and pore size) and the pressure difference, which increases as the average pore size decreases. Higher pressure differences also require membranes with greater mechanical strength. The average pressure applied in microfiltration is 1 to 3 bar (0.1–0.3 MPa) (Ezugbe and Rathial 2020, Giwa and Ogunribido 2012, Mulder 1996).

Since the ionic and atomic radius are in the range of Angstroms (about 10^{-10} m), several orders of magnitude smaller than the average pore size of the microfiltration membrane (micrometers, 10^{-6} m), this process is considered as a pretreatment, since, without the use of a coagulant agent, no soluble heavy metal ion may be removed, unless the materials that compose the membrane act as a sorbent for the metallic ions (Carolin et al. 2017, Hosseini et al. 2016). In Table 2 some filtration kinds are presented, along with some common contaminants, average size, and molecular weights.

Table 2: Filtration spectrum for each kind of contaminant relative to pore size and average molecular weight.

Contaminant type	Colloids	Macromolecules	Ionic species	Small organics
Average molecular weight (Da)	10^4–10^6	10^2–10^4	$< 10^2$	10^1–10^2
Separation process employed	Microfiltration	Ultrafiltration	Nanofiltration	Reverse osmosis
Membrane size pore (nm)	1,000	100	10	1
Pressure difference applied (atm)	1–3	2–7	5–20	30–150

Literature cites the use of microfiltration in wastewater treatment, especially in the removal of heavy metals, in association with coagulation and flotation. Several coagulating agents are proposed and may also be tailored for specific ion capture/reaction. The flotation process increases average particle size, rendering the flakes large enough to be retained by the membrane (Zamboulis et al. 2011, Kyzas and Matis 2018, Sarode et al. 2019).

Hernández et al. (2020) reported a polyvinylidene fluoride–poly(acrylic acid) (PVDF–PAA) microfiltration membrane superficially functionalized with cysteamine (MEA); the authors designed the membrane for specific adsorption of mercury (Hg^{2+}) and silver (Ag^+) ions, reporting removal efficiencies in the range of 94–99% relative to the feed. The authors also cited the potential use of functionalized membranes for specific ion removal, especially regarding industrial wastewaters.

Ibrahim et al. (2019) developed a composite cellulose microfiltration membrane by surface coating with alpha zirconium phosphate. The process presented a removal efficiency of approx. 55% of Cu^{2+}, 60% of Pb^{2+}, 30% of Zn^{2+}, and 15% of Ni^{2+} at pH 7.0; at higher pH values (about 12.0), the removal efficiencies increased to above 90%, with exception of the Pb^{2+} ion, whose removal was about 80%. The coated membranes presented a higher removal efficiency than the pristine cellulose membrane for all metallic ions. The authors attributed this behavior to both an ion-exchange mechanism in the membrane and the coating and also to the formation of the heavy metal hydroxides, which are insoluble and tend to coagulate, generating flakes that are retained by the membrane.

Bhattacharya et al. (2015) used a tubular multichannel ceramic membrane made from an indigenous composition of clay and alumina to treat tannery wastewater contaminated with Cr^{6+}; both the untreated effluent and the permeate from the microfiltration were applied to soil to evaluate the concentration of contaminants. The soil irrigated with effluent presented an average Cr^{6+} content of 15.8 $g \cdot kg^{-1}$, whereas the soil irrigated with the permeate has had an average Cr^{6+} content of 0.03 $g \cdot kg^{-1}$, almost equal to the Cr^{6+} content of the control treatment (0.02 $g \cdot kg^{-1}$).

Zanain and Lovitt (2013) carried out the removal of Ag^+ ions from water using a microfiltration polymeric membrane and dithizone coated alumina as sorbent for cross flow microfiltration, with Ag^+ removal in the range of 56–88%. The authors reported that both cross flow velocity and feed flow rate presented influence in the removal of Ag^+, with higher removals at higher cross flow velocities and flow rates.

Khulbe and Matsuura (2018), in a review work, compiled several studies that employed functionalized and coated microfiltration membranes for the joint absorption/filtration of heavy metal contaminated water. The authors reported the use of functional nanofibrous membranes fabricated by electrospinning. According to the study, these membranes can be used to remove heavy metal ions from contaminated waters through adsorption. Nanofibrous membranes are considered to have high permeation fluxes and lower pressure drop than conventional microfiltration membranes. However, these membranes also possess a high surface-to-volume ratio, whose surface may be modified to remove toxic metal ions through adsorption, with a capability comparable to typical adsorbents.

3. Ultrafiltration

Ultrafiltration, as micro and nanofiltration, also employs a porous membrane to carry out the separation. The operational setup and characteristics of ultrafiltration systems are quite similar to the ones of microfiltration. The main differences between these separation processes are that ultrafiltration works at greater pressures (2 to 7 bar, about two to three times the pressure employed in microfiltration) and the pore size of ultrafiltration membranes are in the range of 1–50 nm, an intermediate size between micro and nanofiltration (Giwa and Ogunribido 2012, Mulder 1996, Nunes and Peinemann 2001).

Ultrafiltration systems are commonly used in the removal/purification of large molecules, especially proteins and polymers, being also capable to retain large viruses. The exact cutoff value is determined by membrane microstructure and average pore size. Small size molecules and ions are permeable, being capable of permeating through the membrane pores (Garba et al. 2019, Mulder 1996).

Although having a much small pore size than microfiltration membranes, ultrafiltration membranes still have pore sizes about 10–500 times larger than typical atoms or ions, still rendering them unable to retain these species by size exclusion (Garba et al. 2019). Like in microfiltration, the use of chemicals to induce partial coagulation, increase of particle size, or the use of high molecular weight sorbents may be strategies to retain heavy metal ions and obtain permeates with low contamination degrees. The use of membranes with adsorptive capacity is also explored; however, these membranes tend to saturate quickly, which hinders their economic feasibility (Huang and Cheng 2020, Yaqub and Lee 2019).

When employing ultrafiltration in the removal of heavy metals from water, literature generally presents the use of a chelating agent or adsorbent. Since the cutoff molecular weight of ultrafiltration membranes lies in the range of 500 kDa, large molecules, such as proteins, polymeric sorbents, and large-sized ligands are retained by the membrane. Due to the interaction and capture of the metallic ions by these agents, the heavy metals are also retained (Garba et al. 2019, Tyagi and Jacob 2020). Figure 3 presents a scheme of removal of metals from wastewater using a complexing (or adsorbent) agent, followed by ultrafiltration.

Several chelating agents are proposed in the literature to be employed in ultrafiltration systems to remove heavy metals. Rezania et al. (2019) reported the use of poly(itaconic acid) in the removal of Pb^{2+}, Sn^{2+}, Cu^{2+}, Zn^{2+}, and Cd^{2+} using

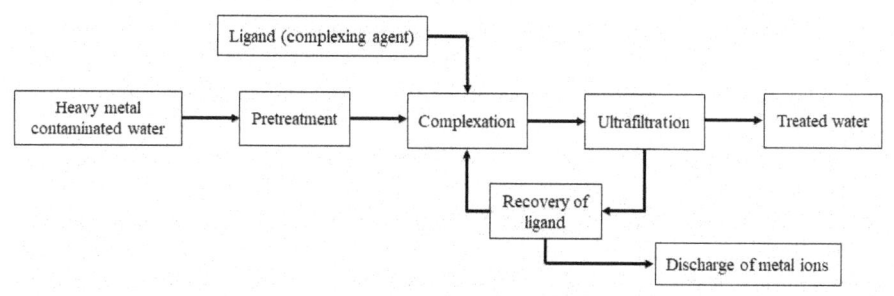

Fig. 3: Scheme of wastewater treatment for heavy metal removal by using a complexing agent, followed by ultrafiltration.

a polyethersulfone membrane with a pore size of 10.4 nm. The authors reported removal efficiencies ranging from 36.5% (Cd^{2+}) up to 83.4% (Pb^{2+}) in basic medium (pH > 7), with a trend of rejection of $Pb^{2+} > Cu^{2+} > Sn^{2+} > Zn^{2+} > Cd^{2+}$. Le and Qiu (2020) used a hybrid coagulation/ultrafiltration process to remove Cd^{2+}, Fe^{2+}, Al^{3+}, Zn^{2+}, and Pb^{2+} from smelter wastewater using copolymer of acrylic acid-maleic acid (PMA) as complexing agent. The hybrid process has had removal efficiencies above 95% for these ions when the wastewater pH was adjusted to 5.5–7.0. Crini et al. (2017) proposed the use of chitosan as a chelating/complexing agent for heavy metal ions due to its biodegradability, high efficiency, easy obtainment, chemical stability, and compatibility with other polymers, and the non-synthetic origin, among other factors.

Along with the use of complexing agents and sorbents, the use of micellar-enhanced ultrafiltration (MEUF) is also proposed to remove heavy metals from water. The use of a surfactant in concentrations above the critical micellar concentration (CMC) allow for the formation of micelles, which interact and traps the metallic ions. The removal can then be carried out successfully by selecting a membrane with appropriate molecular weight cut-off (MWCO) or pore size; the micelles and the heavy metal ions are retained in the retentate stream, whereas the permeate stream is composed by free surfactant molecules and water (Abdullah et al. 2019, Yaqub and Lee 2019). The removal efficiency of several heavy metal ions (such as Cd^{2+}, Ni^{2+}, Pb^{2+}, Cu^{2+}, Zn^{2+}, and Cr^{3+}) by this process is reported as higher than 90%, with some systems presenting efficiency of 99.5% (Aoudia et al. 2003, Yaqub and Lee 2019). Shi et al. (2019) carried out the removal of Cd^{2+} from wastewater by MEUF using sodium dodecyl sulfate as surfactant and commercial polyether sulfone (PES) membranes with a molecular weight cut-off (MWCO) of 10 kDa. The observed results indicated that a concentration polarization of the surfactant may have influenced the overall performance of the process; however, the Cd^{2+} removal efficiency was above 90% for a surfactant concentration of 8 mM and a transmembrane pressure of 0.15 MPa (1.5 bar).

Data on hybrid membranes with addition of nanomaterials to change the overall membrane polarity and/or to interact with metal ions are also described. Abdulkarem et al. (2020) developed a polyethersulfone ultrafiltration membrane with the addition of 0.25 wt.% of α-zirconium phosphate nanoparticles and evaluated its efficiency in the removal of Cd^{2+}, Cu^{2+}, Ni^{2+}, Pb^{2+}, and Zn^{2+}. Despite the large average pore size of the membrane (114 nm), the process presented a removal efficiency in the range of 70.2 ± 1.0 (Cd^{2+}) up to $99.7 \pm 0.2\%$ (Cu^{2+}). The authors attributed the high removal efficiency due to electrostatic effects in the membrane caused by the nanoparticles added.

Zhang et al. (2020) produced β-cyclodextrin-modified/PVDF blend magnetic membranes to remove ions from wastewater directly. The average pore size of the membranes ranged from 30–60 nm. According to the results, the rejection rate of Cu^{2+} and its absorption capacity were 75% and 0.94 mg·g^{-1} after five cycles, respectively; the membrane was best suited to remove Cu^{2+} concentrations below 30 mg·L^{-1} at pH values above 6.0. Arif et al. (2020) prepared a hybrid polyvinylidene fluoride/ titanium dioxide solar active photocatalytic ultrafiltration membrane to be employed in the removal and reduction of Cr^{+6} from tannery wastewater. The authors reported 97.59% rejection and 91.73% reduction values for wastewater pH of 5.55.

4. Nanofiltration

Nanofiltration, despite presenting several characteristics common to micro and ultrafiltration, has some fundamental differences. Due to the extremely small pore size, in the range of 1–2 nm, nanofiltration membranes lay halfway between porous and dense membranes. Since the ionic radius of cations are smaller than 200 pm (0.2 nm), the Knudsen diffusion becomes an important mass transport mechanism (Mazumder et al. 2019, Phillip et al. 2009, Yoshiura et al. 2020).

However, due to the very small pores, for most substances the microfiltration membranes are like dense membranes; in this case, the sorption-diffusion mechanism may also play an important role for molecules larger than the average pore size of the membrane (Phillip et al. 2009, Yoshiura et al. 2020).

Another important difference relative to membrane design is that nanofiltration membranes tend to be highly asymmetric, unlike micro and ultrafiltration membranes, which tend to be more homogeneous, although porous, especially relative to membrane cross section. Nanofiltration membranes are characterized by a compact active layer, followed by a more porous support structure. This construction is used to lower the resistance to mass flow and to allow for a short diffusion path, which increases the separation efficiency (Mazumder et al. 2019).

Other factor that differs in nanofiltration is the pressure differential applied in the membrane; due to the more compact structure and lower pore size, pressures in the range of 5–20 bar may be used to carry out the separation. These high pressures also put great strain in the membrane, which needs to be properly reinforced by specific material or by the use of supports to withstand the pressure differential applied on it (Bliattacharya and Ghosh 2004).

Several kinds of nanofiltration membranes were evaluated/studied for use in the removal of heavy metal ions from many wastewater kinds. It must be kept in mind that some wastewaters may attack chemically some polymers, in which a proper evaluation of the membrane polymer should be carried out. Cuhorka et al. (2020) evaluated the performance of commercial tubular polyamide thin-film nanofiltration membranes in the joint removal of Zn^{2+} as zinc sulfate and zinc nitrate and diclofenac and ibuprofen. The observed results showed a Zn^{2+} rejection above 98% for one of the membranes; in the other two, the rejection percentage of Zn^{2+} was lower when the counterion was nitrate. The authors attributed this behavior as steric effects relative to the bigger volume of the sulfate ion relative to nitrate.

Ali et al. (2020) employed an Alfa Laval-NF99HF flat sheet thin film composite membrane to remove heavy metals (Al^{3+}, As^{3+}, Cr^{3+}, Cu^{2+}, Fe^{3+}, Mg^{2+}, Mn^{2+}, Na^+, and Zn^{2+}) from wastewater of anodizing industry. The authors reported a high influence of the wastewater pH. At alkaline pH conditions, no permeate was obtained, and inacidic pH, the average permeability of the wastewater ranged between 3.1 and 4.1 $L \cdot m^{-2} \cdot h^{-1} \cdot bar^{-1}$, but a low ion rejection (7–13%). On the other hand, neutralized wastewater, after the removal of precipitate, produced high-quality permeate, with a stable permeability of 1 $L \cdot m^{-2} \cdot h^{-1} \cdot bar^{-1}$; the average removal for this feed was complete for all heavy metals and above 80% for Na^+.

Bandehali et al. (2019) developed a hybrid nanofiltration membrane composed by polyetherimide (PEI) with the addition of polyhedral oligomeric silsesquioxane (POSS) functionalized particles to remove Cu^{2+} and Pb^{2+} from wastewater. According to the results, the removal percentages for Cu^{2+} and Pb^{2+} were 85 and 86% for the hybrid membrane, respectively, whereas the rejection of the same ions in a non-hybrid PEI membrane were 40 and 44%, respectively. Barahimi et al. (2020) produced a hybrid $TiO2$/3-cyanopropyltriethoxysilane (CPTES)/Metformin polyethersulfone (PES) nanofiltration membrane with antifouling properties. The membrane was used in the removal of Cu^{2+}, COD content, and dyes from liquorice extraction plant (LEP) wastewater. At a transmembrane pressure of 5 bar and 150 min of process, the Cu^{2+} removal percentage was 98%; the COD and dye removal percentages were 88 and 98%, respectively.

Wang et al. (2019) modified a polyamide nanofiltration membrane by the deposition of a thin film of polyethylenimine; the composite membrane was used to remove Cu^{2+} ions from wastewater. According to the observed results, the rejection of Cu^{2+} increased from 38.9% in the polyamide membrane to 92.5% in the polyethylenimine-coated one. The presence of organic matter in the form of humic acid has not influenced membrane flux and the rejection of Cu^{2+}; however, the performance of the coated membrane was highly dependent on the PEI molecular weight and the preparation method.

Yang et al. (2020) developed a nanocomposite Fe_3O_4@MXene membranes by self-assembly of Fe_3O_4 nanoparticles (NPs) and two-dimensional MXene (Ti_3AlC_2) nanosheets on the surface of cellulose acetate base membrane. The membrane was employed in the removal of Cu^{2+}, Cd^{2+}, and Cr^{6+} from wastewater. The removal percentages of Cu^{2+}, Cd^{2+}, and Cr^{6+} were ~ 63.2%, ~ 64.1% and ~ 70.2%, respectively. The authors attributed this behavior to the introduction of Fe_3O_4, which was considered a good adsorbent and had a synergistic effect with MXene nanosheets in the adsorption of heavy metal ions.

Pino et al. (2020) used a hybrid nanofiltration/solvent extraction process to remove Cu^{2+} from acid mine drainage (AMD) wastewater. It was used as a thin-film composite (TFC) membrane consisting of an active polyamide layer supported on polysulfone and polyester. Ketoxime was selected as the organic extractant. According to the authors, the hybrid NF-SE system was technically feasible, capable of recovering 80% of the water and 97% of the copper from the AMD wastewater.

The authors also commented on the economic feasibility of the process, in which the recovery of copper would compensate for about 10% of the NF operation cost, while recovering water from the AMD may help reduce the consumption of freshwater or desalinated water.

Mu et al. (2020) prepared a polyethersulfone three-channel (TC) capillary membrane incorporated with quaternary ammonium moieties to remove heavy metals from wastewater. The reported results were a removal efficiency of 96.43% for $CuSO_4$, 96.16% for $ZnSO_4$, 91.69% for $Cu(NO_3)_2$, 89.69% for $ZnCl_2$, and 88.37% for $Pb(NO_3)_2$. The stability of the prepared membrane was also tested; the results illustrated that the membrane has a potential applicability to be employed in the removal of heavy metals under certain feed conditions. Gong et al. (2020) also tried a similar approach, preparing a positively charged nanofiltration (NF) membrane by incorporating metal–organic frameworks (MOFs) into polyethyleneimine (PEI) and using trimesic acid (TMA) as a cross-linking system. The novel membrane exhibited a high permeability (2.2 $L·m^{-2}·h^{-1}·bar^{-1}$) and a $NiCl_2$ rejection of 90.9%. Compared to the non-incorporated membrane, the average permeability increase for the modified membrane was 369.2%.

Despite the great potential and several researches employing nanofiltration membranes in the removal of heavy metals in wastewater, its use in industrial/commercial scales is still incipient. Some issues are relative to the difficulties in the production of this kind of membrane and the high intrinsic asymmetry, which may also negatively affect the membrane production in large batches. Further developments in nanofiltration membrane technology are likely to resolve most of these issues, rendering both implantation and maintenance costs considerably lower, increasing the overall economic feasibility of the process (Mulder 1996, Lastra et al. 2004, Touati et al. 2020).

5. Osmosis

Osmosis is a mass transport phenomenon in which solvent flows spontaneously from a less concentrated (hypotonic) solution to a more concentrated (hypertonic) solution through a semi-permeable membrane. In this process, the difference of chemical potential caused by the differences in the concentration of the solutes acts as the driving force, generating an 'osmotic pressure' ($\Delta\pi$). This pressure is higher in the less concentrated solution; thus, a flow of solvent is established, transporting solvent molecules through the membrane to the more concentrated solution, diluting it. When the concentration in both sides equalize, the osmotic pressure also equals and the equilibrium is reached (Mulder 1996, Marbach and Bocquet 2019).

Unlike the micro, ultra, and nanofiltration membranes, which are porous, osmosis membranes are considered as dense (pore size smaller than 1 nm), meaning that the separation does not occur by hydrodynamic sieving, but by sorption and diffusion, a completely different mechanism, which characterizes dense membranes. In sorption and diffusion, the components of the feed current that have more chemical affinity with the membrane are sorbed by the membrane, entering in its structure and diffusing throughout it until reaching the other surface and being collected in the

Table 3: Separation mechanism according to membrane pore size and the corresponding separation process.

Membrane separation process	Membrane pore size range (μm)	Separation mechanism
Particle filtration	> 1000–1	hydrodynamic sieving control
Microfiltration	10–1	hydrodynamic sieving control
Ultrafiltration	1–0.01	hydrodynamic sieving control
Nanofiltration	0.01–0.001	hybrid[1]
Dialysis	0.01–0.001	hybrid[1]
Reverse osmosis	< 0.001	Sorption-diffusion
Pervaporation	< 0.001	Sorption-diffusion
Gas separation	< 0.001	Sorption-diffusion

[1] – Both hydrodynamic sieving control and sorption-diffusion mechanisms occur simultaneously.

permeate stream (Mulder 1996, Nunes and Peinemann 2001). Table 3 presents the transport mechanisms that govern the more common membrane separation processes.

Osmosis may be classified in two groups, according to the solvent flux and the spontaneity of the process: reverse or forward osmosis. In reverse osmosis, hydraulic pressure is applied in the membrane side which has the more concentrated solution (hypertonic medium); the pressure applied forces the solvent to flow to the more diluted solution (hypotonic medium). Since the natural trend is the flow of solvent from a hypotonic to a hypertonic medium, the application of pressure *reverses* the natural flow; thus, the process is called *reverse osmosis* (Mulder 1996). In forward osmosis, the osmotic pressure is the driving force of the process, and, if hydraulic pressure is applied, it may be on both the hypertonic (pressure retarded osmosis) or on the hypotonic medium (pressure enhanced osmosis). Considering that forward osmosis is a spontaneous process, significantly more energy is required for reverse osmosis (Ahmed et al. 2020, Nicoll 2017). In Fig. 4 is presented the relationship between osmotic ($\Delta\pi$) and hydraulic pressure (ΔP) and the kind of osmosis process.

Historically, osmosis systems were first employed in seawater desalination, especially reverse osmosis. Nowadays, several separation processes relay on osmosis, such as water treatment, removal of organic pollutants, concentration/water removal of food products, production of microchips and pharmaceuticals, and in hybrid systems using other separation processes, such as filtration and dialysis (Ang et al. 2020, Francis et al. 2020, Sourirajan 1978, Xu and Ge 2019).

5.1 Reverse osmosis (RO)

In reverse osmosis, a hydraulic pressure gradient (ΔP) is applied to overcome the osmotic pressure ($\Delta\pi$) of the system, reversing the solvent flux through the membrane. Due to the dense nature of the membrane, which is only permeable to the solvent (water), no heavy metal ions pass through the membrane, unless it has an adsorbent character (Bliattacharya and Ghosh 2004, Nicoll 2017).

Fazullin et al. (2019) tested the performance of a polyamide and a polysulfonamide RO membranes in the removal of Fe^{3+}, Cu^{2+}, Ni^{2+}, Zn^{2+}, Cr^{3+}, Pb^{2+}, and Cd^{2+} ions from

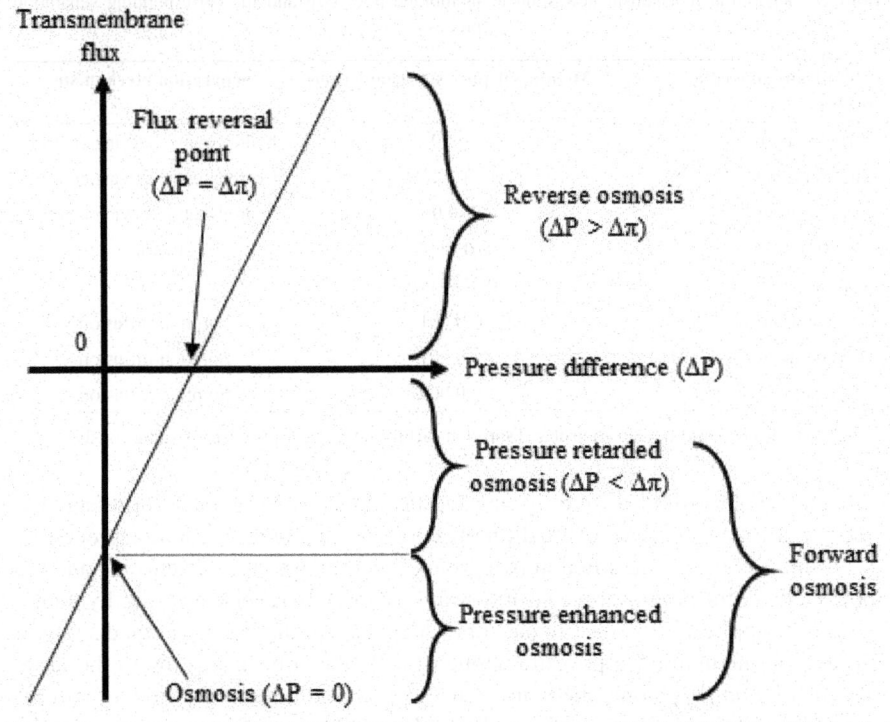

Fig. 4: Relationship between transmembrane flux, osmotic pressure differential ($\Delta\pi$), and hydraulic pressure differential (ΔP), showing the family of osmotic membrane processes.

wastewater of metal processing. The obtained results indicated a removal efficiency ranging between 33% (Cd^{2+}) and 87% (Fe^{3+}). The degree of removal was coincident with the ion hydration energy and with ion charge, with the following trend: $H^+ < Cd^{2+} < Zn^{2+} < Fe^{3+}$. Yoon et al. (2009) used RO to remove chromate (CrO_4^{2-}), arsenate ($HAsO_4^-$), and perchlorate (ClO_4^-) ions from water using a thin-film composite polyamide membrane. The rejection values ranged between 71–99%; according to the authors, the rejection of the ions was mainly governed by electrostatic and steric exclusion.

Bakalár et al. (2009) carried out the RO of Cu^{2+}, Ni^{2+}, and Zn^{2+} ions from water using a polyamide thin-film tubular RO membrane; the influence of the counterion (SO_4^{2-} and NO_3^-) on the removal of the heavy metals was also tested. The observed results showed that the kind of counterion influenced the decrease in transmembrane flux as the separation progressed. The transmembrane pressure also has had a significant influence on the membrane separation; at low transmembrane pressures, the flux dependence on the ion concentration was very low. Ozaki et al. (2002) reported the same trends studying the performance of an ultra-low-pressure RO membrane (ULPROM) of aromatic polyamide in the removal of Ni^{2+}, Cu^{2+}, and Cr^{3+}; the authors also reported rejection values above 95% for all heavy metal ions.

The use of chelating agents to reduce the electrostatic effects is reported. Moshen-Nia et al. (2007) studied the removal of Ni^{2+} and Cu^{2+} using a polyamide

RO spiral membrane and disodium EDTA (Na_2EDTA) as adsorbent. The observed results reported removal values above 98% for the sole RO process. The removal values increased to 99.5% by using 500 mg·L^{-1} Na_2EDTA as chelating agent and carrying out a hybrid chelation/RO process.

Some studies also used sorbent prior to the RO process to reduce the strain on the membrane. Thaçi and Gashi (2019) pretreated wastewater containing Pb^{2+}, Cd^{2+}, Ni^{2+}, Zn^{2+}, Mn^{2+}, and Co^{2+} ions using olive waste, wheat bran, and maize cobs as sorbents; a cellulose acetate RO membrane with coal particles impregnated with 4-nitrobenzene diazonium salt was employed. In general, olive waste presented the highest retention values for all ions studied; the reverse osmosis treatment removed more than 98% of all ions relative to the feed concentrations.

Chung et al. (2014) proposed a hybrid process of RO-ferrite to purify wastewater from electroplating and to recover the metals dissolved in the waste. According to the experiments, after the ferrite reaction, 99.7% of Zn^{2+} was recovered from the second-stage RO concentrate in the form of zinc ferrite. Significant amounts of Cr^{3+} were retained in the concentrate after RO and reaction with ferrite. The authors commented on the economic aspects, proposing a retrofitting option, including a combined RO and ferrite process for an existing wastewater treatment facility of a local metal plating manufacturer.

5.2 Forward osmosis (FO)

Forward osmosis is characterized by a spontaneous mass transfer, in which the osmotic pressure ($\Delta\pi$) acts as the driving force of the separation. Pressure retarded and pressure enhanced osmosis are considered as subclasses of forward osmosis, with difference in the location and intensity of the hydraulic pressure (ΔP), when applied. As with reverse osmosis, forward osmosis membranes are non-permeable to metallic ions, unless the membrane has other characteristics that interact with the ions, such as superficial modifications, addition of adsorbents, charged carriers, among others (Ang et al. 2020, Xu and Ge 2019).

Vital et al. (2018) employed a thin-film composite (TFC) FO membrane to remove Mg, Al, Si, Ca, Mn, Fe, Co, Cu, and Zn from acid mine drainage wastewater. The removal percentages were above 97.0% for all heavy metal ions. However, the authors observed that, when using ammonium bicarbonate (NH_4HCO_3) as draw solution, precipitates formed on the membrane, reducing its efficiency and hindering the FO process. Hussein (2019) used a cellulose triacetate FO membrane to operate through flow and batch FO in the removal of Co^{2+} ions from water. The rejection percentage of Co^{2+} ions dropped from 91.71% after 30 min to 81.19% after 4 h of operation in flow FO process; for the batch FO, the rejection rate dropped from 87.71% after 30 min to 73.19% after 4 h of operation. According to the author, this decrease in rejection percentage was due to the formation of a cobalt layer on the membrane surface; this retarded the back diffusion of the Co^{+2} ions from the membrane surface to the bulk solution. The FO flowing process had a higher efficiency than the FO batch process.

Meng et al. (2020) used an aquaporin FO membrane to remove Sb^{3+} ions from printing and dyeing wastewater; the observed results indicated a Sb^{3+} removal percentage of 99.7% when Cr^{6+} ions were added to the feed current. Naghdali et al. (2020) also used an aquaporin FO membrane to carry out the removal of Cr^{6+} from wastewater; the system was also modelled to compare the experimental and predicted values. Under optimum operational conditions, the predicted removal value for Cr^{6+} was 96.30%, whereas the experimental removal percentage was 97.67%; the authors observed that the data fitted a quadratic equation relative to the experimental design.

The use of hybrid membranes is also reported in the literature. He et al. (2020) developed a polydopamine/metal organic framework thin film nanocomposite (PDA/MOF-TFN) FO membrane to remove Cd^{2+}, Ni^{2+}, and Pb^{2+} ions from water; the results indicated a heavy metal rejection rate of 94.0–99.2% for Ni^{2+}, Cd^{2+}, and Pb^{2+}. Saeedi-Jurkuyeh et al. (2020) developed a thin-film nanocomposite forward osmosis (TFN-FO) membrane and employed it in the removal of Pb^{2+}, Cd^{2+}, and Cr^{3+} from industrial wastewater. According to the authors, the removal rates regarding Pb^{2+}, Cd^{2+}, and Cr^{3+} were 99.9, 99.7, and 98.3%, respectively; the flux recovery ratio was 96.0%.

Qiu and He (2019) used a thin film nanocomposite (TFN) FO membrane with a positively charged and nanofunctional selective layer in the removal of Cu^{2+}, Ni^{2+}, and Pb^{2+} ions from wastewater. The obtained results indicated a removal efficiency of 99.1% for Cu^{2+}, 98.3% for Ni^{2+}, and 97.7% for Pb^{2+}. About 6.2% of Cu^{2+}, 6.9% of Ni^{2+}, and 8.1% of Pb^{2+} ions were adsorbed by the membrane after 24 h, indicating that most of the removal of the heavy metal ions relied on membrane rejection, rather than adsorption.

Despite the large number of studies employing FO in the separation of several kinds of feed mixtures, this process is still largely considered as emergent, with most of its potential unexplored, with limited commercial and industrial applications. However, the much lower energy consumption compared to reverse osmosis (RO) render it an important option, which may be employed in several process that nowadays rely on other separation processes (Ezugbe and Rathial 2020, Francis et al. 2020, Nicoll 2017).

6. Electrodialysis

In electrodialysis, an electrical potential difference is applied and drives the charged chemical species (ions) to the electrode with the inverse charge, i.e., anions will move towards the anode (which is positively charged) whereas cations will move towards the cathode (which is negatively charged). The separation is carried out by a series of alternating cationic and anionic exchange membranes that are placed between the cathode and the anode. The electrically charged membranes allow for the passage of one kind of ions and are used to control the migration of ions; these membranes are electrically conductive. Uncharged species are not affected by the electrical potential difference, thus, are not separated (Gurreri et al. 2020, Hutten 1996, Mulder 1996, Nunes and Peinemann 2001).

In general, cation-selective membranes are composed of polyelectrolytes with negatively charged matter, mainly sulfonic or carboxylic acid groups, which rejects negatively charged ions and allows positively charged ions to flow through. Anion-selective membranes consist mainly of polystyrene with quaternary ammonium cations, which repel the positively charged species, allowing the passage of anions.

Along with the membrane charge, it is also considered that the Donnan exclusion mechanism also plays an important role in the separation itself, especially when large charged particles are present in the feed stream (Hutten 2016, Mulder 1996, Nunes and Peinemann 2001).

High ion concentrations in the feed stream reduce the effectiveness of the Donnan exclusion mechanism and generate a more intense osmotic pressure as the process goes, due to the segregation of ions in each cell following the electrical potential difference. Thus, electrodialysis is more suited, effective and competitive at relatively low ion concentrations in the feed when compared to other processes (Mulder 1996).

Electrodialysis was commercially introduced in the 1950 to desalinate brackish water (Nur-E-Alam et al. 2020). According to the literature, this has already been studied in the removal of heavy metals from wastewater, especially chromium (Cr^{3+} and Cr^{6+}), mainly from tannery wastes. Peng and Guo (2020) commented that electrodialysis is a process with a low energy consumption; however, the high cost of the electrodes and membranes may hinder its economic competitiveness. According to Nur-E-Alam et al. (2020), the use of electrodialysis, both alone and coupled to other processes (coagulation/electrocoagulation), has high removal percentages (above 80%) for chromium; Peng and Guo (2020) reported removal efficiencies in the range of 99.0–99.5% for both Cr^{3+} and Cr^{6+} ions from tannery waste.

Jin et al. (2020) used a hybrid system of electrodialysis coupled to electrodeionization to remove Cr^{6+} ions from synthetic wastewater; the system was composed only by cation exchange membranes and had a removal efficiency above 99.0%. According to the authors, the proposed system may have economic potential to be scaled up to treat large amount of effluent. Santos et al. (2019) employed electrodialysis to remove chromium (VI) as dichromate ($Cr_2O_7^{2-}$) from wastewater produced in a hybrid anaerobic bioreactor; the impact of the concentration of the anion on the process efficiency was also evaluated. At a current of 0.03 A, 75 min of operation, and a starting effluent volume of 76 L with 100 $mg\cdot L^{-1}$ of dichromate, the removal of chromium (VI) was approximately 99.0 ± 0.5%, and the final volume of wastewater concentrated solution was equal to 4 L, with a maximum concentration of chromium (VI) of 570 $mg\cdot L^{-1}$. The authors also cited that more concentrated solutions tend to render the process ineffective after some cycles (about 4–5 cycles for the waste with 100 $mg\cdot L^{-1}$ $Cr_2O_7^{2-}$), both by membrane adsorption and by osmotic effects.

Liu et al. (2020) carried out electrodialysis experiments to remove Fe^{3+}, Zn^{2+}, Ni^{2+}, Cd^{2+}, Cr^{3+}, Cu^{2+}, As^{3+}, and SO_4^{2-} ions from the raffinate stream of the hydrometallurgical processing of copper ore. To help concentrate the metal ions, a bipolar membrane was used. The optimal process conditions were a current density of 3.0 $mA\cdot cm^{-2}$, a volume ratio between the raffinate chamber (RC) and the HMC of 1:15, and a process time of 40 h. At these parameters, the removal efficiencies were 99.3% for Fe^{3+}, 99.1% for Zn^{2+}, 99.0% for Cu^{2+}, 84.9% for Ni^{2+}, 70.6% for Cr^{3+}, 95.8% for Cd^{2+}, and 94.8% for As^{3+}. The authors also observed that 85.9% of SO_4^{2-} in the raffinate stream could be recovered as sulfuric acid (H_2SO_4), also helping to reduce further treatments of the remaining waste.

Abou-Shady (2017) used an electrodialysis system to remove alkaline and alkaline earth metals from wastewater for agricultural use in irrigation. The system removed 99.3% of the Na^+ ions; in the samples that presented heavy metals in very small amounts (trace levels), they were completely removed. In light of the results, the author also cited the potential of this process to treat wastewater for agricultural purposes with joint removal of heavy metals and also alkaline metals and cations that are harmful for plants at high concentrations, as Na^+ and NH_4^+.

Nemati et al. (2019) developed a cation exchange membrane made based on PVC and cation exchange resin with the addition of particles composed by a 2-acrylamido-2-methyl propane sulfonic acid (HAMPS) hydrogel and magnetic nickel ferrite ($NiFe_2O_4$) nanoparticles to remove Pb^{2+}, Ni^{2+}, and Cu^{2+} from wastewater. The modified membrane containing 1.0 wt.% of NiFe2O4-HAMPS particles has the highest removal percentages of Pb^{2+} (~ 98%), Cu^{2+} (~ 48%), and Ni^{2+} (~ 34%); the authors also commented that the energy cost, 6.97 $W \cdot mol^{-1}$, is quite low when compared to other electrodialysis systems that employed other cation exchange membranes.

Jiang et al. (2018) proposed a hybrid coagulation/electrodialysis process to carry out a concomitant removal of heavy metals and organic matter from wastewater. The system employed acetylacetone (acac) as the coagulating agent. The removal efficiency of Cr^{3+} and organic matter reached about 99.4–99.5% and 97.8–99.9%, respectively. The authors also reported the presence of chromium by-products, such as $Cr(acac)_n^{(3-n)+}$, metallo-organics, and oxide nano-powder after post-treatment.

7. Membrane distillation

Membrane distillation is a membrane separation process that employs a porous membrane with a hydrophobic character. In this process, rather than the other membrane processes, where driving force of the separation is a pressure difference (hydraulic, osmotic, vacuum), in membrane distillation, temperature is the driving force. The membrane acts as a barrier to the feed to evaporate; the liquid only evaporates in the areas comprised by the pores of the membrane; hence, the name 'membrane distillation' (Deshmukh et al. 2018, Panagopoulos et al. 2019, Rezaei et al. 2018). Figure 5 presents a simplified scheme of an air-gap membrane distillation system.

Membranes for membrane distillation are commonly manufactured using hydrophobic polymers, such as polyoctylmethylsiloxane (POMS), polydimethylsiloxane (PDMS), polyvinylidene fluoride (PVDF), polytetrafluoroethylene (PTFE), among others; these materials are used to prevent wetting of the membrane by the feed, which will render the process inefficient. The average pore size of this kind of membrane lies in the range of 100–500 nm, somewhat the same as ultrafiltration membranes (Mulder 1996, Rezaei et al. 2018).

Membrane distillation systems are used mainly in water desalination, both seawater and brackish water. However, this process may also be used to remove heavy metal ions from wastewater since these ions are non-volatile and will remain in the retentate fraction; the use of hydrophobic membranes, with non-polar character, also act as a barrier to ion diffusion, due to the charged nature of the ions (mostly cations,

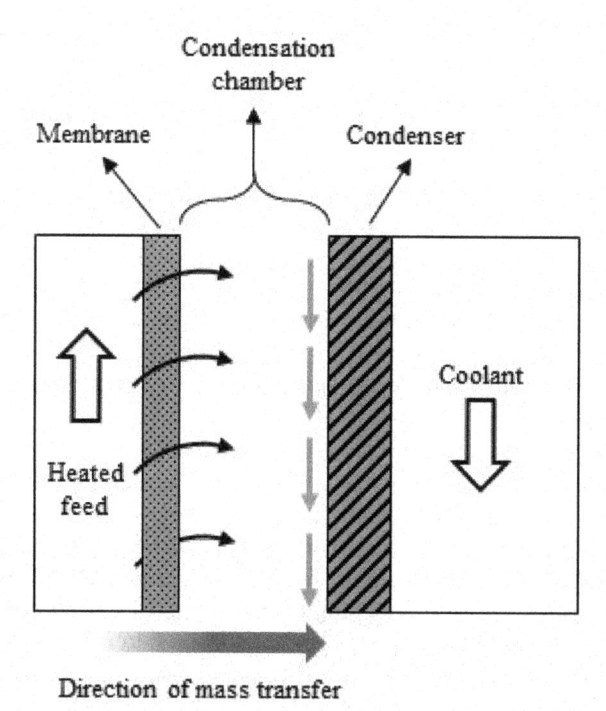

Fig. 5: Scheme of an air-gap membrane distillation system (AG-MD). The heated feed passes parallel to the membrane, some of the solvent (water) evaporates through the membrane pores and, when it gets into contact with the cold plate condenser, it condenses to liquid, being collected as permeate.

but some heavy metals may also be present as anions, such as chromate, dichromate, permanganate, vanadate, among others) (Alkhudhiri et al. 2020, Ghalavand et al. 2014, Rezaei et al. 2018).

Relative to heavy metal removal using membrane distillation, literature has studies of this process in the removal of heavy metal ions from acid rock drainage. Hull and Zodrow (2017) carried out the removal of synthetic acid rock drainage wastewater using a PVDF membrane with an average pore size of 200 nm; the feed current was heated to 90°C and the condenser was kept at 6°C. The authors observed the deposition of solid material (mainly metal oxides and salts) as fouling in the membrane surface in contact with the feed stream; the average rejection was > 99%.

Ryu et al. (2019) also used a hollow-fiber PVDF membrane with an average pore size of 100 nm. The feed was kept at 55.0 ± 0.5°C and the condenser at 22.0 ± 0.5°C; a temperature-treated zeolite was also used as a sorbent to enhance the overall process efficiency. The removal affinity for heavy metal was $Fe^{3+} > Al^{3+} > Zn^{2+} > Cu^{2+} > Ni^{2+}$. The removal of Fe^{3+} and Al^{3+} ions was close to 100% with the 500°C heat treated zeolite, whereas the removal of Zn^{2+}, Cu^{2+}, and Ni^{2+} was in the range of 38–40%. Ryu et al. (2020) carried out a similar study by employing a multi-modified mesoporous silica SBA-15 membrane and a sorbent to enhance the separation efficiency of Cu^{2+}, Ni^{2+}, and Zn^{2+}. The sorbent presented a higher Cu^{2+} adsorption; the adsorption of Ni^{2+} and Zn^{2+} was lower.

Alkhudhiri et al. (2020) used an air-gap membrane distillation (AG-MD) system to remove Hg^{2+}, As^{3+}, and Pb^{2+} ions from synthetic industrial wastewater. The membranes used were of PTFE with average pore sizes of 200, 450, and 1,000 nm; the air-gap was of 5 mm, the feed temperature was 50°C and the condenser was at 10°C. The observed results indicated that bigger pore sizes reduced the overall process efficiency; the removal percentages for the three ions were above 90% for all combinations tested; however, the membrane with the smallest average pore size (200 nm) had the highest removal efficiency (99–100%) for all ions tested. Attia et al. (2017) also employed an AG-MD system to remove Pb^{2+} ions from water. A superhydrophobic electrospun hybrid PVDF membrane with a molecular weight of 275,000 Da and average pore size of 370 nm and superhydrophobic alumina as a carrier was tested. In terms of heavy metal removal, all experiments showed rejection percentages above 99%.

Lou et al. (2019) tested a hybrid membrane-distillation/crystallization system in the removal of Zn^{2+} and Ni^{2+} ions from wastewater. A tubular PTFE membrane was used; the feed was kept at 65°C, whereas the condenser was kept at 10°C. The hybrid system was considered as efficient in the removal of both ions by crystallization in the form of $NiSO_4$ and $ZnSO_4$. Furthermore, the authors commented that the employed membrane had excellent resistance to fouling from highly concentrated solutions.

Jia et al. (2018) used vacuum membrane distillation to remove Co^{2+} ions from radioactive wastewater; a polypropylene (PP) membrane with an average pore size of 180 nm was used. According to the authors, Knudsen diffusion was determined as the main mass transport mechanism; the removal efficiency was 99.67% when the feed concentration of Co^{2+} was 10 mg·L^{-1}. Zou et al. (2020) proposed a hybrid membrane distillation and photocatalysis to carry out the simultaneous removal of heavy metals and organic pollutants from wastewater. The used membrane was of PFTE and the catalyst was a BiOBr film sorted in porous glass sheets. The removal of Ag^+ and 4-chlorophenol (4-CP) was tested. According to the authors, there was an efficient removal of 4-CP and Ag^+ ion via a synergistic effect of photocatalysis and direct contact membrane distillation.

Although regarded as efficient and energy-saving, membrane distillation relies on specific membranes, whose value is quite high. Thus, this kind of separation process is not widely used in industry due to the high implementation and maintenance cost. However, some authors commented on the economic feasibility of this process as a function of further developments in membrane technology, especially the cheapening of the feedstocks and the production methods for this kind of membrane (Alkhudhiri et al. 2020, Attia et al. 2020).

8. Pervaporation

A portmanteau of the terms 'permeation' and 'evaporation', this separation process is still considered emergent, with potential applications in several areas of research and industry. Pervaporation systems employ dense membranes, in which the solution-diffusion model of mass transport applies. Due to the dense character of the membranes, the separation occurs not due to size exclusion, but through chemical affinity, in which both the sorption and diffusion terms influence the overall membrane performance (Jyoti et al. 2015, Silvestre et al. 2019).

Unlike separation processes that employ positive pressures on the feed side and work at atmospheric pressure in the permeate stream, such as micro, ultra, and nanofiltration, and osmosis, most of pervaporation systems work at atmospheric pressure in the feed side and use high vacuum in the permeate side. The use of vacuum helps in establishing a chemical potential gradient between membrane sides and in the desorption of the substances that diffused through the membrane, fostering the mass transport between the streams and increasing overall process efficiency (Mulder 1996, Silvestre et al. 2019). An important issue inherent to the system is the necessity of a very-low temperature condenser in the permeate stream due to the high vacuum, which increases implantation and operational costs and may be difficult to operate/control without process optimization (Dai et al. 2019, Vane 2005). Figure 6 presents a simplified scheme of a typical pervaporation system.

The pervaporation process may be separated in three main groups according to the kind of substance that preferentially is sorbed and diffuses in the membrane: hydrophilic pervaporation, when water is the substance that preferentially permeates the membrane; organophilic pervaporation, when an organic compound permeates the membrane; and target-organophilic pervaporation, when a specific organic compound permeates the membrane from an organic mixture feed. In Fig. 7 is presented a diagram relating each pervaporation kind and their main characteristics.

Due to the nature of the pervaporation process, non-volatile species cannot be separated through sorption-diffusion and collection in the permeate side. However, the concentration of a non-volatile species through removal of the solvent by pervaporation is possible. In this sense, the treatment of heavy-metal contaminated water may be carried out through hydrophilic pervaporation, especially if a high purity water is desired. Nevertheless, it is also important to carry out suitability studies to verify if pervaporation is feasible for this kind of separation (Figoli et al. 2015, Vane 2005).

There are very few literature data on the use of pervaporation in the removal of heavy metals from water and wastewater. Truong et al. (2019) developed a zeolite-chitosan pervaporation membrane and employed it in the pervaporation of Cd, As, Pb, and Cr contaminated wastewater. The authors reported that the membrane acted as a sorbent for the metallic ions, with an affinity degree of Cr > As > Cd > Pb at pH 5.5. The membrane also kept its sorbent capacity even after six regenerations with efficiencies higher than 92%.

Sulaiman et al. (2016) carried out a similar work using a composite pervaporation membrane whose active layer was made of blended polyvinyl alcohol and chitosan crosslinked with tetraethyl orthosilicate (TEOS), supported on porous polysulfone. The removal of Cu^{2+} and Fe^{2+}/Fe^{3+} ions from aqueous solutions was tested. The authors reported that the process removed most of the heavy metal ions from the solution, but the removal was attributed due to the membrane acting as a sorbent, rather than the pervaporation process itself.

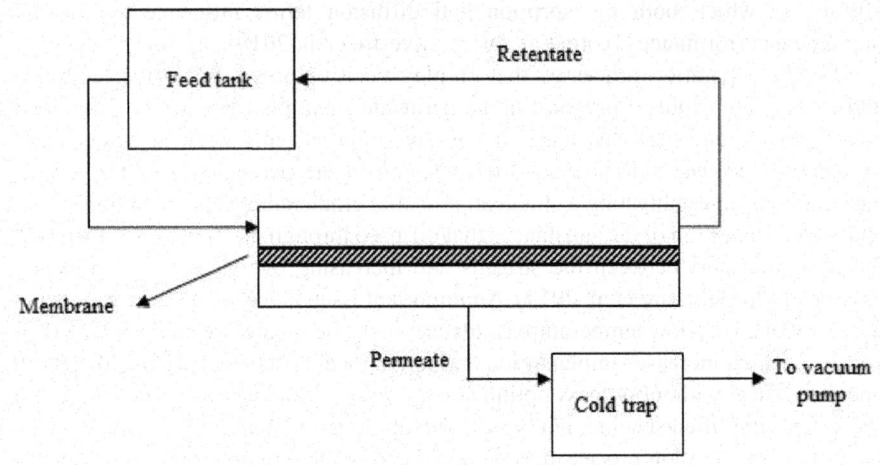

Fig. 6: Simplified scheme of a pervaporation system.

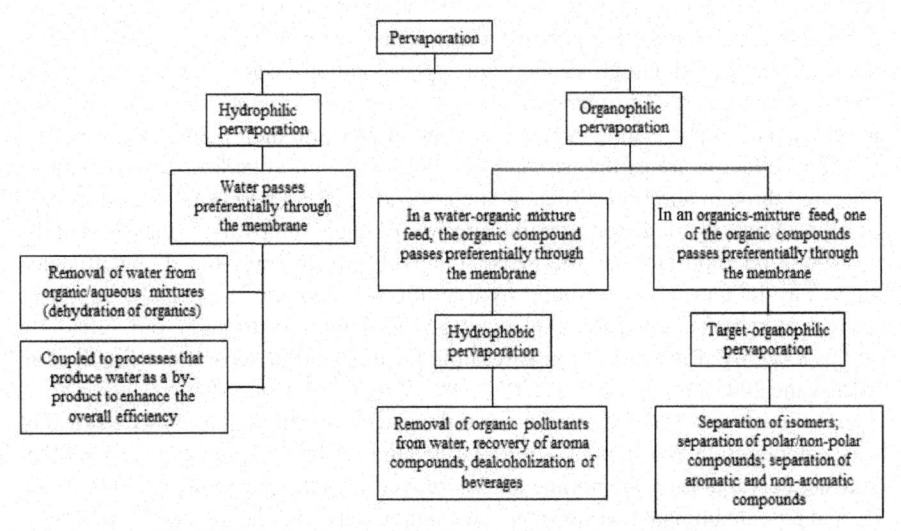

Fig. 7: Diagram of pervaporation kinds and their main characteristics.

Table 4 presents an overview of the reported uses of the membrane separation processes addressed, heavy metals removed, and overall efficiency ranges according to the literature.

9. Future perspectives and concluding remarks

Although being considered well established and reasonably employed methods, several of the previously presented membrane separation methods are still with incipient use in the treatment of heavy metal contaminated wastewater. Literature presents several proofs of concept indicating the technical feasibility of this process;

Table 4: Overview of the applications and results of the membrane separation processes addressed in this chapter.

Membrane separation process	Membrane composition	Feed	Heavy metals removed	Efficiency range (%)	References
Microfiltration	PVDF-PAA Clay/alumina	- synthetic wastewater; - tannery wastewater;	Ag, Cu, Hg, Ni, Pb, Zn	15–99	Bhattacharya et al. (2015), Hernández et al. (2020), Ibrahim et al. (2019), Zanain and Lovitt (2013)
Ultrafiltration	PES PES-hybrid PVDF-β-cyclodextrin PVDF/TiO$_2$	- smelter wastewater; - synthetic wastewater;	Cd, Cu, Pb, Ni, Sn, Zn	36 – > 99	Abdulkarem et al. (2020), Arif et al. (2020), Rezania et al. (2019), Shi et al. (2019), Zhang et al. (2020)
Nanofiltration	PA-TF PE-TF PEI-POSS hybrid PES PA-PETI-TF hybrid CA PS-PE-PA PES PEI-MOF	- synthetic wastewater; - wastewater of anodizing industry; - licorice extraction plant wastewater; - acid mine drainage wastewater;	Al, As, Cd, Cr, Cu, Fe, Mn, Ni, Pb, Zn	63 – >99	Ali et al. (2020), Bandehali et al. (2019), Barahimi et al. (2020), Cuhorka et al. (2020), Gong et al. (2020), Mu et al. (2020), Pino et al. (2020), Yang et al. (2020), Wang et al. (2019)
Reverse osmosis	PA-PSA PA-TF Ar-PA CA	- wastewater of metal processing; - synthetic wastewater; - wastewater from electroplating;	As, Cd, Cr, Cu, Co, Fe, Mn, Ni, Pb, Zn	33 – >99	Balakár et al. (2009), Chung et al. (2014), Fazullin et al. (2019), Moshen-Nia et al. (2020), Ozaki et al. (2002), Thaçi and Gashi (2019), Yoon et al. (2009)
Forward osmosis	CTA Aquaporin PDA/MOF-TFL	- acid mine drainage wastewater; - printing and dyeing wastewater;	Al, Cd, Co, Cr, Cu, Fe, Mn, Ni, Pb, Sb, Zn	73 – >99	He et al. (2020), Hussein (2019), Meng et al. (2020), Naghdali et al. (2020), Vital et al. (2018)

Process	Membrane material	Application	Toxic metals	Removal (%)	References
Electrodialysis	PTFE, PE, PVC	- synthetic wastewater; - wastewater from anaerobic reactor; wastewater of hydrometallurgical processing of copper ore; - wastewater for agricultural use;	As, Cd, Cr, Cu, Fe, Ni, Pb, Zn	34 – 99	Abou-Shady (2017), Jin et al. (2020), Liu et al. (2020), Nemati et al. (2019), Santos et al. (2019)
Membrane distillation	PVDF, PVDF-zeolite, PTFE, PP	- synthetic acid rock drainage; - synthetic industrial wastewater;	Al, As, Co, Cu, Fe, Hg, Ni, Pb, Zn	90 – >99	Alkhudhiri et al. (2020), Attia et al. (2017), Hull and Zodrow (2017), Jia et al. (2018), Lou et al. (2019), Ryu et al. (2019, 2020)
Pervaporation	CS-TEOS-PS, CS-Zeolite	- wastewater;	As, Cd, Cr, Cu, Fe, Pb	>92	Sulaiman et al. (2016), Truong et al. (2019)

PVDF – Polyvinylidene fluoride; PAA – polyacrylic acid; PES – polyethersulfone; PA – polyamide; TF – thin film; PEI – polyetherimide; POSS – polyhedral oligomeric silsesquioxane; PETI – polyethyleneimine; CA – cellulose acetate; PS – polysulfone; PE – polyester; MOF – metal-organic framework; PSA – polysulfonamide; Ar-PA – aromatic polyamide; CTA – cellulose triacetate; PDA – polydopamine; TFL – thin-film layer; PTFE – polytetrafluoroethylene; PE – polyethylene; PVC – polyvinyl chloride; CS – chitosan; TEOS – tetraethyl orthosilicate.

however, depending on the kind and volume of waste to be treated, the economic feasibility of the treatment processes must be evaluated (Chung et al. 2014, Lastra et al. 2004, Touati et al. 2020).

One of the major issues relative to membrane technology is the high cost associated with membrane production and maintenance, especially as the pore size decreases. However, with the development of new fabrication techniques and novel polymers and substrates for membrane production and enhancement of current membranes, there is an inherent trend of cost and price reduction as the methods and applications for membrane processes mature in several areas of industry. This will very probably render these processes more sustainable economically and also render experimental systems feasible in the long run (Ioannou-Ttofa et al. 2017, Passarini et al. 2014, Zubir et al. 2019).

In the last years, membrane technology advanced towards two main fronts. First, the use of biopolymers and green chemistry practices to reduce the environmental impact of membrane production. Several biopolymers, such as starch, chitosan, gelatin, alginate, among others, have specific properties that may be of interest in the removal of contaminants and metallic ions from waste streams, such as chemical resistance, hydrophilicity, capacity to form dense films, metal-adsorption properties, easy production, and the use of less aggressive chemicals in their production/preparation (Ahmed et al. 2020, Ang et al. 2020).

The second trend is the development and use of hybrid and composite membranes; these membranes are shown to have better performance and efficiency than non-hybrid membranes. However, the process efficiency is also highly dependent on the wastewater physical-chemical conditions. Most of the times, the use of a membrane separation process obliges the use of a pretreatment process to reduce fouling and interferences/damage to the membrane, increasing costs; the economic impact of the needed pretreatment must also be evaluated and properly addressed (Abdulkarem et al. 2020, Arif et al. 2020, He et al. 2020, Mulder 1996).

The use of hybrid systems employing a membrane separation process coupled to other unit separation is another approach that may be useful in the treatment of wastewaters; the coupled process may be aimed to remove several kinds of pollutants in a single system or to reduce fouling or other impacts in the membrane process. Literature also comments about the increase in overall process efficiency by the use of hybrid/coupled systems according to the nature of the waste and the physical-chemical properties of the pollutants (Le and Qiu 2020, Moshen-Nia et al. 2007, Pino et al. 2020).

With further development in membrane technology, more efficient and cheaper membranes will become available, rendering the separation processes using membranes as a viable and competitive alternative to classical wastewater treatment processes, especially regarding process time, volume of waste treated, and the quality of the final effluents.

References

Abdulkarem, E., Ibrahim, Y., Naddeo, V., Banat, F., and Hasan, S.W. 2020. Development of Polyethersulfone/α-Zirconium phosphate (PES/α-ZrP) flat-sheet nanocomposite ultrafiltration membranes. Chem. Eng. Res. Design., 161: 206–217.

Abou-Shady, A. 2017. Recycling of polluted wastewater for agriculture purpose using electrodialysis: Perspective for large scale application. Chem. Eng. J., 323: 1–18.

Ahmed, M., Matsumoto, M., Ozaki, A., Thinh, N.V., and Kurosawa, K. 2019. Heavy metal contamination of irrigation water, soil, and vegetables and the difference between dry and wet seasons near a multi-industry zone in Bangladesh. Water, 11: 583.

Ahmed, F.E., Hashaikeh, R., and Hilal, N. 2020. Hybrid technologies: The future of energy efficient desalination – A review. Desalination, 495: 114659.

Ali, A., Nymann, M.C., Christensen, M.L., and Quist-Jensen, C.A. 2020. Industrial wastewater treatment by nanofiltration—a case study on the anodizing industry. Membranes, 10: 85.

Alkhudhiri, A., Hakami, M., Zacharof, M.P., Homod, H.A., and Alsadun, A. 2020. Mercury, arsenic and lead removal by air gap membrane distillation: experimental study. Water, 12: 1574.

Ang, W.L., Mohammad, A.W., Johnson, D., and Hilal, N. 2020. Unlocking the application potential of forward osmosis through integrated/hybrid process. Sci Total Environ., 706: 136047.

Aoudia, M., Allal, N., Djennet, A., and Toumi, L. 2003. Dynamic micellar enhanced ultrafiltration: use of anionic (SDS)-nonionic (NPE) system to remove Cr^{3+} at low surfactant concentration. J. Membr. Sci., 217: 181–192.

Appenroth, K.J. 2009. Definition of "Heavy Metals" and their role in biological systems. pp. 19–29. *In:* Sherameti, I., and Varma, A. [eds.]. Soil Heavy Metals, Springer-Verlag Berlin Heidelberg, Berlin, Germany.

Arif, Z., Sethy, N.K., Mishra, P.K., and Verma, B. 2020. Green approach for the synthesis of ultrafiltration photocatalytic membrane for tannery wastewater: modeling and optimization. Int. J. Environ. Sci. Technol., 17: 3397–3410.

Attia, H., Osman, M.S., Johnson, D.J., Wright, C., and Hilal, N. 2017. Modelling of air gap membrane distillation and its application in heavy metals removal. Desalination, 424: 27–36.

Bai, H., Hu, B., Wang, C., Bao, S., Sai, G., Xu, X., Zhang, S., and Li, Y. 2017. Assessment of radioactive materials and heavy metals in the surface soil around the bayanwula prospective uranium mining area in China. Int. J. Environ. Res. Public. Health, 14: 300.

Baird, C., and Cann, M. 2012. Environental Chemistry. 5th ed. Macmillan Learning, New York.

Balakár, T., Búgel, M., and Gajdošová, L. 2009. Heavy metal removal using reverse osmosis. Acta Mont. Slov., 14: 250–253.

Bandehali, S., Parvizian, F., Moghadassi, A.R., and Hosseini, S.M. 2019. Copper and lead ions removal from water by new PEI based NF membrane modified by functionalized POSS nanoparticles. J. Polym. Res., 26: 211.

Barahimi, V., Taheri, R.A., Mazaheri, A., and Moghimi, H. 2020. Fabrication of a novel antifouling TiO_2/CPTES/metformin-PES nanocomposite membrane for removal of various organic pollutants and heavy metal ions from wastewater. Chem. Pap., 74: 3545–3556.

Bhattacharya, P., Ghosh, S., Swarnakar, S., and Mukhopadhyay, A. 2015. Tannery effluent treatment by microfiltration through ceramic membrane for water reuse: assessment of environmental impacts. Clean Soil Air Water, 43: 633–644.

Bliattacharya, A., and Ghosh, P. 2004. Nanofiltration and reverse osmosis membranes: theory and application in separation of electrolytes. Rev. Chem. Eng., 20: 111–173.

Buonomenna, M.G., G. Golemme, and E. Perrotta. 2012. Membrane operations for industrial applications pp. 543–562. *In:* Nawaz, Z. [ed.]. Advances in Chemical Engineering, InTech Europe, Rijeka, Croatia.

Cainglet, A., Tesfamariam, A., and Heiderscheidt, E. 2020. Organic polyelectrolytes as the sole precipitation agent in municipal wastewater treatment. J. Environ. Manag., 271: 111002.

Carolin, C.F., Kumar, P.S., Saravanan, A., Joshiba, G.J., and Naushad, M. 2017. Efficient techniques for the removal of toxic heavy metals from aquatic environment: A review. J. Environ. Chem. Eng., 5: 2782–2799.

Castellote, M., C. Andrade, and C. Alonso. 2002. Nondestructive decontamination of mortar and concrete by electro-kinetic methods: application to the extraction of radioactive heavy metals. Environ. Sci. Technol., 36: 2256–2261.

Chaoua, S., Boussaa, S., El Gharmali, A., and Boumezzough, A. 2019. Impact of irrigation with wastewater on accumulation of heavy metals in soil and crops in the region of Marrakech in Morocco. J. Saudi Soc. Agric. Sci., 18: 429–436.

Chung, S., Kim, S., Kim, J.O., and Chung, J. 2014. Feasibility of combining reverse osmosis–ferrite process for reclamation of metal plating wastewater and recovery of heavy metals. Ind. Eng. Chem. Res., 53: 15192–15199.

Crini, G., Morin-Crini, N., Fatin-Rouge, N., Déon, S., and Fievet, P. 2017. Metal removal from aqueous media by polymer-assisted ultrafiltration with chitosan. Arab. J. Chem., 10: S3826–S3839.

Cuhorka, J., Wallace, E., and Mikulášek, P. 2020. Removal of micropollutants from water by commercially available nanofiltration membranes. Sci. Total Environ., 720: 137474.

Dai, Y., Li, S., Meng, D., Yang, J., Cui, P., Wang, Y. et al. 2019. Economic and environmental evaluation for purification of diisopropyl ether and isopropyl alcohol via combining distillation and pervaporation membrane. ACS Sust. Chem. Eng., 7: 20170–20179.

Deshmukh, A., Boo, C., Karanikola, V., Lin, S., Straub, A.P., Tong, T., Warsinger, D.M., and Elimelech, M. 2018. Membrane distillation at the water-energy nexus: limits, opportunities, and challenges. Energy Environ. Sci., 11: 1177–1196.

Dhineshkumar, V., and D. Ramasamy. 2017. Review on membrane technology applications in food and dairy processing. Journal of Appl. Biotechnol. Bioeng., 3: 399–407.

Elsaid, K., Kamil, M., Sayed, E.T., Abdelkareem, M.A., Wilberforce, T., and Olabi, A. 2020. Environmental impact of desalination technologies: A review. Sci. Total Environ., 748: 141528.

Ezugbe, E.O., and S. Rathial. 2020. Membrane technologies in wastewater treatment: a review. Membranes., 10: 89.

Fazullin, D.D., Mavrin, G.V., Fazullina, L.I., and Parenkina, A.A. 2019. Removal of heavy metal ions reverse osmosis in wastewater. Biosci. Biotechnol. Res. Comm., 12 (S.I.): 239–244.

Figoli, A., Santoro, S., Galiano, F., and Basile, A. 2015. Pervaporation membranes: preparation, characterization, and application. pp. 19–63. *In:* Basile, A., Figoli, A., and Khayet, M. [eds.]. Pervaporation, Vapour Permeation and Membrane Distillation: Principles and Applications, Woodhead Publishing, Sawston, Cambridge, UK.

Francis, L., Ogunbiyi, O., Saththasivam, J., Lawler, J., and Liu, Z. 2020. A comprehensive review of forward osmosis and niche applications. Environmental Science: Water Res. Technol. 6: 1986–2015.

Garba, M.D., Usman, M., Mazumder, M.A.J., Al-Ahmed, A., and Inamuddin. 2019. Complexing agents for metal removal using ultrafiltration membranes: a review. Environ. Chem. Let., 17: 1195–1208.

Ghalavand, Y., Hatamipour, M.S., and Rahimi, A. 2014. A review on energy consumption of desalination processes. Desalination and Water Treat, 54: 1526–1541.

Giwa, A., and Ogunribido, A. 2012. The applications of membrane operations in the textile industry: a review. Brit. J. Appl. Sci. Technol., 2: 296–310.

Gong, X.Y., Huang, Z.H., Zhang, H., Liu, W.L., Ma, X.H., Xu, Z.L., and Tang, C.Y. 2020. Novel high-flux positively charged composite membrane incorporating titanium-based MOFs for heavy metal removal. Chem. Eng. J., 398: 125706.

Gurreri, L., Tamburini, A., Cipollina, A., and Micale, G. 2020. Electrodialysis applications in wastewater treatment for environmental protection and resources recovery: a systematic review on progress and perspectives. Membranes, 10: 146.

He, M., Wang, L., Lv, Y., Wang, X., Zhu, J., Zhang, Y., and Liu, T. 2020. Novel polydopamine/metal organic framework thin film nanocomposite forward osmosis membrane for salt rejection and heavy metal removal. Chem. Eng. J., 389: 124452.

Hernández, S., Islam, S., Thompson, S., Kearschner, M., Hatakeyama, E., Malekzadeh, N. et al. 2020. Thiol-functionalized membranes for mercury capture from water. Ind. Eng. Chem. Res., 59: 5287–5295.

Hosseini, S.S. Bringas, E., Tan, N.R., Ortiz, I., Ghahramani, M., and Shahmirzadi, M.A.A. 2016. Recent progress in development of high-performance polymeric membranes and materials for metal plating wastewater treatment: A review. J. Water Proc. Eng., 9: 78–110.

Huang, Z.Q., and Cheng, Z.F. 2020. Recent advances in adsorptive membranes for removal of harmful cations. J. Appl. Polym. Sci., 137: 48579.

Hull, E.J., and Zodrow, K.R. 2017. Acid rock drainage treatment using membrane distillation: impacts of chemical-free pretreatment on scale formation, pore wetting, and product water quality. Environ. Sci. Technol., 51: 11928–11934.

Hussein, T.K. 2019. Removal of cobalt ions from wastewater by batch and flowing forward osmosis processes. J. Ecol. Eng., 20: 121–126.

Hutten, I.M. 2016. Filtration mechanisms and theory. pp. 53–107. *In:* Hutten, I.M. [ed.]. Handbook of Nonwoven Filter Media. 2nd ed. Butterworth-Heinemannm, Oxford, UK.

Ibrahim, Y., Abdulkarem, E., Naddeo, V., Banat, F., and Hasan, S.W. 2019. Synthesis of super hydrophilic cellulose-alpha zirconium phosphate ion exchange membrane via surface coating for the removal of heavy metals from wastewater. Sci. Total. Environ., 690: 167–180.

Ioannou-Ttofa, L., Michael-Kordatou, I., Fattas, S.C., Eusebio, A., Ribeiro, B., Rusan, M., Amer, A.R.B., Zuraiqi, S., Waismand, M., Linder, C., Wiesman, Z., Gilron, J., and Fatta-Kassinos, D. 2017. Treatment efficiency and economic feasibility of biological oxidation, membrane filtration and separation processes, and advanced oxidation for the purification and valorization of olive mill wastewater. Water Res., 114: 1–13.

Jia, F., Yin, Y., and Wang, J. 2018. Removal of cobalt ions from simulated radioactive wastewater by vacuum membrane distillation. Prog. Nucl. Energy, 103: 20–27.

Jiang, C, Chen, H., Zhang, Y., Feng, H., Shehzad, M.A., Wang, Y., and Xu. T. 2018. Complexation Electrodialysis as a general method to simultaneously treat wastewaters with metal and organic matter. Chem. Eng. J., 348: 952–959.

Jin, Q, Yao, W., and Chen, X. 2020. Removal of Cr(VI) from wastewater by simplified electrodeionization. Desalin. Water Treat, 183: 301–306.

Jyoti, G., Keshav, A., and Anandkumar, J. 2015. Review on pervaporation: theory, membrane performance, and application to intensification of esterification reaction. J. Eng., 2015: 927068.

Kyzas, G.Z., and Matis, K.A. 2018. Flotation in water and wastewater treatment. Processes, 6: 116.

Khulbe, K.C., and Matsuura, T. 2018. Removal of heavy metals and pollutants by membrane adsorption techniques. Appl. Water Sci., 8: 19.

Lastra, A., Gómez, D., Romero, J., Francisco, J.L., Luque, S., and Álvarez, J.R. 2004. Removal of metal complexes by nanofiltration in a TCF pulp mill: technical and economic feasibility. J. Membr. Sci., 242: 97–105.

Le, H.-S., and Qiu, Y.-R. 2020. Selective separation of Cd(II), Zn(II) and Pb(II) from Pb-Zn smelter wastewater via shear induced dissociation coupling with ultrafiltration. Korean J. Chem. Eng., 37: 784–791.

Liu, Y., Ke, X., Zhu, H., Chen, R., Chen, X., Zheng, X., Jin, Y., and Van der Bruggen, B. 2020. Treatment of raffinate generated via copper ore hydrometallurgical processing using a bipolar membrane electrodialysis system. Chem. Eng. J., 382: 122956.

Lou, X.Y., Xu, Z., Bai, A.P., Resina-Gallego, M., and Ji, Z.G. 2020. Separation and recycling of concentrated heavy metal wastewater by tube membrane distillation integrated with crystallization. Mermbranes, 10: 19.

Maraschin, M., Ferrari, K.F.H., and Carissimi, E. 2020. Acidification and flocculation of sludge from a water treatment plant: New action mechanisms. Sep. Pur. Technol., 252: 117417.

Marbach, S., and Bocquet, L. 2019. Osmosis, from molecular insights to large-scale applications. Chem. Soc. Rev., 48: 3102–3144.

Mazumder, A., Sen, D., and Bhattacharjee, C. 2019. Mass transport through composite asymmetric membranes. Diff. Found., 23: 151–172.

Meng, L., Wu, M., Chen, H., Xi, Y., Huang, M., and Luo, X. 2020. Rejection of antimony in dyeing and printing wastewater by forward osmosis. Sci. Total Environ, 745: 141015.

Moshen-Nia, M., Montazeri, P., and Modarress, H. 2007. Removal of Cu^{2+} and Ni^{2+} from wastewater with a chelating agent and reverse osmosis processes. Desalination, 217: 276–281.

Mu, T., Zhang, H.Z., Sun, J.Y., and Xu, Z.L. 2020. Three-channel capillary nanofiltration membrane with quaternary ammonium incorporated for efficient heavy metals removal. Sep. Pur. Technol., 248: 117133.

Mulder, M. 1996. Basic Principles of Membrane Technology. 2nd. ed. Kluwer Academic Publishers, Dordrecht.

Musale, D.A., and Johnson, B.S. 2008. Method of heavy metal removal from industrial wastewater using submerged ultrafiltration or microfiltration membranes. US Patent 2008/0060999 A1. Filled on 7 September 2006, published on 13 March 2008.

Naghdali, Z., Sahebi, S., Mousazadeh, M., and Jamali, H.A. 2020. Optimization of the forward osmosis process using aquaporin membranes in chromium removal. Chem. Eng. Technol., 43: 298–306.

Nemati, M., Hosseini, S.M., Parvizian, F., Rafiei, N., and Van der Bruggen, B. 2019. Desalination and heavy metal ion removal from water by new ion exchange membrane modified by synthesized $NiFe_2O_4$/HAMPS nanocomposite. Ionics, 25: 3847–3857.

Nicoll, P.G. 2017. Forward osmosis – a brief introduction. Water Today The Magazine, 29: 32–50.

Nunes, S.P., and Peinemann, K.V. [eds.]. 2001. Membrane Technology in the Chemical Industry. Wiley-VCH, Weinheim.

Nur-E-Alam, M., Mia, M.A.S., and Ahmad, F. 2020 An overview of chromium removal techniques from tannery effluent. Appl. Water Sci., 10: 205.

Ozaki, H., Sharma, K., and Saktaywin, W. 2002. Performance of an ultra-low-pressure reverse osmosis membrane (ULPROM) for separating heavy metal: effects of interference parameters. Desalination, 144: 287–294.

Panagopoulos, A., Haralambous, K.J., and Loizidou, M. 2019. Desalination brine disposal methods and treatment technologies - A review. Sci. Total Environ., 693: 133545.

Passarini, K.C., Pereira, M.A., Farias, T.M.B., Calarge, F.A., and Santana, C.C. 2014. Assessment of the viability and sustainability of an integrated waste management system for the city of Campinas (Brazil), by means of ecological cost accounting. J. Clean. Prod., 65: 479–488.

Peng, H., and Guo, J. 2020. Removal of chromium from wastewater by membrane filtration, chemical precipitation, ion exchange, adsorption electrocoagulation, electrochemical reduction, electrodialysis, electrodeionization, photocatalysis and nanotechnology: a review. Environ. Chem. Lett., 18: 2055–2068.

Phillip, W.A., Amendt, M., O'Neill, B.Chen, L., Hillmyer, M.A., and Cussler, E.L. 2009. Diffusion and Flow Across Nanoporous Polydicyclopentadiene-Based Membranes. ACS Appl. Mater. Interf., 1: 472–480.

Pino, L., Beltran, E., Schwarz, A., Ruiz, M.C., and Borquez, R. 2020. Optimization of nanofiltration for treatment of acid mine drainage and copper recovery by solvent extraction. Hydrometallurgy, 195: 105361.

Pivokonský, M., Pivokonská, L., Novotná, K., Čermáková, L., and Klimtová, M. 2020. Occurrence and fate of microplastics at two different drinking water treatment plants within a river catchment. Sci. Total Environ., 741: 140236.

Pourret, O., and Hoursthouse, A. 2019. It's time to replace the term "Heavy Metals" with "Potentially Toxic Elements" when reporting environmental research. Int. J. Environ. Res. Public Health, 16: 4446.

Qiu, M., and He, C. 2019. Efficient removal of heavy metal ions by forward osmosis membrane with a polydopamine modified zeolitic imidazolate framework incorporated selective layer. J. Haz. Mat., 367: 339–347.

Rezaei, M., Warsinger, D.M., Lienhard V, J.H., Duke, M.C., Matsuura, T., and Samhaber, W.M. 2018. Wetting phenomena in membrane distillation: Mechanisms, reversal, and prevention. Water Res., 139: 329–352.

Rezania, H., Vatanpour, V., and Faghani, S. 2019. Poly(itaconic acid)-assisted ultrafiltration of heavy metal ions' removal from wastewater. Iran. Polym. J., 28: 1069–1077.

Ryu, S., Naidu, G., Johir, M.A.H., Choi, Y., Jeong, S., and Vigneswaran, S. 2019. Acid mine drainage treatment by integrated submerged membrane distillation–sorption system. Chemosphere, 218: 955–965.

Ryu, S., Naidu, G., Moon, H., and Vigneswaran, S. 2020. Selective copper recovery by membrane distillation and adsorption system from synthetic acid mine drainage. Chemosphere, 260: 127528.

Saeedi-Jurkuyeh, A., Jafari, A.J., Kalantary, R.R., and Esrafili, A. 2020. A novel synthetic thin-film nanocomposite forward osmosis membrane modified by graphene oxide and polyethylene glycol for heavy metals removal from aqueous solutions. React. Funct. Polym., 146: 104397.

Santos, C.S.L., Reis, M.H.M., Cardoso, V.L., and Resende, M.M. 2019. Electrodialysis for removal of chromium (VI) from effluent: Analysis of concentrated solution saturation. J. Environ. Chem. Eng., 7: 103380.

Sarode, S., Upadhyay, P., Khosa, M.A., Mak, T., Shakir, A., Song, S. et al. 2019. Overview of wastewater treatment methods with special focus on biopolymer chitin-chitosan. Int. J. Biol. Macromol., 121: 1086–1100.

Shi, L., Huang, J., Zhu, L., Shi, Y., Yi, K., and Li, X. 2019. Role of concentration polarization in crossflow micellar enhanced ultrafiltration of cadmium with low surfactant concentration. Chemosphere., 237: 124859.

Silva, L.S., Galindo, I.C.L., Nascimento, C.W.A., Gomes, R.P., Freitas, L., Oliveira, I.A. et al. 2018. Heavy metals in waters used for human consumption and crop irrigation. Amb. Água., 23: e1999.

Silvestre, W.P., Livinalli, N.F., Baldasso, C., and Tessaro, I.C. 2019. Pervaporation in the separation of essential oil components: A review. Trends Food Sci. Technol., 93: 42–52.

Silvestre, W.P., Baldasso, C., and Tessaro, I.C. 2020. Potential of chitosan-based membranes for the separation of essential oil components by target-organophilic pervaporation. Carb. Polym., 247: 116676.

Sourirajan, S. 1978. The science of reverse osmosis - mechanisms, membranes, transport and applications. Pure Appl. Chem., 50: 593–615.

Sulaiman, N.A., Shaari, N.Z.K., and Rahman, N.A. 2016. Removal of Cu (II) and Fe (II) ions through thin film composite (TFC) with hybrid membrane. J. Eng. Sci. Technol., 11: 36–49.

Suwaileh, W., Johnson, D., and Hilal, N. 2020. Membrane desalination and water re-use for agriculture: State of the art and future outlook. Desalination, 491: 114559.

Tchounwou, P.B., Yedjou, C.G., Patlolla, A.K., and Sutton, D.J. 2012. Heavy metal toxicity and the environment. pp. 133–164. *In:* Lush, A. [ed.]. Molecular, Clinical and Environmental Toxicology - Volume 3: Environmental Toxicology, Springer Basel, Basel, Switzerland.

Thaçi, B.S., and Gashi, S.T. 2019. Reverse osmosis removal of heavy metals from wastewater effluents using biowaste materials pretreatment. Pol. J. Environ. Stud., 28: 337–341.

Touati, K., Usman, H.S., Mulligan, C.N., and Rahaman, M.S. 2020. Energetic and economic feasibility of a combined membrane-based process for sustainable water and energy systems. Appl. Energy. 264: 114699.

Truong, T.T.C., Takaomi, K., and Bui, H.M. 2019. Chitosan/zeolite composite membranes for the elimination of trace metal ions in the evacuation permeability process. J. Serb. Chem. Soc., 84: 83–97.

Tunçsiper, B. 2020. Nitrogen removal in an aerobic gravel filtration-sedimentation pond-constructed wetland-overland flow system treating polluted stream waters: Effects of operation parameters. Sci. Total Environ., 746: 140577.

Tyagi, R., and Jacob, J. 2020. Design and synthesis of water-soluble chelating polymeric materials for heavy metal ion sequestration from aqueous waste. React. Funct. Polym., 154: 104687.

Vane, L.M. 2005. A review of pervaporation for product recovery from biomass fermentation processes. J. Chem. Technol. Biotechnol., 80: 603–629.

Vital, B., Bartacek, J., Ortega-Bravo, J.C., and Jeison, D. 2018. Treatment of acid mine drainage by forward osmosis: Heavy metal rejection and reverse flux of draw solution constituents. Chem. Eng. J., 332: 85–91.

Wang, J., Yu, W., Graham, N.J.D., and Jiang, L. 2019. Evaluation of a novel polyamide-polyethylenimine nanofiltration membrane for wastewater treatment: Removal of Cu^{2+} ions. Chem. Eng. J., 392: 123769.

Xu, W., and Ge, Q. 2019. Synthetic polymer materials for forward osmosis (FO) membranes and FO applications: a review. Rev. Chem. Eng., 35: 191–209.

Yang, X., Liu, Y., Hu, S., Yu, F., He, Z., Zeng, G., Feng, Z., and Sengupta, A. 2020. Construction of Fe3O4@MXene composite nanofiltration membrane for heavy metal ions removal from wastewater. Polym. Adv. Technol., DOI: 10.1002/pat.5148.

Yaqub, M., and Lee, S.H. 2019. Heavy metals removal from aqueous solution through micellar enhanced ultrafiltration: A review. Environ. Eng. Res., 24: 363–375.

Yoon, J., Amy, G., Chung, J., Sohn, J., and Yoon, Y. 2009. Removal of toxic ions (chromate, arsenate, and perchlorate) using reverse osmosis, nanofiltration, and ultrafiltration membranes. Chemosphere, 77: 228–235.

Yoshiura, J., Ishii, K., Saito, Y., Nagataki, T., Nagataki, Y., Ikeda, I., and Nomura, M. 2020. Permeation properties of ions through inorganic silica-based membranes. Membranes, 10: 27.

Zamboulis, D., Peleka, E.N., Lazaridis, N.K., and Matis, K.A. 2011. Metal ion separation and recovery from environmental sources using various flotation and sorption techniques. J. Chem. Technol. Biotechnol., 86: 335–344.

Zanain, M., and Lovitt, R. 2013. Removal of silver from wastewater using cross flow microfiltration. Proceedings of the 16th International Conference on Heavy Metals in the Environment. Rome, Italy.

Zhang, R., Li, Y., Zhu, X., Han, Q., Zhang, T., Liu, Y., Zeng, K., and Zhao, C. 2020. Application of β-Cyclodextrin-Modifed/PVDF blend magnetic membranes for direct metal ions removal from wastewater. J. Inorg. Organomet. Polym. Mat., 30: 2692–2707.

Zhiling Wu, Z., Tang, X., and Chen, H. 2021. Seasonal and treatment-process variations in invertebrates in drinking water treatment plants. Front. Environ. Sci. Eng., 15: 62.

Zou, Q., Zhang, Z., Li, H., Pei, W., Ding, M., Zie, Z., Huo, Y., and Li, H. 2020. Synergistic removal of organic pollutant and metal ions in photocatalysis membrane distillation system. Appl. Catal. B-Environ., 264: 118463.

Zubir, M.A., Zahran, M.F.I., Shahruddin, M.Z., Ibrahim, K.A., and Hamid, M.K.A. 2019. Economic, feasibility, and sustainability analysis of energy efficient distillation based separation processes, Chem. Eng. Trans., 72: 109–114.

8

Adsorption as an Efficient Alternative for the Removal of Toxic Metals from Water and Wastewater

Yasmin Vieira,[1] Juliana Machado Nascimento dos Santos,[4] Jeferson S. Piccin,[2] Ádrian Bonilla-Petriciolet[3] and Guilherme Luiz Dotto[1,4,]*

1. Introduction

The progress of humankind into a well-developed and industrial society, alongside city advancement and expansion, culminated in the water pollution (Schwarzenbach et al. 2010, Halder and Islam 2015, Wang and Yang 2016, Rodgers 1970). As the needs for improvement in agricultural methods increased, diverse secondary industries mounted into the present modern society, such as metal plating facilities, mining operations, fertilizer industries, tanneries, batteries, paper industries, and pesticides fabrication (Henderson et al. 1995, Hu et al. 2020, Willow 2020, Heffron et al. 2020). The one thing that these industries have in common is using chemical reactions with or assisted by metals and metal ions; consequently, they all generate metal contaminated wastewaters, which are directly or indirectly discharged into the environment, especially in developing countries (Snyder et al. 2003, Cai et al. 2020, Guo et al. 2020). Unlike many other organic contaminants, heavy metals are not

[1] Department of Chemistry, Federal University of Santa Maria–UFSM, 1000, Roraima Avenue, 97105-900 Santa Maria, RS, Brazil.
[2] Faculty of Engineering and Architecture, University of Passo Fundo–UPF, 99052-900 Passo Fundo, Brazil.
[3] Instituto Tecnológico de Aguascalientes, Aguascalientes, 20256, Mexico.
[4] Chemical Engineering Department, Federal University of Santa Maria–UFSM, 1000, Roraima Avenue, 97105-900 Santa Maria, RS, Brazil.
* Corresponding author: guilherme_dotto@yahoo.com.br

biodegradable (Hanfi et al. 2020, Hoang et al. 2020, Mokarram et al. 2020) and tend to accumulate in living organisms (Khan et al. 2015, Goodyear and McNeill 1999, Turan et al. 2020, Marella et al. 2020). Many metal ions are known to be toxic or carcinogenic (Zhu and Costa 2020, Diaconu et al. 2020, Tchounwou et al. 2012, Snow 1992). Of particular concern in treating industrial wastewaters is zinc, copper, nickel, mercury, cadmium, lead, and chromium (Fu and Wang 2011).

When in excess, zinc and copper can cause stomach aches, skin rash, vomiting, nausea, convulsions, or even death (Singh et al. 2006, Gaetke and Chow 2003, Pawa et al. 2008, Bennett et al. 1997, Ventura et al. 2017). Nickel intoxication presents itself as lung and kidney problems, aside from gastrointestinal distress, pulmonary fibrosis, and dermatitis (Borba et al. 2006). On the other hand, the slightest mercury ingestion can cause impairment of pulmonary and kidney function, chest pain, and shortness of breath (Shah et al. 2020, Baum 1999). Meanwhile, the chronic exposure to cadmium and lead can result in kidney dysfunction, nervous system damage, liver and reproductive system damage, ending up in disrupting basic cellular processes and brain functions (Sharp et al. 1987, Altmann et al. 1993, Saičić 2008, Horiguchi 2007, Yoshiki et al. 1975). Furthermore, chromium presents harm to human physiology in the form of Cr^{6+}, causing severe health problems that vary from simple skin irritation to carcinoma (Kornhauser et al. 2002, Li et al. 2008, Gibb et al. 2000).

There are many methods that can be used to remove heavy metal ions: chemical precipitation (Wang et al. 2005, Matlock et al. 2002), chelation (Flora and Pachauri 2010, Bose et al. 2002), ion-exchange (Rengaraj et al. 2001, Rengaraj and Moon 2002, Peng and Guo 2020), membrane filtration (Hube et al. 2020, Ercarikci and Alanyalioglu 2020), reverse osmosis (Ozaki et al. 2002, Bakalár et al. 2009), electrochemical technologies (Lacasa et al. 2019, Chen 2004), coagulation, flocculation, and flotation (Blöcher et al. 2003, Polat and Erdogan 2007, Sun et al. 2020, Hankins et al. 2006, El Samrani et al. 2008, Kurniawan et al. 2006). The downwards in the use of the methods cited above are their low cost-effectiveness, which is impacted by their high selectivity towards some metal ions. These methods also tend to generate secondary wastes, and thereupon, solve one environmental problem while becoming another (Fu and Wang 2011). In this context, adsorption is proposed as a recognized efficient method for removing heavy metals in low concentrations from water and wastewater. It is economically viable due to the development of several types of low-cost adsorbents, in addition to the fact that the same adsorbent can be reused several times. Furthermore, adsorption is a simple and elegant process capable of enhancing time-consuming effluent treatment steps under its simple separation by filtration or magnetization.

1.1 Adsorption theory

Adsorption is a unit operation involving a fluid, such as water or wastewater, and a solid phase named the adsorbent. In the fluid phase, one or more dissolved contaminants are present. The process aims to transfer the contaminants from the liquid phase to the adsorbent surface, purifying the water (Dotto and McKay 2020). By examining the most recent research, it is clear that adsorption is one of the most attractive techniques, on account of being simple, flexible, cheap, of high performance, and associated with an extensive range of chemical substances separation capability. In this context, adequate mathematical modeling is crucial and well-studied in literature since the elucidation of a determined adsorption process depends on isotherms, thermodynamics, and kinetics (Miraboutalebi et al. 2017, Dotto et al. 2017, Rozumová et al. 2016). Also, operational parameters can be optimized by other valuable tools, such as response surface methodology (RSM), artificial neural networks (ANN), and by General Algebraic Modeling System (GAMS) (Schio et al. 2020, Amosa and Majozi 2016, Yetilmezsoy and Demirel 2008, Arulkumar et al. 2011).

The isotherms represent the relation between the pollutant amount adsorbed and its remaining concentration in the solution under thermodynamic equilibrium conditions (Piccin et al. 2017). The data obtained by the plotting of these curves are crucial for all adsorption systems. In other words, an adsorption isotherm is a curve relating the equilibrium adsorbed concentration, q_e, to the concentration of the solute in the liquid, C_e (Ho and McKay 1998, Sahu and Singh 2018). The relationship between q_e and C_e can be fitted to one or more equilibrium isotherm models, such as Freundlich, Langmuir, Henry, Sips, and others (Atikah et al. 2020, Wang and Guo 2020, Kalra et al. 2018). Thus, using isotherms, it is possible to investigate adsorption information such as the adsorption mechanisms, the maximum adsorption capacity, and adsorbents' properties (Wang and Guo 2020).

In addition to the isotherms, kinetic aspects can provide more details about adsorption mechanisms, characteristics, and application possibilities. For example, contact time to reach the required concentration of adsorbate can be determined by data obtention, making equipment design and operation possible, as well as defining the optimal performance conditions in batch and continuous systems before testing. In batch studies, kinetic curves are a representation given by plots of adsorption capacity versus time. In contrast, in the fixed-bed case, the representation is given by dimensionless pollutant concentration at the column outlet versus time or bed volumes. Among the most representative empirical kinetic models used for batch adsorption, pseudo-first-order, pseudo-second-order, general order, Avrami and Elovich can be cited. The most representative models for fixed-bed systems are bed depth service time, Clark, Thomas, Adams-Bohart, and Yoon-Nelson. Finally, mathematical tools such as RSM, ANN, GAMS, Fuzzy (FIS), and Neuro-Fuzzy (ANFIS) are also used to interpret adsorption behavior in water and wastewater treatment. They are reliable and robust, capable of relating non-linearity between input and output variables from a set of experiments. Using these models, it is possible to optimize a specific response variable as a function of several input variables.

2. Heavy metal adsorption

2.1 Arsenic

Arsenic is one of the 60 most abundant elements in Earth's crust, and its presence occurs in many minerals combined with sulfur and other metals and as a pure element in the form of crystals. Its compounds are used to produce pesticides, treated wood products, herbicides, and insecticides—all of which are declining with the increasing recognition of its toxicity. Arsenic can occur in the environment in several oxidation states (-3, 0, $+3$, and $+5$). In natural waters, it is mostly found in inorganic forms as oxyanions of trivalent arsenite [As(III)] or pentavalent arsenate [As(V)].

Activated carbon produced from oat hulls was investigated for As (V) adsorption in batch reactor tests. The adsorbent surface area was 520 m^2 g^{-1}. Preliminary results indicated that, in this case, the adsorptive capacity of the activated carbon was affected by the initial pH value, which decreased from 3.08 mg g^{-1} in pH 5 to 1.57 mg g^{-1} in pH 8, according to the optimal fit ($R^2 = 0.98$) to Langmuir isotherm model. Besides, a modified linear driving force model, complementary to a pseudo-first-order model, was developed. It was verified that As adsorption onto activated carbon is more influenced by the initial pollutant concentration than by the adsorbent dosage, which led to the conclusion that the process occurs at both slow and fast pace, with transfer coefficients of 2.2×10^{-4} and 1.0×10^{-5} s^{-1}, respectively (Chuang et al. 2005).

Red mud, a bauxite processing residue discarded in alumina production, exhibited adsorption towards the anionic pollutants after an acid treatment step. Tests with raw red mud demonstrated that arsenite (q_{max} = 8.86 μmol g^{-1}) and arsenate (q_{max} = 6.86 μmol g^{-1}) could be removed in pH 9.5 and 1.1–3.2, respectively. Subsequently, heat and acid treatment were employed, and their adsorption efficiency was compared. The acid-treated red mud presented superior removal efficiency. Therefore, it was used for the following experiments. In this study, the optimal pH range was 5.8–7.5 for arsenite, and 1.8–3.5 for arsenate, reaching maximum removals of 96.52% and 87.54%, respectively (C_0 = 133.5 μmol L^{-1}). Considering a contact time of 60 min and 25°C, the optimal dosage of acid-treated red mud was 20 g L^{-1}, which is 5 times less than the necessary in the process with raw red mud. Moreover, the adsorption data followed a first-order rate expression and fit the Langmuir isotherm well (Soner Altundogan et al. 2000, Altundoan et al. 2002).

Hematite and goethite are both naturally obtained iron oxides. Their adsorption efficiency on the sorption of As(V) was verified considering both pH and ionic strength variations. Again, As(V) adsorption was higher in acidic rather than alkaline pH. About 80% of the initial concentration (C_0 = 500 μg L^{-1}) was adsorbed at pH ranging from 5 to 7, which is very interesting, considering that hematite and goethite surface areas were about 1.66 $\pm$ 0.02 and 11.61 $\pm$ 0.19 m^2 g^{-1}, respectively. The dosage of adsorbent was 4 g L^{-1}, 5 times less than the dosage of acid treated red mud and 10 times less than the raw red mud. No effect by ionic strength variation on the adsorption process was observed. In conclusion, the results suggested a formation of an inner sphere surface complex, and both adsorbents are suitable candidates for arsenate removal (Mamindy-Pajany et al. 2009).

Another iron oxide that can be obtained naturally is magnetite. Magnetite is a very interesting oxide because its magnetic character is widely taken advantage of as a recovery particle for adsorbents, such as carbon nanotubes, besides its intrinsic peroxidase-like activity. As a recent trend, carbon nanotubes have been used for adsorption as a new and promising, more efficient alternative to activated carbons. Regarding arsenic adsorption, Wojciechowska and Lendzion-Bieluń (2020) synthesized hybrid $Fe_3O_4/C/T_iO_2$ nanostructures with a surface area of 190 m^2 g^{-1}, which they used then as adsorbents for the removal of As(V) ions from aqueous media. The process efficiency was very high, superior to 90%, considering an initial concentration of 10 mg L^{-1} and pH ranging from 2 to 7. The nanostructures prepared presented a maximum adsorption capacity of 19.34 mg g^{-1}. Liu et al. (2015) aimed for a better understanding of the mechanism of As(V) and As(III) adsorption, and therefore, conducted thermodynamic and spectroscopic studies, which consisted of macroscopic adsorption experiments associated with thermodynamic calculation, alongside micro-spectroscopic characterization using synchrotron-radiation-based X-ray absorption spectroscopy (XAS) and X-ray photoelectron spectroscopy (XPS). Results suggested a predominant form of bidentate binuclear corner-sharing complexes for As(V), while a tridentate hexanuclear corner-sharing complexation occurred for As(III) on the magnetite nanoparticles surfaces. Both macroscopic and microscopic data were coherent, leading to the conclusively identified formation of inner-sphere complexes between As and the magnetite surfaces. Moreover, the adsorbed's complex redox transformation as exposed to air was noticed to occur simultaneously with magnetite oxidation. According to previous studies reported elsewhere, the oxidation of As(III) and magnetite is to be expected. Still, the observed As(V) reduction was a surprise, mostly due to the reactive Fe(II) role.

Giménez et al. (2007) compared the sorption of As(III) and As(V) on different natural iron oxides: hematite, magnetite, and goethite. The solids' surface areas were determined as 0.381 ± 0.002 m^2 g^{-1} for hematite, 2.009 ± 0.004 m^2 g^{-1} for goethite, and 0.890 ± 0.002 m^2 g^{-1} for magnetite. The three iron oxides' sorption kinetics shows that equilibrium is reached faster for goethite and magnetite than for hematite. However, natural hematite presents the highest sorption capacity, especially at acidic pH. The equilibrium data were fitted with a non-competitive Langmuir isotherm. The most relevant aspect observed was the trend on arsenic sorption towards pH: when pH increased, sorption decreased, and vice-versa. The authors also compared the natural minerals studied in this work to their synthetic versions, which had similar sorption capacities for arsenic. This observation is interesting because it indicates that the experimental methodologies applied are free of artifact effects and that the data found in this work and elsewhere is authentic. Furthermore, the similar sorption capacities found for the natural solids used in this work and the synthesized materials used by different authors could indicate that the sorption mechanism is similar in both kinds of solids. In addition to goethite, hematite, magnetite, Mamindy-Pajany et al. (2011) studied zero-valent iron under different physical-chemical conditions. The reversibility of the adsorption process was also studied using chlorides and phosphates as competing ions. The modeling of isotherms by the Langmuir model suggested a monolayer type of adsorption. Diversely to Giménez et al. (2007) magnetite, and goethite, Mamindy-Pajany et al. (2011) found that the adsorption rate increases in goethite, hematite, and magnetite, and zero-valent iron. However, desorption experiments pointed that arsenic is strongly adsorbed onto hematite then to zero-valent iron. Therefore, among the adsorbents studied, hematite appears to be, as well, the most suitable for removing arsenate in natural medium, mostly because it is effective over large ranges of pH and arsenic concentration.

2.2 Cadmium

Cadmium is a transition metal present in Ni-Cd batteries, playing the role of covering components made of iron and steel. It is also widely used in the formation of alloys, manufacture of photovoltaic cells, and capacitors/semiconductors. In addition to the microelectronics and electroplating sectors, cadmium acts as plastic stabilizer, an important component of phosphate fertilizers and pigments for the textile industry (Alyasi et al. 2020, Kavand et al. 2020, Uddin 2017). Even though the presence of cadmium in the environment may be from a geogenic origin, because this metal naturally exists in soils and rocks, the anthropogenic actions still are the major cause for soil and water pollution, caused mostly by runoff from mishandled waste batteries, galvanized pipes corrosion, mine drainage operation or wastewater overflow from mineral processing (Purkayastha et al. 2014, Soliman and Moustafa 2020).

Cadmium is capable of forming water-soluble complexes with anions and also with dissolved organic matter. Therefore, this metal can be considered of great environmental mobility (Kozyatnyk et al. 2016). It is typically found in an aqueous solution in its divalent cationic state, Cd^{2+}, which tends to remain in solution if pH values are below 6.5, under oxygenated conditions. These Cd^{2+} ions may accumulate in the environment and the human body, primarily through the food chain. Cd^{2+} ions easily replace Ca^{2+} ions, promoting a significant bone loss in contaminated individuals, which is a huge evidence of cadmium toxicity (Pyrzynska 2019, Kubier et al. 2019, Kavand et al. 2020). Then, to ensure population safety, the World Health Organization (WHO) states that cadmium concentration should not exceed 3 µg L^{-1} in drinking/domestic waters (WHO 2011).

A brief survey identified at least three episodes of river waters, from different locations, with high concentrations of cadmium after improper dumping of mining effluents. In 2005, the Beijiang River's Cadmium spill left millions of people without water in Guangdong province in China (Chan 2005, Asia News 2005). In 2012, another environmental disaster occurred at Liu River in Guangxi province, also in China, which forced authorities to apply around 300 tons of caustic soda and polyaluminum chloride each day to flocculate the cadmium and favor its removal (Huang 2012). Still, the greatest cadmium contamination ever recorded, with clinical repercussions on the nearby population that fed on rice irrigated by Jinzu River waters, occurred in Japan during the year of 1950 (Aoshima 2016). Thousands of middle-aged and older adults complained of severe pain and developed bone fractures with minimal strength. This set of symptoms, caused by kidney failure and bone demineralization resulting from chronic cadmium poisoning, became known as the Itai-Itai (Ouch-Ouch) disease in 1968 (Suwazono et al. 2019).

The environmental issues related to cadmium pollution still happen in 2019. Chinese researchers concluded that the cadmium levels found nowadays are comparable to those that triggered the Itai-Itai disease in the Japanese population, mostly by evaluating the impact of Cd contaminated soils on the food security of locally produced rice (Wang et al. 2019). To prevent scenarios like these from occurring again, the development of technologies applied to effluents' treatment is a key factor. Precipitation technique is by far the most used (75% of electroplating facilities) and consists of an insoluble precipitate formation, compounded by reagents and metal contaminant removed by filtration or sedimentation. However, adsorption, especially biosorption, has focused on many types of research related to the treatment of effluents contaminated by heavy metals. High removal efficiency combined with simplicity and low operating cost makes the adsorption a promising alternative to those companies that must treat effluents without increasing operational costs (Purkayastha et al. 2014).

An example of bioadsorption using marine material that presents low toxicity risk is given by Satya et al. (2020), where, in this case, the cyanobacterium *Aphanothece* sp., carrier of a cell membrane rich in functional groups and polysaccharides, capable of interacting with heavy metals, had its performance as an adsorbent evaluated. When placed in a packed column to predict parameters and rupture curve for scale-up purposes, the proposed adsorbent managed to remove about 90% of cadmium in the aqueous phase, with a maximum capacity of 8.20 mg g^{-1}, under 0.60 L h^{-1} and 4.85 mg L^{-1} as flow rate and Cd inlet concentration, respectively. Only after five adsorption-regeneration cycles, using diluted hydrochloric acid solution as regeneration agent, the authors observed an adsorption efficiency loss of column only around 21% that, together with other characteristics, justify the spotlight on biosorption operations focused on metals removal (Satya et al. 2020).

Xie et al. went further in their studies. In addition to providing construction of three-dimensional g-C$_3$N$_4$/attapulgite hybrids to remove cadmium from a liquid effluent, the authors reused cadmium containing-adsorbent as a catalyst to treat effluents contaminated by tetracycline antibiotic through photocatalysis. Graphitic carbon nitride (g-C$_3$N$_4$), a multilayer polymeric material, was used to functionalize attapulgite (ATP), a natural clay mineral. Electrons from g-C$_3$N$_4$ nitrogen atoms acted as a molecular structure for metal ions attraction, and their coordination capacity provided a stable ions adsorption. The g-C$_3$N$_4$/ATP hybrid adsorbent, with abundant surface hydroxyl and tri-s-triazine unit, showed greater cadmium adsorption capacity than pure attapulgite, 61.1 and 34.7 mg g^{-1}, respectively (Xie et al. 2020).

When nano chitosan was used as an adsorbent, Alyasi et al. (2020) considered that cadmium adsorption embraced two types of mechanisms, noticeable in Freundlich's multi-stage isotherm: (i) Cd^{2+} ions chelation with the pair of free electrons in chitosan amino groups through dative bonds, where the electron pair and metal ions act as a base and acid of Lewis, respectively, and (ii) surface complexation of metal ions by sharing electron pairs of oxygen atoms. The Freundlich model fitting to experimental data provided that the amino-chitosan-cadmium chelation capacity was 1.749 mmol of Cd, and the surface complexation capacity was 0.184 mmol g^{-1} (Alyasi et al. 2020).

In real effluents, where metal removal is more complex, cadmium ions are unlikely to be the only contaminants founded. For this reason, studies that evaluate the influence of coexisting ions in adsorption operation are so important. Kavand et al., for example, observed that the coexistence of lead and cadmium ions in solution caused a significant reduction in the adsorption capacity of activated carbon. In the single system, the equilibrium adsorption capacity for Pb^{2+} and Cd^{2+} was 119.32 and 117.86 mg g^{-1}. These values decreased, respectively, to 97.63 and 94.43 mg g^{-1} when a multiple system was considered. The higher the atomic weight and the greater electronegativity of Pb^{2+} compared to Cd^{2+} ions, as well as the lower polarization power of Cd, results in weaker carbon-cadmium interaction forces than carbon-lead interactions, which can explain the higher adsorption capacity for Pb when it coexists with Cd (Kavand et al. 2020).

Competitive behavior between Cd^{2+}, Ni^{2+}, and NH_4^+ ions for adsorption on fresh and artificially aged biochars was studied by Deng et al. (2020). They highlighted the existence of cation exchange mechanisms, mineral co-precipitation, and surface complexation with oxygen-containing functional groups. Cation exchange occurs when the cations K^+, Ca^{2+}, Na^+ and Mg^{2+}, retained in biochar by electrostatic attraction, are replaced by the ions to be adsorbed. The surface complexation mechanism is characterized by complex formation between biochar-oxygenated groups and the adsorbate, when carboxyl and hydroxyl release H^+ ions. On the other hand, mineral co-precipitation occurs when the biochar presents many minerals trapped at its surface. These minerals precipitate or co-precipitate the Cd^{2+}, Ni^{2+}, and NH_4^+ ions. The SEM images showed white granular crystals, identified as cadmium mineral phase ($CdCO_3$). Regarding cadmium and nickel ions adsorption, cation exchange mechanisms and co-precipitation were dominant, although competition between them favored other potential mechanisms (Deng et al. 2020).

2.3 Chromium

Chromium is one of the most worldwide used metals, but still, its behavior is very misunderstood. Despite being a necessary macronutrient in its trivalent state, it is subject to several discussions because it can exist in toxic forms, such as hexavalent, which can be hazardous to the human health and the environment. The chemistry of chromium includes redox transformations, precipitation and dissolution, and adsorption and desorption reactions. The most prominent threat regarding chromium is that it exists concomitantly with manganese oxides that can transform Cr(III) to Cr(VI). In its trivalent state, chromium plays the role of maintaining proper carbohydrate and lipid metabolism at a molecular level; therefore, it is known to be a micronutrient for mammals for four decades. Cr(III) readily forms hydroxide complexes, compounds such as $Cr(OH)_3$ and $(Cr, Fe)(OH)_3$ in aquatic environments, which present amphoteric solubility behavior. Their relatively low solubilities limit Cr(III) concentrations to less than the drinking water limit over much of the pH range of environmental interest. Hence, Cr(III) is not an environmental problem at all. However, Cr(VI) induces sensitization of the respiratory tract and is a potent sensitizer of the skin; it can induce mutations *in vitro* and *in vivo* and cause cancer in experiments carried out with both animals and humans. Therefore, it presents a threat to public health.

Hexavalent chromium in groundwater has generally been assumed anthropogenic contamination since it is used in several industrial applications, including electroplating, tannery industries, industrial water-cooling, paper pulp production, petroleum refining, and others. The carcinogenicity of different hexavalent chromium compounds differs markedly. Studies show that strontium chromate ($SrCrO_4$) is more carcinogenic by far than any other chromium compound. Also, calcium chromate

(CaCrO$_4$) and zinc chromate (ZnCrO$_4$) are potentially toxic and carcinogenic. The good news is that these chromates are relatively insoluble in water. In most soil environments, the iron oxides are naturally present as the most important adsorbents for aqueous Cr(VI) species. Still, natural depuration is not enough to solve this contamination problem, and remediation strategies need to be developed to restrain Cr(VI)'s way into the environment.

Sharma and Forster (1994) verified four organic wastes as potential adsorbents: sugar cane bagasse, sawdust, sugar beet pulp, and maize cob. The only process these materials underwent was drying for 24 h in an oven at 110°C, as well as sieving and homogenization. The adsorption capacities of sorbents at varying pH were examined. Regression analysis of the Langmuir parameters resulted in high correlation coefficients, indicating a strong positive relationship. The highest maximum adsorption capacities attained were 39.7, 17.2, 13.4, and 13.8 mg g^{-1} for sawdust, sugar beet pulp, sugar cane bagasse, and maize cob. All values were found at about pH 2, except maize cob, which presented its q_{max} at pH 1-5. These materials are abundantly available throughout the world and are considered sub-products of other industries, mostly treated as waste. Therefore, applying them in their raw form as adsorbents is very interesting from an ecological perspective.

The initiative of using activated charcoal for Cr(VI) removal can be combined with the reuse of agricultural waste. Mohan et al. (2005) proposed developing low-cost activated carbons derived from agricultural waste. They developed activated carbon derived from coconut fibers (FAC, S$_{BET}$ = 343 m^2 g^{-1}), activated carbon derived from coconut shells (SAC, S$_{BET}$ = 378 m^2 g^{-1}), activated carbon derived from acid-treated coconut fibers (ATFAC, S$_{BET}$ = 512 m^2 g^{-1}), and activated carbon derived from acid-treated coconut shells (ATSAC, S$_{BET}$ = 380 m^2 g^{-1}). Adsorption results indicated that the Langmuir isotherm model fits the data better than the Freundlich model, and the data was more correlated in a nonlinear form than conversely. The kinetic studies conducted suggested that the adsorption of Cr(VI) follows pseudo-second-order rate kinetics. Overall, the removal of Cr(VI) in concentrations ranging from 1 to 100 mg L^{-1} increased in the order of SAC < ATFAC < ATSAC < FAC, which reached q_{max} values of 9.53 < 9.86 < 11.51 < 21.75 mg g^{-1}, respectively.

Di Natale et al. (2015) performed dynamic tests on Cr(VI) adsorption by commercially obtained granular activated carbons in terms of breakthrough curves of a lab-scale fixed-bed column, considering different pH, inlet concentration, and flow rate. The results allowed the determination of the intercorrelation between the main process parameters. Both the adsorption isotherms and the breakthrough curves behaved towards non-linear and unconventional trends. The chromium C_0 varied between 5–50 mg L^{-1}, while the sorbent concentration spanned from 2 to 10 g L^{-1}. Optimal experimental conditions were found to be reached at T = 25°C, pH 6-7, Q = 6.5 L h^{-1}, C$_0$ = 50 mg L^{-1}, and adsorbent dosage of 10 g. In these conditions, granular activated carbon reached an average loading of about 1.6 ± 0.15 mg g^{-1} of chromium. Furthermore, the experimental results pointed out that speciation played a key role in the adsorption process, mostly by reducing reactions.

Duranoğlu et al. (2012) presented another way of obtaining activated carbon for chromium adsorption: deriving it from acrylonitrile-divinylbenzene. Their choice was based on controlling the polymer-derived adsorbent's structure by choosing acrylonitrile-divinylbenzene as the precursor material. In general, polymer-derived carbons tend to present a considerably high adsorption capacity due to high specific surface area. Besides, they are more stable and mechanically resistant than other types of carbonaceous materials, combined with good hydrodynamic properties, making them good candidates for use in fixed bed/fluidized bed type separation applications. The developed adsorbent present a surface area of 579 $m^2 g^{-1}$ and attained an adsorption capacity of 101.2 mg g^{-1} at pH 2. Still, this proposal has limitations, and the adsorbent regeneration needs to be improved.

In light of the recent trends towards investigating multiple applications of graphene, Wang et al. (2020) used ammonium thiocyanate to functionalize graphene oxide-supported nanoscale zero-valent iron for both the adsorption and reduction of Cr(VI). This composite presented itself as more chemically stable than nanoscale zero-valent iron alone, which is very important since the authors aimed to reduce Cr(VI) to Cr(III). Moreover, the observed Cr(VI) removal efficiency by the prepared composite was higher than that of bare ammonium thiocyanate graphene oxide and pristine nanoscale zero-valent iron, reaching up to 94% removal against 86% and 88%, respectively.

2.4 Cobalt

A silvery-gray metal with high hardness, the elemental cobalt is naturally present in the environment in some rock and soils, in plants, and in animals that consume cobalt-containing food. It is common in solid-state that cobalt be associated with other chemical elements such as oxygen, nitrogen, and sulfur, forming nitrates, sulphates, carbonates, and cobalt oxides. In an aqueous medium, it is usually found in its ionic form and small amounts. The cobalt applicability list is large and quite diverse. When used as dye, it provides blue color to glasses, ceramics, and paints, where it can also act as a catalyst and drying agent. Metal alloys, which hold cobalt as a component, are used to manufacture magnets, batteries, cutting tools, motors, and orthopedic prostheses that replace worn-out hip and knee joints (Agency for Toxic Substances and Disease Registry 2004, Liu et al. 2019, Dehaine et al. 2021).

Among the 26 cobalt element isotopes already known, only one of them is stable, the ^{59}Co. All the rest of the 25 are radioactive, capable of ionizing radiation production (Szymanski and Thoennessen 2010). The ^{60}Co isotope, for example, is a source of gamma rays widely used in radiation therapy treatment for cancer patients and the sterilization of surgical equipment. The ^{58}Co isotope can be produced when a neutron source reaches the nickel-metal that makes up nuclear reactors. Therefore, ^{58}Co isotope is one of the contaminants to be considered in the water of nuclear reactor cooling systems. Although there are differences among them, such as applicability and decay period, all known isotopes of cobalt exhibit the same chemical behavior in the environment and human body, with beneficial or deleterious effects, depending on the considered amount (WHO 2006, Saleh et al. 2020, Agency for Toxic Substances and Disease Registry 2004).

Cyanocobalamin, for example, also known as vitamin B12, is a cobalt biochemical compound essential to the central nervous system and red blood cell formation in humans (Jägerstad and Arkbåge 2003). However, in 1960, many beer consumers felt severe nausea after drinking, and some of them died after cardiac complications. Some breweries used cobalt salts as foam stabilizers that took the cobalt concentration over the range of 0.04–0.14 mg cobalt per kg of body weight (Alexander 1972). Like other contaminants, the amount of cobalt present in the air and water is regulated by the responsible public institutions, especially near industrial units, where employees and nearby communities are constantly exposed and risk developing lung, kidney, and liver diseases (WHO 2006). In this context, the simplicity, efficiency, and robustness of the adsorption systems make the adsorption a strong ally for companies required to treat wastewater to comply with current legislation standards, but also to ensure population health security (Zhang et al. 2020a, Gunjate et al. 2020).

In the scientific literature, studies on cobalt adsorption focus mainly at mechanisms' understanding and in finding best operational condition, employing conventional adsorbents and alternative/conventional modified adsorbents. Kara et al. (2003), for example, tried to identify the Co(II) adsorption mechanism by sepiolite, a clay mineral with ion exchange properties. This adsorption study was the first to propose a relationship between the replacement of Mg (II) ions, naturally present in the material, by Co(II) ions. The authors observed that, for each Co(II) adsorbed, an equivalent amount of Mg(II) migrates from the sepiolite surface to the aqueous phase, mainly with pH values between 5 and 8.2. This proportion has not been verified above this pH range since the hydrolysis of Co(II) ions occurs as insoluble complexes. Although the activated sepiolite had a greater specific surface area when compared to natural sepiolite, the collapse of Mg(II) ions or the formation of an amorphous layer during acid activation, followed by heat treatment, reduced its ability to release ions Mg(II) and, consequently, adsorb the ions of interest (Kara et al. 2003).

Kyzas et al. prepared three charcoal types from potato peels, activated with H_3PO_4, to remove Co(II) from synthetic wastewaters (PoP400, PoP600, and PoP800, where 400, 600, and 800 indicated the activation temperatures in °C). The results presented by the authors demonstrated that the activation step, in addition to the significant increase in specific surface area of PoP400 and PoP600, provided the introduction of functional groups, especially aliphatic ones, capable of promoting complexation sites for Co(II) at the adsorbent surface. The complexation, together with the cation exchange and electrostatic interaction, were considered the main adsorption mechanisms. Co(II) ions' maximum coal adsorption capacity was quite high, with values of 373 and 405 mg g^{-1}, for PoP400 and PoP600, respectively (Kyzas et al. 2016).

In the age of automation and search for new ways of reducing fossil fuel emissions, electrical devices' development grows fast, emphasizing the automotive industry with a high level of cobalt consumption. In this context, it is possible to highlight the adsorption as effluent treatment technology and as a very efficient method for recovery of cobalt, an essential metal in electrical components. By the year 2025, it is estimated that at least 11 million electric vehicles will be manufactured and sold. These cars will then represent about 50% of the new cars sold fleet in 2040 (Haji and Slocum 2019). The lithium-ion batteries used in electric vehicles hold a large cobalt amount. This scenario may represent this technology's bottleneck since the batteries' demand, projected for 2040, would exhaust current cobalt land reserves in just 9 years. In the Tesla model car battery, for example, there are about 4.8 kg of cobalt. In the Volkswagen electric model, this amount reaches 24 kg (Benchmark Mineral Intelligence 2018, Bloomberg 2018, Haji and Slocum 2019).

It is already known that cobalt is much more abundant in the oceans than in the Earth's crust. The marine reserves contain around 50 million tons of dissolved cobalt, while terrestrial reserves contain only 7 million tons of this element (Diallo et al. 2015, United States Geological Survey 2018). Given this information, Haji and Slocum present a new and ambitious opportunity to obtain cobalt based on estimates, using deactivated oil platforms to support industrial cobalt adsorption systems. With high adsorption capacity adsorbents, adequately located along the water column, the authors concluded that it is possible to obtain between 521 and 4,172 tons of cobalt annually in each of the 76 offshore platforms evaluated. Therefore, the cobalt recovery/adsorption is, in addition to an environmental issue, a strategic question (Haji and Slocum 2019).

2.5 *Copper*

Copper is abundant in nature, being found in a variety of rocks and minerals. It is one of the essential micronutrients necessary for numerous metabolic processes of the most diverse life forms. At least thirty Cu-containing enzymes are known, all of which function as redox catalysts or dioxygen carriers. It can exist in three oxidation states: zero-valent (Cu^0, a solid metal); Cu(I), cuprous ion; Cu(II), cupric ion.

The anthropogenic activities by which Cu enters soils and sediments include smelting, mining, metal plating, steelworks, refineries, domestic waste emission, and the application of fertilizers, sewage sludge, algicides, fungicides, and molluscicides. At concentrations of 100 to 200 $\mu L\ L^{-1}$, Cu can disturb most kinds of plants' metabolic processes and growth. Even though copper tends to bind to proteins, it can be released and, when free, it acts as a catalyst in the formation of highly detrimental hydroxyl radicals. In humans, chronic Cu exposure leads to toxicity primarily in the liver because it tends to first deposit after entering the bloodstream. Therefore, toxicity is typically manifested by cirrhosis development accompanied by the damage to renal tubules, brain, and other organs. These symptoms can progress to coma, hepatic necrosis, vascular collapse, and eventually death.

Chen et al. (2003) modified commercially available activated carbon with 1.0 M citric acid (resulting in 431 $m^2\ g^{-1}$), followed by reaction with 1.0 M sodium hydroxide (resulting in 448 $m^2\ g^{-1}$), which was optional. The surface modification reduced specific surface area (originally 631 $m^2\ g^{-1}$) and the point of zero charge of the material. Despite the reduction in surface area, it became more homogeneous. The maximum adsorption capacity was 14.92 mg g^{-1}, which was 140% higher than the unmodified carbon. Adsorption kinetic mechanisms responded well to the intraparticle diffusion model. Imamoglu and Tekir (2008) developed an activated carbon from a new precursor hazelnut husks with zinc chloride activation. Its surface area was found to be 1092 $m^2\ g^{-1}$. Studies were conducted in batch by investigating initial pH, contact time, adsorbent dosage, and Cu concentrations. Equilibrium data were fitted to both Freundlich and Langmuir isotherms well, but Langmuir presented better correlation coefficients. The maximum adsorption capacity obtained was 6.645 mg g^{-1}, respectively. Even though it is lower than the q_{max} value for the activated carbon modified with citric acid, the importance of Imamoglu and Tekir's (2008) work lies in transforming a residue to a promising adsorbent.

Papandreou et al. (2007) used fired coal fly ash, a solid by-product produced in power plants worldwide. Their preparation method consisted of shaping the coal fly ash into pellets with a diameter in-between 3–8 mm. Due to the cementitious properties of fly ashes, the authors' most promising stabilization technique was encapsulation of metal saturated pellets in concrete structures. Despite the fly ash pellets presenting a low surface area of 10.20 m^2 g^{-1}, the material reached a maximum adsorption capacity of 20.92 mg g^{-1} of adsorbent. The adversity of this experiment was the contact time needed, of about 2 months.

Aydin et al. (2008) invested in the line of low-cost adsorbents as a replacement for current methods, most of which are costly. Different adsorbents were obtained from shells of lentil, wheat, and rice. The preparation process consisted of air drying at 110°C for 24 h, proceeded by sieving (0.6 mm). Equilibrium adsorption level was determined as a function of the solution pH (2–6), temperature (20°C), contact time (3 h), initial adsorbate concentration (100 mg L^{-1}), and adsorbent doses (20 g L^{-1}). Langmuir model was used, and the maximum adsorption capacities for lentil, wheat, and rice shells adsorbents at 20°C were 8.977, 7.391, and 1.854 mg g^{-1}, respectively. Yu et al. (2000) used locally available sawdust, a byproduct of the world industry, directly for adsorption experiments without any treatment. Maximum adsorption capacity was 1.79 mg g^{-1}, similar to the one obtained by rice shells in Aydin et al.'s (2008) study. Many byproducts have little or no economic value, are still produced in large quantities, and often present a disposal problem. If these wastes could be used as an absorbent, both the environment and the economy would benefit. Contaminated streams would be cleaned, and a new market would be opened for the residues. Therefore, results demonstrate that low-cost adsorbents can be economically viable and effective in Cu(II) removal.

Chen et al. (2011) produced biochars—adsorbents obtained by pyrolysis of biomass under oxygen-limited conditions—of hardwood and corn straw. Adsorption equilibrium data fit well Langmuir isotherm models, with q_{max} values of 6.79 and 12.52 mg g^{-1} for corn straw and hardwood, respectively. Even though the values can be considered low, they are surprisingly high when the surface area of the materials obtained is considered, of about 0.43 13.08 m^2 g^{-1} obtained for hardwood and corn straw biochars. Furthermore, the thermodynamic analysis pointed to an endothermic process, which did not occur spontaneously. On the other hand, chitosan, the most abundant natural biopolymer with relatively strong mechanical strength, was used by Li and Bai (2005) in the construction of chitosan–cellulose hydrogel beads. Both the chitosan–cellulose and the cross-linked chitosan–cellulose beads developed presented high adsorption capacities for Cu adsorption, with q$_{max}$ values around and 53.2 mg g^{-1}. Still, the adsorption process was highly pH-dependent. Chitosan is a very interesting material because it can be cheaply obtained from chitin, which conveniently is the second most naturally abundant biopolymer, alongside cellulose, that combines very well together. Both can be readily attained from seafood-processing wastes.

At last, regarding copper adsorption in both single and binary compound systems, with Pb, Sellaoui et al. (2017) investigated treated Sea Mango fruits. The obtained fruit was dried, pulverized until particle size reached 80 to 100 mesh, and then delignified. Two different powders were obtained, hence: sea mango and delignified sea mango powder. The surface area of pristine sea mango powder was $3.28 \, m^2 g^{-1}$, while the surface area of the delignified sea mango was $15.74 \, m^2 g^{-1}$. For single- and binary-compound adsorption, each metal's initial concentration was set at $200 \, mg \, L^{-1}$. As theoretically expected, a significant reduction in the adsorption capacity in the binary system was observed. This trend is very important because the applications developed must attend to environmental needs, and, in the environment, neither pollutant is alone.

2.6 *Manganese*

Manganese is a transition metal and presents different oxidation states varying between $+2, +3, +4, +6,$ and $+7$. In the environment, the elemental manganese, in metal form, does not naturally occur but as a component of minerals such as manganese carbonate, dioxide, silicate, borate, etc. The manganese oxidation states Mn(II) and Mn(VII) are soluble in water. However, Mn(VII) of oxidizing permanganate ions is not formed in most natural waters. Mn(II) mostly predominates at 4–7 pH range, Mn(III) is stable only as a complex, Mn(V) and Mn(VI) are not stable in neutral solutions. In terms of availability, manganese lags only behind iron among heavy metals, representing around 0.1% of Earth's crust composition. Its most popular use is on iron-manganese alloys composition applied on steel manufacturing, where manganese acts like deoxidizer, sulfur control content, and improves alloy hardness and strength. Manganese is also an important component of dry-cell batteries, fireworks, matches, fertilizer, varnish, fungicides, and food additive (Patil et al. 2016, U.S. Environmental Protection Agency 2003, El-Aassar and Mohamed 2021).

For biological systems, the Mn(II) and Mn(III) oxidation states are the most relevant. In humans, manganese is required for proper iron metabolism and brain function. However, when Mn concentration exceeds $0.1 \, mg \, L^{-1}$ in drinking waters, poisoning symptoms tend to appear, such as fatigability, cephalalgia, sleep disturbances, muscular pain, masklike face, reduced coordination, hallucinations, and mental irritability, mostly in children and infants. In addition to health issues, manganese in water leads to rusted pipes, clogged plumbing, and taste in beverages. Whereas manganese releases into waterways result mainly from anthropogenic action through mining and industry activities, manganese water contamination has become a serious issue since its application tends to increase. According to World Steel Association 2019 report, the global steel demand increased by 1.3% from 2018. In 2020, an increase of 1% was estimated, reaching 1.752 tons. To reduce potential environmental impacts resulted from Mn increased demand, the adsorption process with flexible design and operation is fully capable of providing an effluent free of odor, color, or sludge (Agency for Toxic Substances and Disease Registry 2012, Rudi et al. 2020, Michalke and Fernsebner 2014, Youngwilai et al. 2020, Soliman and Moustafa 2020).

Rudi et al. (2020) provided an overview of the manganese adsorption using agricultural waste adsorbents and the adsorbate-adsorbent interaction mechanisms. For all the residues studied (flowers, peels, seeds, even fibers of fruits and grains), the manganese adsorption tended to occur through two main mechanisms: ions exchange and complex formation. Despite the high removal efficiencies reported, the authors indicated a prior deep investigation when evaluating the costs related to agricultural waste employability in manganese adsorption, taking into account the availability and the chemical or physical procedures commonly used to improve waste structural characteristics and its adsorption capacity for manganese ions. The most commonly used physical treatments are heating, freezing, drying, autoclaving, and lyophilization. Chemical treatments and removing existing impurities also promote the insertion of superficial functional groups capable of interacting with manganese. It substitutes hydrogen ions for contaminants or even donating pairs of electrons to form complexes (Rudi et al. 2020).

Sugarcane bagasse (SCB) is such an example of agricultural waste applied to manganese adsorption. This material is a by-product of sugar production with high cellulose, pentosan, and lignin concentrations that provide carboxylic, carbonyl, amine, and hydroxyl groups as adsorptive sites for manganese ions. When investigating the adsorption of Mn(II) by sugarcane bagasse (SCB) treated with HCl and commercial activated carbon for comparison purposes, Esfandiar et al. identified that the modified SCB was able to remove superior Mn(II) quantities than untreated SCB, even activated carbon. The performed analysis results suggested that carboxyl and hydroxyl groups were involved in Mn(II)/SCB binding. Moreover, the analysis indicated that electrostatic attraction and hydrogen bond formation between SCB treated surface and Mn(II) dominated the adsorption process (Esfandiar et al. 2014).

Compared to activated carbon, the most widely used adsorbent for metal removal, the acquisition costs of biochar tend to be about 6 times less. The high surface area, the functional group's diversity, and its microporous structure are probably the factors that make biochar an efficient material for removing manganese. Idrees et al. (2018) produced biochar from poultry and farmyard manure as an alternative to reducing the associated environmental issues to manure, applying as an organic soil amendment and providing a solution for manganese contaminated water. It was possible to observe that percentage removal increased up to 80% at less acidic pH, possibly due to low H^+ ion concentration that competes with Mn^{2+} ions for negatively charged sites at biochar surface. Adsorbent structural analyses after adsorption indicated the formation of manganese compounds with carbonates and phosphates, suggesting that the adsorption presents a chemical nature (Idrees et al. 2018).

Youngwilai et al. (2020) also evaluated the potential for manganese removal, but by biochar from the wood vinegar production. After modifying the biochar surface through chemical treatment with hydrogen peroxide, the authors immobilized manganese-oxidizing bacterium, forming a biologically activated carbon that, in addition to adsorbing the metal, was able to oxidize the dissolved manganese (Mn^{2+}) to particulate Mn^{3+} or Mn^{4+}. Thus, the biochar produced also played the role of a biofilter. Although Mn^{2+} ions and immobilized cells occupied the same type of active site, the immobilization provides a significant enhancement of manganese removal performance, reaching almost 80% (Youngwilai et al. 2020).

The manganese adsorption in a continuous system was evaluated by Li et al. (2020) and El-Aassar and Mohamed (2021). The red mud, an alkaline and mineral by-product from alumina production by the Bayer process, was prepared and immobilized into a sodium alginate matrix gel. At pH 6, the maximum adsorption capacity reached 56.81 mg g^{-1} in a batch experiment. The estimated removal efficiency was about 3 times higher in the column, where 40 mL h^{-1} and 1.57 h were the volume flow and the retention time considered, respectively (Li et al. 2020). El-Aassar and Mohamed (2021) used anthracite modified by microwave to adsorb Mn(VII) ions considering batch and continuous systems. The maximum adsorption capacity, obtained in the first cycle, was 35.7 mg g^{-1} at pH 3. At each column regeneration cycle, using H_2SO_4 diluted solution, the percentage of Mn (VII) removal was reduced, dropping from 96 to 95, 94, 93, and 89%, after each of the 5 consecutive cycles were evaluated, respectively (El-Aassar and Mohamed 2021).

The adsorption of coexisting ions with Mn(II) was investigated by Goher et al. (2015). Two different adsorbents were evaluated on iron, aluminum, and manganese adsorption: granular activated carbon (GAC) and Amberlite IR-120H (AIR-120H), a strong cationic exchanger. At pH 5, the removal efficiency was higher for iron and aluminum contaminants, while 7 was the most appropriate pH value for manganese adsorption. The removal percentage of each metal was higher, but the values indicated that coexisting metals might cause a decrease in manganese adsorption on both adsorbents. The removal tendency was $Al^{+3} > Fe^{+2} > Mn^{+2}$ with 99.2, 99.02 and 79.05 and 99.55, 99.42, and 96.65% of metal removal with GAC AIR-120H, respectively (Goher et al. 2015).

2.7 *Lead*

Lead is a very dense metal; it is also soft and malleable. Its toxicity was recognized in the 19th century, and since then, its use has been banished from many applications. However, in many countries, some paints and bullets containing lead are still sold. In the human body, lead acts as a neurotoxin, accumulating in soft tissues and bones, damaging the nervous system, interfering with biological enzymes' function, leading to neurological disorders, mostly brain damage and behavior problems. It is presented as two main oxidation states: +4 and +2.

Nonetheless, only a few inorganic Pb(IV) compounds are known because they are only formed in highly oxidizing solutions and do not normally exist under standard conditions. On the other hand, Pb(II) compounds are more likely to be found. Hence, their existence is something almost characteristic in the inorganic chemistry of lead. It can form compounds even with strong oxidizing agents, such as fluorine and chlorine, forming PbF_2 and $PbCl_2$. In its ionic form, Pb^{2+} is usually colorless, and it tends to hydrolyze, forming Pb(OH)+ partially, and later $[Pb_4(OH)_4]^{4+}$.

Nowadays, the largest use of lead is in lead-acid battery production. Even though it does not undergo direct contact with humans, there are still toxicity concerns. Even so, the people who work in battery production plants can be exposed to form lead dust, which can unwind in harmful effects after inhalation. Even though these batteries have lower energy density and lower charge-discharge efficiency than lithium-ion batteries, they are still significantly cheaper. Thus, its production perseveres. Overall, lead has entered the environment from industrial production, incineration, and recycling. For example, Tetraethyllead was mixed with gasoline at the beginning of the 1920s as a patented octane rating booster that allowed engine compression to be raised substantially. Only after the 1970s phasing out had begun, which eventually led to the banning of tetraethyl-lead addition in automotive fuel. The most interesting part is that it was not specifically banned to combat poisoning or because its adverse effects were discovered. It was found that it interfered with the newest smog-reducing catalytic converters, soon to be incorporated in the market. Because of this negligence, its impacts are still felt nowadays: lead was buried in many disposal attempts, only to end soon being mobilized in soils and sediments in post-industrial and urban areas.

Sekar et al. (2004) studied the removal of lead onto coconut-shell carbon. The adsorbent was prepared with coconut shells collected from oil industries located in India. The shells were crushed, washed with deionized water, and dried. They were then mixed with concentrated sulphuric acid, and the solution rested for 24 h, with occasional stirring. The acid-carbonized coconut shells were then washed, sieved, and used in further experiments. At pH 4.5, the highest Pb^{2+} adsorption was noted, reaching 26.50 mg g^{-1}. Interestingly, Langmuir, Freundlich, and Tempkin isotherm models described well the equilibrium data. Despite being simple, this method of preparation proved to be effective, resulting in an efficient adsorbate with a surface area of 265.96 m^2 g^{-1} obtained by simple acid carbonization of coconut residue.

Goel et al. (2005) investigated in batch and column studies coconut shell-based granulated activated carbon adsorption, considering a pristine (1000 m^2 g^{-1}) and a sulfur enriched version of the material (900 m^2 g^{-1}). More efficient adsorption was obtained in column mode adsorption studies and the sulfur enriched carbon. Breakthrough curves were plotted regarding continuous-flow column operation, considering operating parameters such as hydraulic loading rate (3.0 to 10.5 $m^3h^{-1} m^{-2}$), bed height (0.3 to 0.5 m), and feed concentrations (2.0 6.0 mg L^{-1}). In this study, Bohart–Adams modeling provided an objective framework to the subjective interpretations, illuminating the path towards upscaling the adsorption system and designing the adsorption process at the pilot plant scale level. Besides consideration towards the differences between batch and column mode, factors as the adsorbent's selectivity are very important. Therefore, Peng et al. (2014) verified lead adsorption behavior in Two-Dimensional Titanium Carbide with an activated hydroxyl group. They prepared an alk-MXene, $Ti_3C_2(OH/ONa)_xF_{2-x})$, by chemical exfoliation followed by alkalization intercalation. Surprisingly, adsorption equilibrium was achieved in 120 s. In adsorption tests, it exhibited preferential Pb(II) sorption behavior when competing with other cations, such as Ca(II) and Mg(II), even when they coexisted at high concentration levels, hence presenting high selectivity towards Pb(II) adsorption.

Jin and Bai (2002) investigated chitosan/polyvinyl alcohol hydrogel beads in batch adsorption experiments, considering pH values ranging from 2 to 7.6. They noticed that lead adsorption on the chitosan composite was strongly pH-dependent, with its maximum uptake capacity at around pH 4, while its minimum was observed at around pH 6. The dominant mechanisms were complexation, ion exchange, and electrostatic interaction. As observed, the relative importance of each of these mechanisms varied alongside pH values. Finally, to explore the effect of inorganic minerals on the activation process and lead adsorption onto sludge-based biochar, Zhang et al. (2020) verified the influence of pre-deashing the adsorbate with hydrochloric or hydrofluoric acid, followed by potassium acetate activation. An increase in pore parameters was observed, which subsequently promoted a more effective activation by potassium acetate addition. After the acid pretreatment, the surface area increased from 583.36 $m^2 g^{-1}$ to 718.70 and 991.55 $m^2 g^{-1}$ for hydrochloric and hydrofluoric acid, respectively. The biochars that underwent the acid pretreatment showed better lead adsorption capacities: 16.70 and 49.47 mg g^{-1} versus 7.56 and 38.49 mg g^{-1}.

3. Conclusion

This chapter highlighted the removal of arsenic, cadmium, chromium, cobalt, copper, manganese, and lead by adsorption methods. Different adsorbents were employed and enhanced over the years, by studying the different adsorption mechanisms involved in each process, such as physical adsorption, electrostatic interaction, surface complexation, and chemical interaction between the surface functional groups and the metal ions. Moreover, very important operational parameters were evaluated, such as pH, adsorbent dosage, time, ionic strength, temperature, and surface charge. The feasibility of reuse through regeneration by desorption of metal ions was also highlighted. Hence, it can be seen that adsorption applicability grows larger each year, followed by the development of more efficient adsorbents, taking into consideration its cost-effectiveness.

References

Agency for Toxic Substances and Disease Registry, ATSDR. 2004. Public Health Statement: Cobalt.

Agency for Toxic Substances and Disease Registry, ATSDR. 2012. Toxicological Profile for Manganese.

Alexander, Carl S. 1972. Cobalt-Beer Cardiomyopathy. The American Journal of Medicine 53(4): 395–417. doi: 10.1016/0002-9343(72)90136-2.

Altmann, Lilo, Frank Weinsberg, Karolina Sveinsson, Hellmuth Lilienthal, Herbert Wiegand, and Gerhard Winneke. 1993. Impairment of long-term potentiation and learning following chronic lead exposure. Toxicology Letters, 66(1). Elsevier: 105–112. doi:10.1016/0378-4274(93)90085-C.

Altundoan, H. Soner, Sema Altundoan, Fikret Tümen, and Memnune Bildik. 2002. Arsenic adsorption from aqueous solutions by activated red mud. Waste Management, 22(3). Pergamon: 357–363. doi: 10.1016/S0956-053X(01)00041-1.

Alyasi, Haya, Hamish Mackey, and Gordon McKay. 2020. Novel model analysis for multimechanistic adsorption processes: case study: cadmium on nanochitosan. Separation and Purification Technology, October, 117925. doi: 10.1016/j.seppur.2020.117925.

Amosa, Mutiu Kolade, and Thokozani Majozi. 2016. GAMS supported optimization and predictability study of a multi-objective adsorption process with conflicting regions of optimal operating conditions. Computers and Chemical Engineering, 94 (November). Elsevier Ltd: 354–361. doi: 10.1016/j.compchemeng.2016.08.014.

Aoshima, Keiko. 2016. Itai-itai disease: renal tubular osteomalacia induced by environmental exposure to cadmium—historical review and perspectives. Soil Science and Plant Nutrition, 62(4): 319–326. doi: 10.1080/00380768.2016.1159116.

Arulkumar, M., P. Sathishkumar, and T. Palvannan. 2011. Optimization of orange G dye adsorption by activated carbon of *Thespesia populnea* pods using response surface methodology. Journal of Hazardous Materials, 186(1). Elsevier: 827–834. doi:10.1016/j.jhazmat.2010.11.067.

Asia News. 2005. Cadmium Spill in the Beijiang River Leaves Millions of People without Water.

Atikah, Nur, Abdul Salim, Mohd Hafiz Puteh, Hairul Khamidun, Mohamad Ali Fulazzaky, Noorul Hudai Abdullah, Abdull Rahim et al. 2020. Interpretation of Isotherm Models for Adsorption of Ammonium onto Granular Activated Carbon. doi: 10.33263/BRIAC112.92279241.

Aydin, Haluk, Yasemin Bulut, and Çiğdem Yerlikaya. 2008. Removal of copper (II) from aqueous solution by adsorption onto low-cost adsorbents. Journal of Environmental Management, 87(1). Academic Press: 37–45. doi: 10.1016/j.jenvman.2007.01.005.

Bakalár, Tomáš, Milan Búgel, and Lucia Gajdošová. 2009. Heavy Metal Removal Using Reverse Osmosis. Acta Montanistica Slovaca Ročník. Vol. 14.

Baum, Carl R. 1999. Treatment of mercury intoxication. Current Opinion in Pediatrics, 11(3): 265–268. doi: 10.1097/00008480-199906000-00018.

Benchmark Mineral Intelligence, BMI. 2018. Panasonic Reduces Tesla's Cobalt Consumption by 60% in 6 Years…but Cobalt Supply Challenges Remain.

Bennett, Daniel R., Curtis J. Baird, Kwok Ming Chan, Peter F. Crookes, Cedric G. Bremner, Michael M. Gottlieb, and Wesley Y. Naritoku. 1997. Zinc toxicity following massive coin ingestion. American Journal of Forensic Medicine and Pathology, 18(2). Am J Forensic Med Pathol: 148–153. doi: 10.1097/00000433-199706000-00008.

Blöcher, C., J. Dorda, V. Mavrov, H. Chmiel, N.K. Lazaridis, and K.A. Matis. 2003. Hybrid flotation - membrane filtration process for the removal of heavy metal ions from wastewater. Water Research, 37(16). Elsevier Ltd: 4018–4026. doi:10.1016/S0043-1354(03)00314-2.

Bloomberg, News. 2018. Tesla's Battery Tweaks Won't Solve World's Cobalt Conundrum.

Borba, C.E., R. Guirardello, E.A. Silva, M.T. Veit, and C.R.G. Tavares. 2006. Removal of Nickel(II) ions from aqueous solution by biosorption in a fixed bed column: experimental and theoretical breakthrough curves. Biochemical Engineering Journal, 30(2). Elsevier: 184–191. doi:10.1016/j.bej.2006.04.001.

Bose, Purnendu, M. Aparna Bose, and Sunil Kumar. 2002. Critical evaluation of treatment strategies involving adsorption and chelation for wastewater containing copper, zinc and cyanide. Advances in Environmental Research, 7(1). Pergamon: 179–195. doi:10.1016/S1093-0191(01)00125-3.

Cai, Hao, Yadong Mei, Junhong Chen, Zhenhui Wu, Lan Lan, and Di Zhu. 2020. An analysis of the relation between water pollution and economic growth in china by considering the contemporaneous correlation of water pollutants. Journal of Cleaner Production, 276 (December). Elsevier Ltd: 122783. doi: 10.1016/j.jclepro.2020.122783.

Chan, John. 2005. Chemical Spill Pollutes Water Supply in North-Eastern China.

Chen, Guohua. 2004. Electrochemical technologies in wastewater treatment. Separation and Purification Technology, 38(1). Elsevier: 11–41. doi:10.1016/j.seppur.2003.10.006.

Chen, J. Paul, Shunnian Wu, and Kai Hau Chong. 2003. Surface modification of a granular activated carbon by citric acid for enhancement of copper adsorption. Carbon, 41(10). Pergamon: 1979–1986. doi:10.1016/S0008-6223(03)00197-0.

Chen, Xincai, Guangcun Chen, Linggui Chen, Yingxu Chen, Johannes Lehmann, Murray B. McBride, and Anthony G. Hay. 2011. Adsorption of copper and zinc by biochars produced from pyrolysis of hardwood and corn straw in aqueous solution. Bioresource Technology, 102(19): 8877–8884. doi: 10.1016/j.biortech.2011.06.078.

Chuang, C.L., M. Fan, M. Xu, R.C. Brown, S. Sung, B. Saha and C.P. Huang. 2005. Adsorption of Arsenic(V) by activated carbon prepared from oat hulls. Chemosphere, 61(4). Elsevier Ltd: 478–483. doi: 10.1016/j.chemosphere.2005.03.012.

Dehaine, Quentin, Laurens T. Tijsseling, Hylke J. Glass, Tuomo Törmänen, and Alan R. Butcher. 2021. Geometallurgy of cobalt ores: a review. Minerals Engineering, 160(January): 106656. doi: 10.1016/j.mineng.2020.106656.

Deng, Yiyi, Shuang Huang, Caiqin Dong, Zhuowen Meng, and Xiugui Wang. 2020. Competitive adsorption behaviour and mechanisms of cadmium, nickel and ammonium from aqueous solution by fresh and ageing rice straw biochars. Bioresource Technology, 303 (May): 122853. doi: 10.1016/j.biortech.2020.122853.

Di Natale, F., A. Erto, A. Lancia, and D. Musmarra. 2015. Equilibrium and dynamic study on hexavalent chromium adsorption onto activated carbon. Journal of Hazardous Materials 281 (January). Elsevier: 47–55. doi: 10.1016/j.jhazmat.2014.07.072.

Diaconu, Mariana, Lucian Vasile Pavel, Raluca Maria Hlihor, Mihaela Rosca, Daniela Ionela Fertu, Markus Lenz, Philippe Xavier Corvini, and Maria Gavrilescu. 2020. Characterization of heavy metal toxicity in some plants and microorganisms—a preliminary approach for environmental bioremediation. New Biotechnology, 56 (May). Elsevier B.V.: 130–139. doi:10.1016/j.nbt.2020.01.003.

Diallo, Mamadou S., Madhusudhana Rao Kotte, and Manki Cho. 2015. Mining critical metals and elements from seawater: opportunities and challenges. Environmental Science & Technology, 49(16): 9390–9399. doi: 10.1021/acs.est.5b00463.

Dotto, Guilherme Luiz, Nina Paula Gonçalves Salau, Jeferson Steffanello Piccin, Tito Roberto Sant'Anna Cadaval, and Luiz Antonio Almeida De Pinto. 2017. Adsorption kinetics in liquid phase: modeling for discontinuous and continuous systems. In Adsorption Processes for Water Treatment and Purification, 53–76. Cham: Springer International Publishing. doi: 10.1007/978-3-319-58136-1_3.

Dotto, Guilherme L., and Gordon McKay. 2020. Current scenario and challenges in adsorption for water treatment. Journal of Environmental Chemical Engineering, 8(4). Elsevier Ltd: 103988. doi: 10.1016/j.jece.2020.103988.

Duranoğlu, Dilek, Andrzej W. Trochimczuk, and Ulker Beker. 2012. Kinetics and thermodynamics of hexavalent chromium adsorption onto activated carbon derived from acrylonitrile-divinylbenzene copolymer. Chemical Engineering Journal, 187 (April). Elsevier: 193–202. doi: 10.1016/j.cej.2012.01.120.

El-Aassar, M.R., and F.M. Mohamed. 2021. Characterization valorized anthracite and its application in manganese (VII) adsorption from aqueous solution; batch and column studies. Microporous and Mesoporous Materials, 310 (January): 110641. doi: 10.1016/j.micromeso.2020.110641.

El Samrani, A.G., B.S. Lartiges, and F. Villiéras. 2008. Chemical coagulation of combined sewer overflow: heavy metal removal and treatment optimization. Water Research, 42 (4-5). Elsevier Ltd: 951–960. doi: 10.1016/j.watres.2007.09.009.

Ercarikci, Elif, and Murat Alanyalioglu. 2020. Dual-functional graphene-based flexible material for membrane filtration and electrochemical sensing of heavy metal ions. IEEE Sensors Journal, September. Institute of Electrical and Electronics Engineers (IEEE), 1–1. doi: 10.1109/jsen.2020.3021988.

Esfandiar, Narges, Bahram Nasernejad, and Taghi Ebadi. 2014. Removal of Mn(II) from groundwater by sugarcane bagasse and activated carbon (a Comparative Study): application of response surface methodology (RSM). Journal of Industrial and Engineering Chemistry, 20(5): 3726–3736. doi: 10.1016/j.jiec.2013.12.072.

Flora, Swaran J.S., and Vidhu Pachauri. 2010. Chelation in metal intoxication. International Journal of Environmental Research and Public Health, 7(7). Molecular Diversity Preservation International: 2745–2788. doi: 10.3390/ijerph7072745.

Fu, Fenglian, and Qi Wang. 2011. Removal of heavy metal ions from wastewaters: a review. Journal of Environmental Management. Academic Press. doi: 10.1016/j.jenvman.2010.11.011.

Gaetke, Lisa M., and Ching Kuang Chow. 2003. Copper toxicity, oxidative stress, and antioxidant nutrients. Toxicology. Elsevier Ireland Ltd. doi: 10.1016/S0300-483X(03)00159-8.

Gibb, Herman J., Peter S.J. Lees, Paul F. Pinsky, and Brian C. Rooney. 2000. Lung cancer among workers in chromium chemical production. American Journal of Industrial Medicine, 38(2). John Wiley & Sons, Ltd: 115–126. doi: 10.1002/1097-0274(200008)38:2<115::AID-AJIM1>3.0.CO;2-Y.

Giménez, Javier, María Martínez, Joan de Pablo, Miquel Rovira, and Lara Duro. 2007. Arsenic sorption onto natural hematite, magnetite, and goethite. Journal of Hazardous Materials, 141(3). Elsevier: 575–580. doi: 10.1016/j.jhazmat.2006.07.020.

Goel, Jyotsna, Krishna Kadirvelu, Chitra Rajagopal, and Vinod Kumar Garg. 2005. Removal of Lead(II) by adsorption using treated granular activated carbon: batch and column studies. Journal of Hazardous Materials, 125(1–3): 211–220. doi:10.1016/j.jhazmat.2005.05.032.

Goher, Mohamed E., Ali M. Hassan, Ibrahim A. Abdel-Moniem, Ayman H. Fahmy, Mohamed H. Abdo, and Seliem M. El-sayed. 2015. Removal of aluminum, iron and manganese ions from industrial wastes using granular activated carbon and amberlite IR-120H. The Egyptian Journal of Aquatic Research, 41(2): 155–164. doi:10.1016/j.ejar.2015.04.002.

Goodyear, K.L., and S. McNeill. 1999. Bioaccumulation of heavy metals by aquatic macro-invertebrates of different feeding guilds: a review. Science of the Total Environment. Elsevier. doi:10.1016/S0048-9697(99)00051-0.

Gunjate, J.K., Y.K. Meshram, R.U. Khope, and R.S. Awachat. 2020. Adsorption based recovery of cobalt using chemically modified activated carbon. Materials Today: Proceedings, 29: 1150–1155. doi: 10.1016/j.matpr.2020.05.388.

Guo, Chuanbo, Yushun Chen, Wentong Xia, Xiao Qu, Hui Yuan, Songguang Xie, and Lian Shin Lin. 2020. "Eutrophication and heavy metal pollution patterns in the water suppling lakes of china's south-to-north water diversion project. Science of the Total Environment, 711 (April). Elsevier B.V.: 134543. doi: 10.1016/j.scitotenv.2019.134543.

Haji, Maha N., and Alexander H. Slocum. 2019. An offshore solution to cobalt shortages via adsorption-based harvesting from seawater. Renewable and Sustainable Energy Reviews, 105 (May): 301–309. doi: 10.1016/j.rser.2019.01.058.

Halder, Joshua Nizel, and M. Nazrul Islam. 2015. Water Pollution and its Impact on the Human Health, 2(1). doi: 10.15764/EH.2015.01005.

Hanfi, Mohamed Y., Mostafa Y.A. Mostafa, and Michael V. Zhukovsky. 2020. Heavy metal contamination in urban surface sediments: sources, distribution, contamination control, and remediation. Environmental Monitoring and Assessment. Springer. doi: 10.1007/s10661-019-7947-5.

Hankins, Nicholas P., Na Lu, and Nidal Hilal. 2006. Enhanced removal of heavy metal ions bound to humic acid by polyelectrolyte flocculation. Separation and Purification Technology 51(1). Elsevier: 48–56. doi: 10.1016/j.seppur.2005.12.022.

Heffron, Raphael, Marc Fabian Körner, Jonathan Wagner, Martin Weibelzahl, and Gilbert Fridgen. 2020. Industrial demand-side flexibility: a key element of a just energy transition and industrial development. Applied Energy, 269 (July). Elsevier Ltd: 115026. doi: 10.1016/j.apenergy.2020.115026.

Henderson, V., A. Kuncoro, and M. Turner. 1995. Industrial development in cities. Journal of Political Economy, 103(5). The University of Chicago Press : 1067–1090. doi: 10.1086/262013.

Ho, Y.S., and G. McKay. 1998. A Comparison of chemisorption kinetic models applied to pollutant removal on various sorbents. Process Safety and Environmental Protection, 76(4): 332–340. doi: 10.1205/095758298529696.

Hoang, Hong Giang, Chitsan Lin, Huu Tuan Tran, Chow Feng Chiang, Xuan Thanh Bui, Nicholas Kiprotich Cheruiyot, Chien Chuan Shern, and Chia Wei Lee. 2020. Heavy metal contamination trends in surface water and sediments of a river in a highly-industrialized region. Environmental Technology and Innovation, 20 (November). Elsevier B.V.: 101043. doi: 10.1016/j.eti.2020.101043.

Horiguchi, Hyogo. 2007. Anemia induced by cadmium intoxication. Japanese Journal of Hygiene, 62(3): 888–904. doi: 10.1265/jjh.62.888.

Hu, Wanqiu, Jinping Tian, and Lujun Chen. 2020. An industrial structure adjustment model to facilitate high-quality development of an eco-industrial park. Science of The Total Environment, October. Elsevier BV, 142502. doi: 10.1016/j.scitotenv.2020.142502.

Huang, Cecily. 2012. Cadmium Spill Threatens Water Supplies of Major Chinese City.

Hube, Selina, Majid Eskafi, Kolbrún Fríða Hrafnkelsdóttir, Björg Bjarnadóttir, Margrét Ásta Bjarnadóttir, Snærós Axelsdóttir, and Bing Wu. 2020. Direct membrane filtration for wastewater treatment and resource recovery: a review. Science of the Total Environment. Elsevier B.V. doi: 10.1016/j.scitotenv.2019.136375.

Idrees, Muhammad, Saima Batool, Hidayat Ullah, Qaiser Hussain, Mohammad I. Al-Wabel, Mahtab Ahmad, Amjad Hussain, Muhammad Riaz, Yong Sik Ok, and Jie Kong. 2018. Adsorption and thermodynamic mechanisms of manganese removal from aqueous media by biowaste-derived biochars. Journal of Molecular Liquids, 266 (September): 373–380. doi: 10.1016/j.molliq.2018.06.049.

Imamoglu, Mustafa, and Oktay Tekir. 2008. Removal of Copper (II) and Lead (II) ions from aqueous solutions by adsorption on activated carbon from a new precursor hazelnut husks. Desalination, 228 (1–3). Elsevier: 108–113. doi: 10.1016/j.desal.2007.08.011.

Jägerstad, M., and K. Arkbåge. 2003. COBALAMINS | Properties and determination. pp. 1419–1427. In: Encyclopedia of Food Sciences and Nutrition. Elsevier. doi: 10.1016/B0-12-227055-X/00257-1.

Jin, Li, and Renbi Bai. 2002. Mechanisms of lead adsorption on chitosan/pva hydrogel beads. Langmuir, 18(25): 9765–9770. doi: 10.1021/la025917l.

Kalra, Akanksha, Pejman Hadi, Hamish R. Mackey, Tareq Al Ansari, and Gordon Mckay. 2018. Sorption of Heavy Metal Ions onto E-Waste-Derived Ion-Exchange Material-Selecting the Optimum Isotherm. doi: 10.5004/dwt.2018.23038.

Kara, M., Yuzer, H., Sabah, E., and Celik, M.S. 2003. Adsorption of cobalt from aqueous solutions onto sepiolite. Water Research, 37(1): 224–232. doi: 10.1016/S0043-1354(02)00265-8.

Kavand, Mohammad, Parisa Eslami, and Laleh Razeh. 2020. The adsorption of cadmium and lead ions from the synthesis wastewater with the activated carbon: optimization of the single and binary systems. Journal of Water Process Engineering, 34 (April): 101151. doi: 10.1016/j.jwpe.2020.101151.

Khan, Anwarzeb, Sardar Khan, Muhammad Amjad Khan, Zahir Qamar, and Muhammad Waqas. 2015. The uptake and bioaccumulation of heavy metals by food plants, their effects on plants nutrients, and associated health risk: a review. Environmental Science and Pollution Research, 22(18). Springer Verlag: 13772–13799. doi: 10.1007/s11356-015-4881-0.

Kiran Marella, Thomas, Abhishek Saxena, and Archana Tiwari. 2020. Diatom mediated heavy metal remediation: a review. Bioresource Technology. Elsevier Ltd. doi: 10.1016/j.biortech.2020.123068.

Kornhauser, Carlos, Katarzyna Wróbel, Kazimierz Wróbel, Juan Manuel Malacara, Laura Eugenia Nava, Leobardo Gómez, and Rita González. 2002. Possible Adverse Effect of Chromium in Occupational Exposure of Tannery Workers. Industrial Health. Vol. 40.

Kozyatnyk, Ivan, Sylvain Bouchet, Erik Björn, and Peter Haglund. 2016. Fractionation and size-distribution of metal and metalloid contaminants in a polluted groundwater rich in dissolved organic matter. Journal of Hazardous Materials, 318(November): 194–202. doi: 10.1016/j.jhazmat.2016.07.024.

Kubier, Andreas, Richard T. Wilkin, and Thomas Pichler. 2019. Cadmium in soils and groundwater: a review. Applied Geochemistry, 108(September): 104388. doi:10.1016/j.apgeochem.2019.104388.

Kurniawan, Tonni Agustiono, Gilbert Y.S. Chan, Wai Hung Lo, and Sandhya Babel. 2006. Physico-chemical treatment techniques for wastewater laden with heavy metals. Chemical Engineering Journal, 118(1–2). Elsevier: 83–98. doi: 10.1016/j.cej.2006.01.015.

Kyzas, George Z., Eleni A. Deliyanni, and Kostas A. Matis. 2016. Activated carbons produced by pyrolysis of waste potato peels: cobalt ions removal by adsorption. Colloids and Surfaces A: Physicochemical and Engineering Aspects, 490 (February): 74–83. doi: 10.1016/j.colsurfa.2015.11.038.

Lacasa, E., S. Cotillas, C. Saez, J. Lobato, P. Cañizares, and M.A. Rodrigo. 2019. Environmental applications of electrochemical technology. what is needed to enable full-scale applications? Current Opinion in Electrochemistry. Elsevier B.V. doi: 10.1016/j.coelec.2019.07.002.

Li, Nan, and Renbi Bai. 2005. Copper adsorption on chitosan-cellulose hydrogel beads: behaviors and mechanisms. Separation and Purification Technology, 42(3): 237–247. doi: 10.1016/j.seppur.2004.08.002.

Li, Yan, Xijin Xu, Junxiao Liu, Kusheng Wu, Chengwu Gu, Guo Shao, Songjian Chen, Gangjian Chen, and Xia Huo. 2008. The hazard of chromium exposure to neonates in guiyu of China. Science of the Total Environment, 403(1–3). Elsevier: 99–104. doi: 10.1016/j.scitotenv.2008.05.033.

Li, Yongchao, He Huang, Zheng Xu, Hongqing Ma, and Yifei Guo. 2020. Mechanism Study on Manganese(II) removal from acid mine wastewater using red mud and its application to a lab-scale column. Journal of Cleaner Production, 253(April): 119955. doi:10.1016/j.jclepro.2020.119955.

Liu, Chen, Donglin Zhao, Kehua Zhang, Han Xuan, Ahmed Alsaedi, Tasawar Hayat, and Changlun Chen. 2019. Fabrication of Si/Ti–based amino-functionalized hybrids and their adsorption towards Cobalt(II). Journal of Molecular Liquids, 289(September): 111051. doi: 10.1016/j.molliq.2019.111051.

Liu, Cheng Hua, Ya Hui Chuang, Tsan Yao Chen, Yuan Tian, Hui Li, Ming Kuang Wang, and Wei Zhang. 2015. Mechanism of arsenic adsorption on magnetite nanoparticles from water: thermodynamic and spectroscopic studies. Environmental Science and Technology, 49(13). American Chemical Society: 7726–7734. doi: 10.1021/acs.est.5b00381.

Mamindy-Pajany, Yannick, Charlotte Hurel, Nicolas Marmier, and Michèle Roméo. 2009. Arsenic adsorption onto hematite and goethite. Comptes Rendus Chimie, 12(8). No longer published by Elsevier: 876–881. doi: 10.1016/j.crci.2008.10.012.

Mamindy-Pajany, Yannick, Charlotte Hurel, Nicolas Marmier, and Michèle Roméo. 2011. Arsenic (V) adsorption from aqueous solution onto goethite, hematite, magnetite and zero-valent iron: effects of ph, concentration and reversibility. Desalination, 281(1). Elsevier: 93–99. doi: 10.1016/j.desal.2011.07.046.

Matlock, Matthew M., Brock S. Howerton, and David A. Atwood. 2002. Chemical precipitation of heavy metals from acid mine drainage. Water Research, 36(19). Elsevier Ltd: 4757–4764. doi: 10.1016/S0043-1354(02)00149-5.

Michalke, Bernhard, and Katharina Fernsebner. 2014. New insights into manganese toxicity and speciation. Journal of Trace Elements in Medicine and Biology, 28(2): 106–116. doi:10.1016/j.jtemb.2013.08.005.

Miraboutalebi, Seyed Mohammadreza, Soheil Kordmirza Nikouzad, Mohammad Peydayesh, Nima Allahgholi, Leila Vafajoo, and Gordon McKay. 2017. Methylene blue adsorption via maize silk powder: kinetic, equilibrium, thermodynamic studies and residual error analysis. Process Safety and Environmental Protection, 106 (February). Institution of Chemical Engineers: 191–202. doi: 10.1016/j.psep.2017.01.010.

Mohan, Dinesh, Kunwar P. Singh, and Vinod K. Singh. 2005. Removal of hexavalent chromium from aqueous solution using low-cost activated carbons derived from agricultural waste materials and activated carbon fabric cloth. Industrial and Engineering Chemistry Research, 44(4). American Chemical Society, 1027–1042. doi: 10.1021/ie0400898.

Mokarram, Marzieh, Ali Saber, and Vahideh Sheykhi. 2020. Effects of heavy metal contamination on river water quality due to release of industrial effluents. Journal of Cleaner Production, 277(December). Elsevier Ltd: 123380. doi: 10.1016/j.jclepro.2020.123380.

Ozaki, Hiroaki, Kusumakar Sharma, and Wilasinee Saktaywin. 2002. Performance of an ultra-low-pressure reverse osmosis membrane (ULPROM) for separating heavy metal: effects of interference parameters. Desalination, 144 (1–3). Elsevier: 287–294. doi: 10.1016/S0011-9164(02)00329-6.

Papandreou, A., C. J. Stournaras, and D. Panias. 2007. Copper and cadmium adsorption on pellets made from fired coal fly ash. Journal of Hazardous Materials, 148(3). Elsevier: 538–547. doi: 10.1016/j.jhazmat.2007.03.020.

Patil, Deepti S., Sanjay M. Chavan, and John U. Kennedy Oubagaranadin. 2016. A review of technologies for manganese removal from wastewaters. Journal of Environmental Chemical Engineering, 4(1): 468–487. doi: 10.1016/j.jece.2015.11.028.

Pawa, Swati, Ahmad J. Khalifa, Murray N. Ehrinpreis, Charles A. Schiffer, and Firdous A. Siddiqui. 2008. Zinc toxicity from massive and prolonged coin ingestion in an adult. American Journal of the Medical Sciences, 336(5). Lippincott Williams and Wilkins: 430–433. doi: 10.1097/MAJ.0b013e31815f2c05.

Peng, Hao, and Jing Guo. 2020. Removal of chromium from wastewater by membrane filtration, chemical precipitation, ion exchange, adsorption electrocoagulation, electrochemical reduction, electrodialysis, electrodeionization, photocatalysis and nanotechnology: a review. Environmental Chemistry Letters. Springer. doi: 10.1007/s10311-020-01058-x.

Piccin, Jeferson Steffanello, Tito Roberto Sant Anna Cadaval, Luiz Antonio Almeida De Pinto, and Guilherme Luiz Dotto. 2017. Adsorption isotherms in liquid phase: experimental, modeling, and interpretations. In Adsorption Processes for Water Treatment and Purification, 19–51. Springer International Publishing. doi: 10.1007/978-3-319-58136-1_2.

Polat, H., and D. Erdogan. 2007. Heavy metal removal from waste waters by ion flotation. Journal of Hazardous Materials, 148(1-2). Elsevier: 267–273. doi: 10.1016/j.jhazmat.2007.02.013.

Purkayastha, Debasree, Umesh Mishra, and Swarup Biswas. 2014. A comprehensive review on Cd(II) removal from aqueous solution. Journal of Water Process Engineering, 2 (June): 105–128. doi: 10.1016/j.jwpe.2014.05.009.

Pyrzynska, Krystyna. 2019. Removal of cadmium from wastewaters with low-cost adsorbents. Journal of Environmental Chemical Engineering, 7(1): 102795. doi: 10.1016/j.jece.2018.11.040.

Rengaraj, S., and Seung Hyeon Moon. 2002. Kinetics of Adsorption of Co(II) removal from water and wastewater by ion exchange resins. Water Research, 36(7). Elsevier Ltd: 1783–1793. doi: 10.1016/S0043-1354(01)00380-3.

Rengaraj, S., Kyeong Ho Yeon, and Seung Hyeon Moon. 2001. Removal of chromium from water and wastewater by ion exchange resins. Journal of Hazardous Materials, 87 (1–3). Elsevier: 273–287. doi: 10.1016/S0304-3894(01)00291-6.

Rodgers, William H. Jr. 1970. Industrial water pollution and the refuse act: a second chance for water quality. University of Pennsylvania Law Review 119. https://heinonline.org/HOL/Page?handle=hein.journals/pnlr119&id=775&div=&collection=.

Rozumová, Lucia, Ondřej Životský, Jana Seidlerová, Oldřich Motyka, Ivo Šafařík, and Mirka Šafaříková. 2016. Magnetically modified peanut husks as an effective sorbent of heavy metals. Journal of Environmental Chemical Engineering, 4(1). Elsevier Ltd: 549–555. doi: 10.1016/j.jece.2015.10.039.

Rudi, Nurul Nadia, Mimi Suliza Muhamad, Lee Te Chuan, Janifal Alipal, Suhair Omar, Nuramidah Hamidon, Nor Hazren Abdul Hamid, Norshuhaila Mohamed Sunar, Roslinda Ali, and Hasnida Harun. 2020. Evolution of adsorption process for manganese removal in water via agricultural waste adsorbents. Heliyon, 6(9): e05049. doi: 10.1016/j.heliyon.2020.e05049.

Sahu, Omprakash, and Nagender Singh. 2018. Significance of bioadsorption process on textile industry wastewater. pp. 367–416. *In*: The Impact and Prospects of Green Chemistry for Textile Technology. Elsevier. doi: 10.1016/B978-0-08-102491-1.00013-7.

Saičić, Zorica S. 2008. Effect of Chronic Cadmium Exposure on Antioxidant Defense System in Some Tissues of Rats: Protective Effect of Selenium. https://www.researchgate.net/publication/267030334.

Saleh, Hosam M., Helal R. Moussa, Fathy A. El-Saied, Maher Dawoud, El Said A. Nouh, and Reda S. Abdel Wahed. 2020. Adsorption of cesium and cobalt onto dried myriophyllum spicatum l. from radio-contaminated water: experimental and theoretical study. Progress in Nuclear Energy, 125 (July): 103393. doi: 10.1016/j.pnucene.2020.103393.

Satya, Awalina, Ardiyan Harimawan, Gadis Sri Haryani, M.A.H. Johir, Luong N. Nguyen, Long D. Nghiem, Saravanamuthu Vigneswaran, Huu Hao Ngo, and Tjandra Setiadi. 2020. Fixed-bed adsorption performance and empirical modeling of cadmium removal using adsorbent prepared from the cyanobacterium aphanothece sp cultivar. Environmental Technology & Innovation, October, 101194. doi: 10.1016/j.eti.2020.101194.

Schio, R.R., N.P.G. Salau, E.S. Mallmann, and G.L. Dotto. 2020. Modeling of fixed-bed dye adsorption using response surface methodology and artificial neural network. Chemical Engineering Communications, April. Taylor and Francis Ltd., 1–12. doi: 10.1080/00986445.2020.1746655.

Schwarzenbach, René P., Thomas Egli, Thomas B. Hofstetter, Urs Von Gunten, and Bernhard Wehrli. 2010. Global water pollution and human health. doi: 10.1146/annurev-environ-100809-125342.

Sekar, M., V. Sakthi, and S. Rengaraj. 2004. Kinetics and equilibrium adsorption study of lead(ii) onto activated carbon prepared from coconut shell. Journal of Colloid and Interface Science, 279(2): 307–313. doi: 10.1016/j.jcis.2004.06.042.

Shah, Ekta, Ian Butler, Pedro Mancias, Kelly Block, and Carlos A. Pérez. 2020. Mercury-induced neurotoxicity and neuroinflammation: a role for heavy metal intoxication in the pathogenesis of autoimmune diseases of the nervous system (912). Neurology, 94 (15 Supplement).

Sharma, D.C., and C.F. Forster. 1994. A preliminary examination into the adsorption of hexavalent chromium using low-cost adsorbents. Bioresource Technology, 47(3). Elsevier: 257–264. doi: 10.1016/0960-8524(94)90189-9.

Sharp, Dan S., Charles E. Becker, and Allan H. Smith. 1987. Chronic low-level lead exposure: its role in the pathogenesis of hypertension. Medical Toxicology and Adverse Drug Experience, 2(3). Springer: 210–232. doi: 10.1007/BF03259865.

Singh, Rashim Pal, Sandeep Kumar, Ritambra Nada, and Rajendra Prasad. 2006. Evaluation of copper toxicity in isolated human peripheral blood mononuclear cells and it's attenuation by zinc: *ex vivo*. Molecular and Cellular Biochemistry, 282 (1–2). Springer: 13–21. doi: 10.1007/s11010-006-1168-2.

Snow, Elizabeth T. 1992. Metal carcinogenesis: mechanistic implications. Pharmacology and Therapeutics. Pergamon. doi: 10.1016/0163-7258(92)90043-Y.

Snyder, Shane A., Paul Westerhoff, Yeomin Yoon, and David L. Sedlak. 2003. Pharmaceuticals, personal care products, and endocrine disruptors in water: implications for the water industry. Environmental Engineering Science. Mary Ann Liebert Inc. doi: 10.1089/109287503768335931.

Soliman, N.K., and A.F. Moustafa. 2020. Industrial solid waste for heavy metals adsorption features and challenges; a review. Journal of Materials Research and Technology, 9(5): 10235–10253. doi: 10.1016/j.jmrt.2020.07.045.

Soner Altundogan, H., Sema Altundogan, Fikret Tümen, and Memnune Bildik. 2000. Arsenic removal from aqueous solutions by adsorption on red mud. *In*: Waste Management, 20: 761–767. Elsevier Science Ltd. doi: 10.1016/S0956-053X(00)00031-3.

Sun, Yongjun, Shengbao Zhou, Shu Yuan Pan, Sichen Zhu, Yang Yu, and Huaili Zheng. 2020. Performance evaluation and optimization of flocculation process for removing heavy metal. Chemical Engineering Journal, 385 (April). Elsevier B.V.: 123911. doi: 10.1016/j.cej.2019.123911.

Suwazono, Yasushi, Yuuka Watanabe, Kazuhiro Nogawa, and Koji Nogawa. 2019. Itai-Itai Disease. pp. 712–719. *In*: Encyclopedia of Environmental Health. Elsevier. doi: 10.1016/B978-0-12-409548-9.11657-4.

Szymanski, T., and M. Thoennessen. 2010. Discovery of the cobalt isotopes. Atomic Data and Nuclear Data Tables, 96(6): 848–854. doi: 10.1016/j.adt.2010.06.006.

Tchounwou, Paul B., Clement G. Yedjou, Anita K. Patlolla, and Dwayne J. Sutton. 2012. Heavy metal toxicity and the environment. EXS. Springer, Basel. doi: 10.1007/978-3-7643-8340-4_6.

Turan, Funda, Meltem Eken, Gul Ozyilmaz, Serpil Karan, and Haluk Uluca. 2020. Heavy metal bioaccumulation, oxidative stress and genotoxicity in african catfish clarias gariepinus from orontes river. Ecotoxicology, July. Springer, 1–16. doi: 10.1007/s10646-020-02253-w.

U.S. Environmental Protection Agency, USEPA. 2003. Contaminant Candidate List Regulatory Determination Support Document for Manganese.

Uddin, Mohammad Kashif. 2017. A review on the adsorption of heavy metals by clay minerals, with special focus on the past decade. Chemical Engineering Journal. Elsevier B.V. doi: 10.1016/j.cej.2016.09.029.

United States Geological Survey, USGS. 2018. Mineral Commodity Summaries 2018: Cobalt.

Ventura, Francesco, Sara Candosin, Rosario Barranco, Alessandro Bonsignore, Luisa Andrello, Luca Tajana, and Antonio Osculati. 2017. A fatal case of coin battery ingestion in an 18-month-old child: case report and literature review. American Journal of Forensic Medicine and Pathology, 38(1). Lippincott Williams and Wilkins: 43–46. doi: 10.1097/PAF.0000000000000297.

Wang, Jianlong, and Xuan Guo. 2020. Adsorption isotherm models: classification, physical meaning, application and solving method. Chemosphere. Elsevier Ltd. doi: 10.1016/j.chemosphere.2020.127279.

Wang, Lawrence K., David A. Vaccari, Yan Li, and Nazih K. Shammas. 2005. Chemical precipitation. pp. 141–197. *In*: Physicochemical Treatment Processes. Humana Press. doi: 10.1385/1-59259-820-x:141.

Wang, Peng, Hongping Chen, Peter M. Kopittke, and Fang-Jie Zhao. 2019. Cadmium contamination in agricultural soils of china and the impact on food safety. Environmental Pollution, 249 (June): 1038–1048. doi: 10.1016/j.envpol.2019.03.063.

Wang, Qing, and Zhiming Yang. 2016. Industrial water pollution, water environment treatment, and health risks in China. Environmental Pollution 218 (November). Elsevier Ltd: 358–365. doi: 10.1016/j.envpol.2016.07.011.

Wang, Yangyang, Donglin Zhao, Shaojie Feng, Yan Chen, and Rong Xie. 2020. Ammonium thiocyanate functionalized graphene oxide-supported nanoscale zero-valent iron for adsorption and reduction of Cr(VI). Journal of Colloid and Interface Science, 580 (November). Academic Press Inc.: 345–353. doi: 10.1016/j.jcis.2020.07.016.

WHO, World Health Organization. 2006. Concise International Chemical Assessment Document 69 - COBALT AND INORGANIC COBALT COMPOUNDS. WHO Library Cataloguing-in-Publication Data.

WHO, World Health Organization. 2011. Guidelines for Drinking-Water Quality. 4th ed. WHO Library Cataloguing-in-Publication Data.

Willow, Anna J. 2020. Embrace it, accept it, or fight like hell: understanding diverse responses to extractive industrial development. Environment, Development and Sustainability, 22(7). Springer: 7075–7096. doi: 10.1007/s10668-019-00529-8.

Wojciechowska, Agnieszka, and Zofia Lendzion-Bieluń. 2020. Synthesis and characterization of magnetic nanomaterials with adsorptive properties of arsenic ions. Molecules, 25(18). MDPI AG: 4117. doi: 10.3390/molecules25184117.

Xie, Hongtian, Jingnan Zhang, Ding Wang, Jie Liu, Lidong Wang, and Huining Xiao. 2020. Construction of Three-Dimensional g-C3N4/Attapulgite Hybrids for Cd(II) adsorption and the reutilization of waste adsorbent. Applied Surface Science, 504 (February): 144456. doi: 10.1016/j.apsusc.2019.144456.

Yetilmezsoy, Kaan, and Sevgi Demirel. 2008. Artificial Neural Network (ANN) approach for modeling of Pb(II) adsorption from aqueous solution by antep pistachio (*Pistacia Vera* L.) Shells. Journal of Hazardous Materials, 153(3). Elsevier: 1288–1300. doi: 10.1016/j.jhazmat.2007.09.092.

Yoshiki, Shusaku, Masami Kimura, and Michiko Suzuki. 1975. Bone and kidney lesions in experimental cadmium intoxication. Archives of Environmental Health, 30(11). Taylor & Francis Group : 559–562. doi:10.1080/00039896.1975.10666776.

Youngwilai, Atcharaporn, Pinit Kidkhunthod, Nichada Jearanaikoon, Jitrin Chaiprapa, Nontipa Supanchaiyamat, Andrew J. Hunt, Yuvarat Ngernyen, Thunyalux Ratpukdi, Eakalak Khan, and

Sumana Siripattanakul-Ratpukdi. 2020. Simultaneous manganese adsorption and biotransformation by *Streptomyces violarus* Strain SBP1 cell-immobilized biochar. Science of The Total Environment, 713 (April): 136708. doi: 10.1016/j.scitotenv.2020.136708.

Yu, Bin, Y. Zhang, Alka Shukla, Shyam S. Shukla, and Kenneth L. Dorris. 2000. The removal of heavy metal from aqueous solutions by sawdust adsorption - removal of copper. Journal of Hazardous Materials, 80 (1–3). Elsevier Science Publishers B.V.: 33–42. doi: 10.1016/S0304-3894(00)00278-8.

Zhang, Junjie, Jingai Shao, Qianzheng Jin, Xiong Zhang, Haiping Yang, Yingquan Chen, Shihong Zhang, and Hanping Chen. 2020b. Effect of deashing on activation process and lead adsorption capacities of sludge-based biochar. Science of the Total Environment, 716 (May). Elsevier B.V.: 137016. doi: 10.1016/j.scitotenv.2020.137016.

Zhang, Qian, Shuting Zhuang, and Jianlong Wang. 2020a. Biosorptive removal of Cobalt(II) from Aqueous solutions using magnetic cyanoethyl chitosan beads. Journal of Environmental Chemical Engineering, 8(6): 104531. doi: 10.1016/j.jece.2020.104531.

Zhu, Yusha, and Max Costa. 2020. Metals and molecular carcinogenesis. Carcinogenesis, 41(9). Oxford University Press (OUP): 1161–1172. doi: 10.1093/carcin/bgaa076.

9

Biosorption of Toxic Metals from Multicomponent Systems and Wastewaters

Heloisa Pereira de Sá Costa,[1] Giani de Vargas Brião,[1] Talles Barcelos da Costa,[1] Cléophée Gourmand,[2] Caroline Bertagnolli,[2] Meuris Gurgel Carlos da Silva[1] and Melissa Gurgel Adeodato Vieira[1,]*

1. Introduction

The rapid expansion of industries such as chemical, textile, electronics, pharmaceutical, metallurgical industries and nuclear power plants drastically increases the quantity and diversity of pollutants releases in water, air and soil. Among those pollutants, metallic species are a priority. Indeed, their toxicity even at low concentration causes their accumulation, which could have undesired consequences on living organisms. Moreover, the population growth, urbanization intensification and the climate change increase the need for fresh and clean water. While processes for water depollution exist, they are not fully efficient, especially at low concentrations and lead to the presence of micro metallic pollutants in water. In the search of low cost, efficient, environmentally friendly processes, biosorption stood out. Based on the use of natural materials, it has become popular in the past few years due to its simplicity, low cost and efficiency. This section aims to introduce biosorption, present different biosorbent materials such as algae, fungi, residual biomass and agro-industrial waste and their application in metal removal. Kinetic and equilibrium aspects as well as desorption, sorbent elimination and selectivity will be addressed.

[1] University of Campinas, School of Chemical Engineering, Department of Processes and Products Design, Albert Einstein Avenue, 500, 13083-852, Campinas, São Paulo, Brazil.
[2] Université de Strasbourg, CNRS, IPHC UMR 7178, F-67000 Strasbourg, France.
* Corresponding author: melissag@unicamp.br

1.1 Contamination by toxic metals and treatment technologies

Metals are natural constituents of the Earth's crust and can be easily liberated in the environment. Naturally, processes such as erosion, volcanic activity and others control the flow of metals between the hydrosphere, lithosphere, atmosphere and biosphere. However, the increase of human activity and industry severely modified those flows, having consequences on the speciation and behavior of the elements (Callender 2003). This increase of anthropogenic activity also contributed to the augmentation of toxic metals emissions in air, water and soil (Callender 2003). Water pollution is one of the main environmental problems nowadays (Pyrzynska 2019). Indeed, large quantities of aqueous effluents are generated by industries, such as mineral extraction (As, Cd, Co, Cr, Cu, Mn, Ni, Pb, Zn), energy and fuel production (Cd, Cr, Hg, Pb), surface finishing and electroplating (Ag, Au, Cd, Co, Cr, Cu, In, Ni, Pb, Pd, Pt, Sn, Zn), aerospace (Cd, Co, Cu, Fe, Pb, Zn), metallurgy (Al, Cr, Cu, Ni, Sn, Zn, W, Ta, Ti), metal molding and coating, electronics and batteries (As, Cd, Co, Cr, Hg, In, Li, Ni, rare earth elements, Sb, Tl), nuclear industry (Am, Cs, Ra, Th, U), milling and mining (Cd, Cr, Cu, Fe, Ni, Zn), pharmaceutical industries (As, Cd, Co, Cr, Cu, Fe, Hg, Mn, Ni, Pb, Zn), fertilizer industries (As, Cd, Cr, Cu, Pb), tanneries (Cr), pulp and paper industries (Cd, Co, Cr, Cu, Fe, Hg, Mn, Ni, Pb, Zn) and agriculture (As, Cd, Cr, Cu, Ni, Pb, Zn) (Volesky and Holan 1995, Volesky 2001, Wang and Chen 2009, Fu and Wang 2011, Carolin et al. 2017, Wu et al. 2017, Gupta et al. 2018, Kumari and Tripathi 2019, Beni and Esmaeili 2020). Among the main water pollutants, toxic metals represent a challenge. Because of their solubility, their oxidation-reduction properties and their ability to form complexes, metal removal plays a major concern (Carolin et al. 2017).

The most commonly found metals in wastewaters are As, Cd, Cr, Cu, Pb, Hg, Ni and Zn (He and Chen 2014, Carolin et al. 2017, Freitas et al. 2019). Table 1 hereafter presents the current regulations of effluent and drinking water regarding metals established by European Union (European Council 2010, 2020), World Health Organization (World Health Organization 2008) and United States Environmental Protection Agency (United States Environmental Protection Agency 2018, Gao et al. 2019). Those values should appear as target levels for biosorption processes.

Metals tend to accumulate and are not degraded into less harmful compounds. Most of them are toxic, sometimes carcinogenic. They can affect different organs, respiratory, cardiovascular and nervous systems, provoke nauseas, anemia and induce diseases (Kurniawan et al. 2006, Fu and Wang 2011, Koedrith et al. 2013, Carolin et al. 2017). Therefore, the development of treatment processes able to remove micro pollutants from water is of prime importance. .

Wastewater treatment presents significant variation according to the countries. On average, in high-income countries, about 70% of wastewater is treated. This number goes down to 38–28% in middle-income countries and 8% only in low-income countries. Generally, the wastewater treatment plants do not removal efficiently the traces of metals (Rojas and Horcajada 2020). Conventional methods for metal removal include physical-chemical processes such as chemical precipitation, filtration, ion exchange, membrane processes, electrochemical treatment, evaporation, solvent extraction, coagulation-flocculation, flotation and adsorption (Volesky 1987, 2001,

Table 1: Current regulations on effluents and drinking water composition for the eight most common metals found in water.

Contaminant	Directive EU 2010/75 (effluents) (mg/L)	Directive EU 2020/2184 (drinking water) (mg/L)	WHO (drinking water) (mg/L)	USEPA (drinking water) MCL* (mg/L)
Arsenic	0.15	0.010	0.010	0.050
Cadmium	0.05	0.005	0.003	0.010
Total Chromium	0.50	0.025	0.050	0.050
Copper	0.50	2.000	2.000	0.250
Mercury	0.03	0.001	0.006	0.200
Nickel	0.50	0.020	0.070	0.800
Lead	0.20	0.005	0.010	0.006
Zinc	1.50	-	3.000	0.00003

*maximum contaminant level

Veglio' and Beolchini 1997, Kurniawan et al. 2006, Wang and Chen 2009, Fu and Wang 2011, Bădescu et al. 2018, Pyrzynska 2019). Those traditional methods often have a high cost, need maintenance, are not fully efficient, lack of selectivity, require high energy and reagents, do not benefit from low concentration of metal, generate secondary pollution (sludge) and are very expensive at the industrial scale (Volesky 1987, 2001, Wang and Chen 2009, Bădescu et al. 2018, Freitas et al. 2019, Beni and Esmaeili 2020, Golnaraghi Ghomi et al. 2020, Qin et al. 2020).

Table 2 presents the most common processes as well as their advantages and drawbacks (Volesky 2001, Kurniawan et al. 2006, Fu and Wang 2011, Carolin et al. 2017b). Among those, adsorption is one of the most attractive technique since it presents a high efficiency even at low concentration, is simple and allows an easy recovery of metals. However, commonly used adsorbents are usually very expensive. The most commonly found is activated carbon but other materials such as zeolites, silica, metal–organic frameworks, clay, nanomaterials, etc., can also be used (Fu and Wang 2011, Carolin et al. 2017, Da'na 2017, Gupta et al. 2018, Gao et al. 2019, Qin et al. 2020). The use of nature-based sorbent is therefore of great interest to reduce the cost and availability issues.

1.2 Biosorption of Toxic Metals

One of the promising alternatives to conventional methods for metal removal is biosorption. The removal of low concentrations of metals by biosorption has been investigated since the 1980s (Çolak et al. 2009).

1.2.1 Definition

Biosorption is a subcategory of adsorption. It is defined as the treatment of a contamination using biological material or microorganisms (Volesky and Holan 1995, Avery et al. 1993, Wang and Chen 2009, Michalak et al. 2013, Bădescu et al. 2018, Chojnacka and Mikulewicz 2019, Beni and Esmaeili 2020, Golnaraghi Ghomi et al.

Table 2: Main metal removal techniques (Volesky 2001, Kurniawan et al. 2006, Fu and Wang 2011, Carolin et al. 2017b).

Technique	Method	Advantages	Drawbacks
Chemical precipitation	Use of chemicals (hydroxide, sulphide, chelating agents) to precipitate metals and isolate them	Cheap, simple	Followed by filtration or sedimentation, use of reagents, formation of secondary pollution, control of the environment, adapted to high concentrations
Ion exchange	Exchange of cations between an ion-exchanger resin (sulfonic or carboxylic acid) and metals in the wastewater	High treatment capacity, fast kinetics, high removal efficiency	High operational cost, remove limited metal ions, use of synthetic resins, needs a pretreatment to eliminate particles in suspension
Membrane processes	Separation of metals according to their size and charge with a membrane	High efficiency, easy operation, space saving	Expensive, high energy consumption, complex, risk of membrane fouling
Electrochemical treatment	Use of electrochemistry to precipitate the metal on electrodes	Low use of chemicals, efficient, can be combined to other techniques, metal recovery	High investment, high energy consumption, high maintenance
Coagulation-flocculation/ flotation	Coagulation: destabilization of colloids neutralizing the forces that keep apart; Flocculation: action of polymer to bind particles into large agglomerates. Flotation: separation of solids and dispersed liquids using bubble attachment	Cost effective, better removal of small particles, can treat a high range of concentrations	Generation of sludge, high use of chemicals, followed by filtration or sedimentation, impossible metal recovery
Adsorption	Mass transfer of the metals from liquid phase to a solid which binds the metal physically or chemically	Easy, less sludge, selectivity, flexibility	Reusability of the adsorbent, expensive materials

2020). It is a non-metabolically controlled process. Indeed, biosorption is based on the rapid and reversible binding of desired substances to non-living microorganisms (Michalak et al. 2013, Freitas et al. 2019). When living organisms are used, the process is named bioaccumulation (Volesky 2007). The main differences between biosorption and bioaccumulation lies in the living nature of the biosorbent. Indeed, a living cell has to be kept under suitable conditions, needs metabolic energy and does not have much environmental flexibility (Beni and Esmaeili 2020). Moreover, the renderability is very limited leading to an overall higher cost. Nevertheless, selectivity is believed to be higher with living organisms (Chen et al. 2017).

In comparison to conventional processes and conventional adsorption, biosorption presents a low cost and a high efficiency without the need of any additional nutrient while minimizing the production of sludge (Michalak et al. 2013, Bădescu et al. 2018). It can be used at concentrations below 100 ppm where other treatment methods appear ineffective (Schiewer and Volesky 1995, Michalak et al. 2013, Freitas et al. 2019).

Even if pollutant removal and wastewater treatment are the main uses of biosorption, the process can also be used in biomass enrichment by trace metals and for the recovery of valuable metals (Michalak et al. 2013, Freitas et al. 2019). Another interesting application of biosorption would be in analytical chemistry for the development of green techniques using preconcentration by biosorption for speciation in order to reduce the use of acidic and organic solvents (Chojnacka and Mikulewicz 2019).

1.2.2 Biosorbents

While the metal binding from aqueous solutions by living microorganisms was known since the 18th and 19th century, it is only during the 20th century that biosorbents started being studied (Freitas et al. 2019). A material can be considered biosorbent whenever it is extracted from biological sources, which includes all kinds of plant, animal and microbial biomass and their derivatives (Volesky and Holan 1995, Volesky 1987, Vegliò and Beolchini 1997, Wang and Chen 2009, Michalak et al. 2013). The main types of biosorbents are bacteria, algae, fungi and yeasts, biopolymers, active sludge, animal and fruit skins, plant residues, agricultural and industrial waste biomass and biological compounds (Volesky and Holan 1995, Volesky 1987, Beni and Esmaeili 2020, Golnaraghi Ghomi et al. 2020, Gupta et al. 2018, Michalak et al. 2013, Vegliò and Beolchini 1997, Wang and Chen 2009). With biosorption being a surface-driven process, functional groups on the surface are the metal binding sites. The most commonly present groups in biosorbents are carbonyls, amines, phosphorus or sulfur-based groups, carboxylic, hydroxyls and amides (Beni and Esmaeili 2020, Gupta et al. 2018, Michalak et al. 2013, Pyrzynska 2019, Wang and Chen 2009). Table 3 sums up the different biosorbent groups and their properties. A commercial biosorbent should follow the criteria hereafter (Wang and Chen 2009, Yang and Volesky 1999):

- High biosorption capacity
- Suitable and fast kinetics
- Good size, shape
- Good physical properties
- Fast, cheap and high-performing biosorbent separation
- Strong mechanical strength, thermal stability, mechanical resistance
- Availability in nature in large quantities
- Cost effective preparation methods
- High versatility, regenerability and reusability
- Effortless regeneration
- Metal selectivity

Table 3: Main families of biosorbents (Volesky and Holan 1995, Wang and Chen 2009, Gupta et al. 2018, Golnaraghi Ghomi et al. 2020).

Biosorbent	Contain	Functional groups	Examples	Advantages	Drawbacks
Algae	Cellulose, alginate, carrageenan	$-COOH$, $-OH$, $-NH_2$, $-SH$, $-C=O$, $-SO_4^-$, $-CONH$	Red, brown and green seaweed, diatoms	Very abundant, high adsorption capacity, biocompatible	Variable size and shape, need light and appropriate conditions to grow, low selectivity, difficult separation
Fungi	Chitin	imidazole, $-PO_4^{2-}$, $-OH$, $-NH_2$, $-SH$, $-SO_4^-$	Molds, mushrooms, yeasts	Inexpensive and easy to grow, nonpathogenic, various structure and morphologies	Not always selective, wide range of structures to choose from, not always efficient/selective
Bacteria	Peptidoglycan, teichoic acids	$-COOH$, $-OH$, $-NH_2$, phosphonates	*Cellulosimicrobium, Arthrobacter, Citrobacter, Escherichia, Pseudomonas*	Most abundant microorganism, very stable, small size	Low selectivity, non-reusable, living organism
Virus	Nucleic acid and proteins	$-COOH$, $-OH$, $-NH_2$, $-SH$, $-C=O$	Phages	Small particle size, high availability	Not widespread, possibly pathogenic, living organism
Biowaste	Polysaccharides, lignocelluloses, amino acids, carbohydrates	$-COOH$, $-OH$, $-NH_2$, $-SH$, $-C=O$, $-SO_4^-$, $-CONH$, $-PO_4^{2-}$, ...	Fruit peels, shells and stones, rice husk, tea, straw	Chemisorption and ion exchange, usable in raw form, multi functionality, inexpensive, hydrophilicity	Limited porosity
Plant and animal-based	Cellulose, lignin, chitosan	$-COOH$, $-OH$, $-NH_2$, aromatic amines, $-C=O$, $-phenols$, $-ethers$	Pollen, seeds, leaves, roots, scales, egg shells, bones	Economic, renewable, abundant	Often need treatment/modification to increase performances
Modified/ hybrid biosorbent	All of the above	All of the above	Alginate encapsulated sorbents, PEI-grafted sorbents, nano sorbents	Increase adsorption capacities and selectivity, multi-functional, control of the nature and density of functional groups, combine the properties of the biomass and the second material (porosity, size, functional groups)	Necessity of modification/ functionalization, risk of biosorbent degradation, can lead to secondary pollution

1.2.3 Preparation and modification of biosorbents

To enhance selectivity and improve sorbent separation and adsorption capacities, a surface modification is performed prior to the sorption. This modification can be both physical and chemical modifications which aim to remove surface impurities, change the surface charge, the particle size and increase the number of binding sites (Veglio' and Beolchini 1997, Freitas et al. 2019, Beni and Esmaeili 2020, Qin et al. 2020). In biosorption studies, the following preparation steps are often found (Beni and Esmaeili 2020): collection of the raw sorbent; washing/rinsing; drying; modification to improve properties; shaping/optimization of the geometry; storage.

Table 4 hereafter presents the main sorbent modifications found in biosorption studies with a few examples. Other biosorbent modification methods mentioned in the literature include the combination of biomasses, the interaction of biomasses and microflora. Howbeit, they are not widely used, do not always give full results nor present sufficient adsorption capacities, limiting, therefore, their commercial application (Qin et al. 2020).

1.2.4 Biosorption Mechanisms

Biosorption mechanisms are one of the main limiting factors for industrial applications. Indeed, they are not fully known and combine several physicochemical interactions, mostly on the sorbent surface. Those interactions can differ depending on the functional groups and their conformation and can be a combination of ion exchange, (micro)precipitation, physical and chemical adsorption (Volesky 2001, Veglio' and Beolchini 1997, Michalak et al. 2013, Robalds et al. 2016, Gupta et al. 2018, Freitas et al. 2019, Golnaraghi Ghomi et al. 2020). Robalds et al. proposed in 2016 a new classification of biosorption mechanisms, therefore replacing the obsolete one given by Vegliò and Beolchini in 1997.

Most of the uncertainty of biosorption mechanisms lies in the complexity of the surface chemistry and complex formation. Two steps can however be presented. First, the ion in solution moves and reaches the sorbent surface. Then, the dissolved compound is transferred from the surface to the internal areas of the sorbent to bind to active sites. This second step is the slowest (Vishan et al. 2019, Beni and Esmaeili 2020). Kinetic models commonly used in biosorption studies include pseudo first order, pseudo second order, Weber and Morris, Boyd, external mass transfer resistance and Elovich model (Veglio' and Beolchini 1997, Michalak et al. 2013, Gupta et al. 2018, Beni and Esmaeili 2020, Golnaraghi Ghomi et al. 2020).

According to Srivastava et al., kinetic models for biosorption contain four steps (Srivastava et al. 2015). The metal ion is transferred from the solution to the boundary layer, then to the adsorbent surface, into the adsorbent sites and finishes being bounded to those active sites. Adsorption isotherms are modeled using Langmuir, Freundlich, Dubinin-Radushkevich, Elovich, Jovanovic, Temkin, Halsey, Flory-Huggins, Redlich-Peterson, Sips, Koble-Corrigan, Radke-Prausnitz, Toth, Fritz and Fritz-Schlunder isotherms (Veglio' and Beolchini 1997, Gavrilescu 2004, Wang and Chen 2009, Michalak et al. 2013, Saadi et al. 2015, Gupta et al. 2018, Al-Ghouti and Da'ana 2020, Beni and Esmaeili 2020, Golnaraghi Ghomi et al. 2020).

Table 4: Main modification methods for biosorbents (Veglio' and Beolchini 1997, Qin et al. 2020).

Chemical modification

Modification	Objectives	Method	Remarks	Examples	References
	Remove residual products, expose binding sites	Wash with acidic reagents (HNO_3, H_2SO_4, HCl)	Quick, simple, possible loss of biomass quality	*Ecklonia* sp. with HNO_3, H_2SO_4, and HCl for Cr(VI) removal	(Park, et al. 2005)
	Provide binding sites, eliminate organic compounds, improve affinity	Wash with alkaline reagents (NaOH, $Ca(OH)_2$)	Quick, simple, possible loss of biomass that dissolves	*Cyanobacterium* with NaOH to improve selectivity for Cd(II)	(Nagase et al. 2005)
Washing biomass	Protects biosorbent stability	Wash with inorganic salts (NaCl, $NaNO_3$)	Quick, simple, less effect, removed metal has to have more affinity than the inorganic ion	Seaweed treatment with $CaCl_2$ and Na-EDTA for Cu(II) and Cr removal	(Murphy, et al. 2009)
	Remove protein and lipid fraction on the surface, expose binding sites	Wash with organic solvents (ethanol, acetone)	Quick, simple, possible decrease of efficiency with unadapted solvents	*Sargassum muticum* with ethanol, methanol or acetone for Hg(II) removal	(Carro et al. 2013)
Biological surface functionalization	Modify the type and density of functional groups	Chemical treatment or long-chain polymers grafting (polyethylenimine, epichlorohydrin, acrylic acid)	Higher efficiency but creates secondary pollution, residual chemicals	PEI-modified *Penicillium chrysogenum* for Cu(II), *Pb(II) and Ni(II)*	(Deng and Ting 2005)
Intracellular resistance modification	Artificial regulation of proteins and enzymes to increase resistance, tolerance and metal-accumulation	Genetically engineered microorganisms, keeping biomass in a harsh environment	Promising but is complex, unpredictable, use exogenous chemicals	Zn-induced phytochelatin synthesis in *Dunaliella tertiolecta*	(Tsuji et al. 2002)
Intracellular autogenous nanomaterial	Increase efficiency by combining the biosorbent properties with microorganism-generated nanoparticles	Enhance nanoparticle production by breaking cell ions balance	Increase efficiency, no secondary pollution, not very widespread	*Saccharomyces cerevisiae* cell nanoparticles with $CaCO_3$	(Ma et al. 2011)

Immobilization

Combination with organic polymers	Obtain a material with a mechanical strength, a right size, rigidity and porosity	Trap cells in a polymeric network (Na-Alginate, Agar, Polyvinyl alcohol)	Simple, easier separation, low resistance, can damage the biosorbent	*Halomonas BVR 1 strain* immobilized in sodium alginate for Pb(II) adsorption	(Manasi et al. 2014)
Combination with inorganic materials	Obtain a material with a mechanical strength, a right size, rigidity and porosity	Adsorption of the biomass on activated carbon, zeolites, magnetite	Conservation of biomass properties but easy to fall off,	*Pseudomonas aeruginosa* immobilized on activated carbon for heavy metals removal	(Orhan et al. 2006)
Combination with natural organic materials	Obtain a material with a mechanical strength, a right size, rigidity and porosity	Adsorption of biomass and growth on straw, lignocellulose, agricultural waste	Recyclable, contaminant free, not widespread	Fungal hyphae entrapment in papaya wood for heavy metals removal	(Iqbal and Saeed 2006)
Combination with new nanomaterials	Give nanomaterial properties absent from traditional biosorbents	Various physical and chemical preparations	More efficient and economic, caution required because of nanoparticles	*Phanerochaete chrysosporium* immobilized on nano-Fe_3O_4 for Pb(II) removal	(Xu et al. 2012)

1.2.5 Parameters influencing biosorption mechanisms

Several parameters have an influence on a biosorption process. The main one is the nature of biosorbents (Tsezos 1985, Beni and Esmaeili 2020). Indeed, different biosorbents present different functional groups and different binding ability/ selectivity.

pH is also one of the main parameters (Tsezos 1985, Guibal et al. 1992, Kapoor 1995, Gavrilescu 2004, Gupta et al. 2006, Schiewer and Balaria 2009, Freitas et al. 2019, Pyrzynska 2019, Beni and Esmaeili 2020, Golnaraghi Ghomi et al. 2020). Indeed, when working with metal species, speciation is given by the pH. pH also governs the attractiveness of sites, ionization degree, surface charge of the adsorbent, adsorption mechanisms, biosorbent separation, competition with the adsorption of H^+ and can separate bonding sites (Sibi 2016, Beni and Esmaeili 2020).

Biosorption processes also depend a lot on sorbent dosage (Modak and Natarajan 1995, Gupta et al. 2006, Beni and Esmaeili 2020, Golnaraghi Ghomi et al. 2020). If the dosage is above the optimal quantity, diminutions of capacities can be observed mainly due to the variation of viscosity.

To reduce costs of biosorption processes, biosorption often runs at room temperature. However, temperature also affects biosorption processes (Modak and Natarajan 1995, Cruz et al. 2004). First of all because metal stability and solubility depend on the temperature (Freitas et al. 2019). The same goes for the ligands and therefore their ability to bind can change. Temperature also affects inherent parameters of the biosorbent, mainly the chemical composition, effect which is enhanced if the material is a living organism. Finally, temperature can also affect the kinetics and quality of the sorption process.

The influence of the ionic strength and the competition with other ions, stirring time and speed, particle size, physicochemical properties of the metal ions, initial metal concentration and contact time are sometimes mentioned (Schiewer and Volesky 1997, Schiewer and Balaria 2009, Gavrilescu 2004, Michalak et al. 2013, Pyrzynska 2019). Other parameters specific to biosorption such as culture medium, biosorbent growth and preprocessing are also of great importance (Golnaraghi Ghomi et al. 2020).

1.2.6 Desorption and regeneration

There are several main drawbacks to the industrialization of biosorption. The treatment capacity, the sorbent separation, the pretreatment costs, the lack of knowledge regarding the sorption mechanisms are all factors that prevent the wide use of biosorption for water treatment (Bădescu et al. 2018, Golnaraghi Ghomi et al. 2020). Bădescu et al. (2018) mention the elimination of spent biosorbents and their valorization as one of the main factors hindering the industrial development of biosorption. They cited three ways for exhausted sorbent valorization:

(1) Regeneration of the sorbent and reuse in cycles

Desorption is the treatment of a spent sorbent by an eluent which can be mineral acids (HCl, HNO_3, H_2SO_4) alkalis (KOH, $NaOH$), inorganic salts (sodium, potassium or calcium chlorides/nitrates) or chelating agents (citric acid, EDTA) (Lata et al. 2015, Bădescu et al. 2018, Chatterjee and Abraham 2019, Freitas et al. 2019).

The main advantages of desorption are the recovery of most metal ions, the regeneration and reuse of the sorbent and the decrease of the overall cost of the process through the reuse (Lata et al. 2015, Bădescu et al. 2017, Chatterjee and Abraham 2019). However, the incomplete desorption (only weakly bonded metal is desorbed) and the degradation of material due to the use of highly concentrated acids limit the efficiency and applications of desorption for biosorbents.

(2) Use of exhausted biosorbents as soil fertilizers

Metal-loaded sorbents can be used for soil enrichment in agricultural areas poor in essential metal elements and nutrients. Indeed, metal-biomasses not only contain metal elements but also a lot of nutrients such as nitrogen or phosphorus. In this case, metals would be directly desorbed in soils to avoid a regeneration step (Cole et al. 2017). While this alternative looks promising, it is actually very difficult to apply it. Indeed, it cannot be applied to all materials and all metals. To use sorbents as fertilizers, three conditions have to be met (Bădescu et al. 2017). First, the adsorbed metal needs to be an essential element as described in Section 1.1. Then, industrial effluents composition has to be known in detail and cannot contain toxic species. Finally, the biomass has to be easily degradable without leading to secondary pollution. Using this method would allow the common resolution of two environmental problems (removal of metal ions and soil enrichment), without desorption eluent, but, large amounts of biomasses are necessary and an uncontrolled release or an unadapted sorbent could easily lead to soil pollution (Bădescu et al. 2018).

(3) Pyrolysis under well-defined conditions

An advantage of pyrolysis compared to the use of biosorbent as fertilizers is that is can be adapted to a large scale while staying economically viable. Pyrolysis is the thermal decomposition of organic matter conducted through the calcination of the biomass under a low oxygen atmosphere at a temperature below 800°C, which depends on the type of biomass (Kan et al. 2016, Bădescu et al. 2018). Pyrolysis can be applied directly to a dried and grinded sorbent without further treatment step. The thermal decomposition leads to the formation of a gas mixture (H_2, CO, CO_2, CH_4, NO_x and C_2H_y) whose composition depends on the biomass and the calcination conditions, residual heat and biochar, containing mostly carbon, organic/inorganic impurities, reduced metals and metal oxides. Part of the gas mixture could then be reused to produce electricity (CO_2 and NO_x) and the other part to produce biofuels.

The residual heat could be reused to heat the production area or in the pyrolysis process (Bridgwater 2012) and the biochar can be reused as sorbent, catalyst, additive, precursor for synthesis of carbon-based materials or as an amendment for soils (Agrafioti et al. 2014, Bădescu et al. 2018, Mosa et al. 2018). Pyrolysis presents several advantages because unlike the use of sorbents as fertilizers, it is a single step process which can be applied to all types of sorbents, all types of effluents and all type of metals and can value all reaction "products". However, it still presents limiting factors such as the high cost of equipment, the presence of hazardous species in the recovered biochar and the attention needed to avoid releasing large quantities of greenhouse gases (Bădescu et al. 2018).

2. Application of biosorbents for the removal of toxic metals from multicomponent systems and wastewaters

Toxic metals release in wastewater occurs mainly by natural and human sources. The natural factors include volcanic activities, soil erosion, and aerosols particulate while the human factors include metal finishing and electroplating processes, mining extraction operations, textile industries, and nuclear power (Akpor 2014). This section addresses the use of biosorbents for wastewater remediation, focused on the electroplating wastewater, acid mine drainage water, textile and tannery effluents, and residual water from nuclear power plants.

2.1 Electroplating wastewater

Electroplating is a simple metal finishing technique widely used for various purposes in automotive, electronic, medical, and aerospace industries. The purpose of the galvanizing process is to improve characteristics of a particular material or metal through electrolytic deposition of a specific plating metal on its surface. The process can enhance or incorporate various properties to the material, e.g., corrosion protection and increased surface hardness (Liu et al. 2016).

Large volumes of contaminated wastewater are generated mainly in the washing and rinsing steps. The composition of these effluents depends directly on the metals used in the electroplating process, with chromium, zinc, copper, cadmium, nickel, and lead being the most commonly applied. In addition to these, precious metals such as gold and silver are also used. According to Dermentzis et al. (2011), in the electroplating process only 30 to 40% of the metals are effectively used, the rest being removed mainly in the washing and rinse step, so the effluents from these steps can contain up to 1000 mg/L of toxic metals used in the process. Since these metals present highly dangerous risks for living organisms, they need to be properly removed before the effluent discharge into the environment.

Electroplating wastewater treatments include chemical precipitation, electrochemical methods, ion exchange and bio/adsorption (Naja and Volesky 2017). Biosorption proves to be efficient for the removal of several metals frequently present in electroplating effluents; in the literature, there are several articles reporting satisfactory removals in single-ion systems of Cr, Zn, Cu, Ni, Cd and other toxic metals through adsorption using biomass as sorbent material (Moino et al. 2017, Freitas et al. 2018, Nishikawa et al. 2018, Cardoso et al. 2020). Regarding the application of the process to the treatment of real wastewaters, since they have a much more complex composition than synthetic effluents, a reduced number of studies are found on the subject. Table 5 summarizes studies from the last decade that present different biosorbents applied to the treatment of real effluents from electroplating industries/workshops.

Zinicovscaia and coworkers (2020) evaluated the use of the yeast biomass *Saccharomyces cerevisiae* in a biosorption system for the decontamination of an effluent predominantly contaminated with zinc, generated from an electroplating industry in Dubna region, Russia. In addition to zinc, the effluent also contained nickel, strontium, copper, and barium in its composition. After one hour of process,

Table 5: Electroplating wastewater treatment through biosorption processes using different biosorbents.

Biosorbent	Metals	Operational conditions			Adsorption Capacity (mg/g)	Total Removal (%)	References
		pH	Dosage (g/L)	Initial Concentration (mg/L)			
Yeast *Saccharomyces cerevisiae*	Zn	6	10–20	49.84	-	85.0	(Zinicovscaia et al. 2020)
Penicillium sp. biomass	Ni	6	7.5	639	63.6	74.6	(Sundararaju et al. 2020)
Escherichia coli biomass	Ni	6	20	2.86	26.45	100	(Kwak et al. 2011)
Modified palm oil Empty Fruit Bunch	Cu Fe Zn Cr Ni	6	20	24.22 0.399 0.749 0.435 5.97	-	98.34 98.74 76.70 97.47 96.35	(Rahman et al. 2020)
Spirogyra biomass	Cu	6	60	194.0	160.6	82.8	(Ilyas et al. 2018)
Modified *Spirogyra* biomass	Cu	6	60	194.0	187.0	96.4	(Ilyas et al. 2018)
Modified *Eriobotrya japonica* biomass	Ni	6	8	12.48	-	92.4	(Salem and Awwad 2014)

the authors noted that the total removal was less than that obtained in tests using synthetic effluents. This effect is commonly observed in studies with real effluents and can be explained, for instance, by the difference in the concentration of metallic ions or even by the aqueous matrix complexity. Aiming to improve the uptake of Zn(II) ions, the authors implemented a subsequent scheme of adding biosorbent to the effluent. In optimized conditions, first 20 g of biomaterial were added to the system; after one hour, the mixture was filtered and more 10 g of yeast was added. Using this methodology of adding the biosorbent in two different cycles, up to 85% of the zinc present in the effluent was removed; in addition to this, other metals such as nickel were completely removed.

The biomass obtained from the filamentous fungus *Penicillium* sp. was the biosorbent selected by Sundararaju et al. (2020) to remove nickel present in a galvanizing effluent from an industry located in Jaihindpuram, India. Using the Response Surface Methodology, the researchers observed that the contact time, pH and biomass dosage have a positive effect on the system, in addition to being able to determine the optimized values of these operating conditions in order to improve the removal of Ni (II) ions. The percentage of removal presented in this study (74.6%) is relatively lower in relation to others reported in Table 5; this may be linked, mostly, to the initial concentration of nickel in the wastewater, which suggests that some type of pretreatment of the solution or even some modification in the biosorbent, which makes it more effective at higher concentrations, could substantially upgrade the results of the process.

The removal of nickel in real wastewater was also studied by Kwak and collaborators (2011). The authors chose to use an *E. colli* biomass, residual from fermentative processes, as a biosorbent for nickel uptake in a real aqueous matrix; a 100% removal of the metal was achieved for these operating conditions. In addition, in adsorption/desorption cycles in batch mode, the biomass showed superior performance compared to Amberlite IRN-150 resin, maintaining over 3 cycles a percentage of adsorption and elution around 95% and 80%, respectively. The low initial concentration of Ni(II) present in the real effluent and the high dosage of biomass used directly influences the high percentages of removal obtained by the researchers.

Residues from agro-industrial processes are also frequently investigated as biosorbents for the treatment of electroplating industrial wastewaters. Rahman et al. (2020) investigated the waste generated during the palm oil production process, known as empty fruit bunch (EFB), as a viable biosorbent for the uptake of various toxic metals present in the effluent of an electroplating workshop. The biosorbent was modified through chemical treatment with acrylonitrile, being converted to the ligand poly (amidoxime). The synthesized polymer demonstrated satisfactory removal efficiencies for the wide range of metals present in the effluent, e.g., copper, iron, zinc, chromium, and nickel, which indicates that, in this adsorbent material, the effect of competitiveness between metal ions is minor.

Aiming at the removal of Ni(II) present in the effluent from the rinsing step of an electroplating industry, Salem and Awwad (2014) examined the use of loquat bark wastes as a sorbent material in a finite bath system. Chemical modifications were made to the biosorbent through treatment with sodium hydroxide, aiming to improve its removal efficiency. Under optimized conditions, the system was able to remove about 92% of Ni(II) ions present in the medium. Considering that the initial metal concentration in the effluent was 12.48 mg/L, the removal percentage achieved was relatively satisfactory, which may indicate that the presence of sulfates and chlorides in the effluent does not interfere with the potential for removal from the system.

Algae have a prominent role in biosorption studies of toxic metals (Pozdniakova et al. 2016). Biomasses derived from algae demonstrate great potential for the treatment of industrial wastewaters. Its potential for the treatment of electroplating effluents was demonstrated by Ilyas and associates (2018), who investigated the removal of Cu(II) in real wastewater using the biomass of the green alga *Spirogyra*. The researchers evaluated the removal efficiency of both natural algae and modified-algae, with chemical treatments to enhance its surface and functional groups properties. From Table 5, it is possible to notice, under the same operational conditions, a substantial improvement in the results obtained with the algae treated with sulfuric acid. The remarkable adsorptive capacity of this material demonstrates that it has great potential in the large-scale treatment of real effluents polluted by toxic metals, making essential the investigation regarding the performance of the biosorbent in continuous systems.

Continuous or dynamic systems are often studied in biosorption processes because this configuration is the most viable for industrial scale applications

(Rangabhashiyam et al. 2016). In recent years, studies on biosorption of electroplating effluents in a continuous system have been increasing, indicating the trend and a closer step to the expansion of the process. Fixed bed column configuration is the most explored technique. Barquilha et al. (2019) investigated the application of brown algae *Sargassum* sp. for the continuous treatment of an effluent from the rinsing and washing steps of an electroplating industry; the effluent contained a high concentration of Ni(II) ions, making it necessary to accomplish a pre-treatment step via chemical precipitation so that reasonable results were obtained. Nevertheless, the biosorbent showed a good metal selectivity, especially for nickel and copper. On the other hand, Suganya et al. (2020) evaluated the removal of chromium present in real wastewater through biosorption using as sorbent natural and modified *Eucalyptus camaldulensis* seeds. Chromium was present in the effluent with a concentration of 190 mg/L, the highest among the metals present in the composition. The system was able to remove 45% Ni(II) in about three hours of process. From the examples shown, it is possible to note that there is a need to deepen the optimizations in continuous systems for the treatment of real electroplating effluents, seeking to achieve removal percentages above 90%; in addition, studies on the elution of saturated fixed bed columns are necessary for subsequent column reuse and possible metal recovery.

The treatment of electroplating wastewaters by biosorption is well-established in batch system, bringing up optimized conditions and remarkable results. The research in dynamic system has been expanding in the past decade, seeking in several ways to optimize the process in this configuration. The studies compiled in this section provide some important points about the biosorption of real electroplating effluents, such as the optimized pH of the process. Unanimously, in batch mode, all authors adopted pH 6 for their experiments, some performed optimization tests, others based on the metallic speciation of the metal of interest, but it is important to note how this pH can be used for several toxic metals found in real effluent. In general, all have relatively high removal rates (> 70%), proving that biosorption is a highly viable methodology for the treatment of effluents in this category.

2.2 *Acid mine drainage water*

Acid Mine Drainage (AMD) water as a result of human activities can be attributed in large part to the oxidative decomposition of exposed pyrite (iron sulfide, FeS_2) by water and oxygen (Bwapwa et al. 2017). With the formation of net acidity, effluent water becomes increasingly laden with Fe, Mn, Al, Zn, Cu, Ni, Pb, As, and Cd as the main contaminants. Diverse biomaterials have been applied to remediate this kind of wastewater (Table 6), such as algae, agro-industrial wastes, and biopolymers.

Algae have been successfully applied to retain residual metals from AMD considering they can remove metals through bioaccumulation and/or biosorption (Bwapwa et al. 2017). In the study of Choi (2015), the adsorptive performance of a hybrid system containing sericite beads and microalgae *Chlorella* sp. were evaluated. The acid mine drainage was collected from a Korean mine. In that case, the AMD was rich in Fe, Cu, Zn, Mn, As, and Cd, and was an acid media (pH = 2.41). The proposed remediation system consisted of a fixed-bed filled with sericite, silica feedstock, to neutralize the medium pH and remove partially the toxic

Table 6: AMD wastewater treatment through biosorption processes using different biosorbents.

Biosorbent	Metals	Operational Conditions			Adsorption Capacity (mg/g)	Total Removal (%)	References
		pH	Dosage (g/L)	Initial concentration (mg/L)			
Microalgae *Chlorella* sp. in a bioreactor	Fe	2.41	-	137.5	-	98.9	(Choi 2015)
	Cu			22.8		99.2	
	Zn			19.8		99.0	
	Mn			10.4		97.9	
	As			0.45		97.8	
	Cd			0.27		99.3	
Microalgae *Scenedesmus* sp. in a nutrient-rich media	Fe	2.33	0.65	611.4	-	100.0	(dos Santos et al. 2020)
	Al			269.4		99.9	
	Zn			62.7		99.9	
	Mn			37.9		98.8	
	Pb			0.41		95.1	
	As			0.85		85.8	
Olive oil solid waste	Fe	1.0	0.2	409	2181.5	84.9	(İlay et al. 2019)
	Al			273	764.06	44.8	
	Cd			0.033	0.07	38.7	
	B			11.7	23.7	32.4	
	Ti			0.037	0.12	55.3	
Cassava peels	Ca	7.0	-	95.0	-	65	(Pondja et al. 2017)
	Mg			0.021		34	
	Co			0.046		58	
	Hg			42.0		42	
	Mn			1.31		70	
Exhausted brewer's yeast (*Saccharomyces cerevisiae*)	Zn	3.3 - 4.5	-	107	-	76	(Ramírez-paredes et al. 2011)
	Cu			2		70	
	Mn			2		62	
	Ni			6		48	
	Al			7		30	
Chicken Egg-shell	Cd	2.25 - 2.51	-	0.39	-	18.9	(Zhang et al. 2017)
	Pb			1.20		77.1	
	Cu			6.30		55.7	
	Fe			200.3		62.4	
Ca-alginate beads	Cu	2.65	-	-	-	86	(Park and Lee 2017)
	Cd					60	

metals, and a posterior bioreactor contained *Chlorella* sp. that effectively remediate the wastewater, simultaneously to the growth of the microalgae. The total efficiency of the system varied from 97.8 for arsenic to 99.3 % for cadmium, indicating that this approach is a useful alternative to the treatment of AMD.

In a batch mode, in turn, a treatment process composed of precipitation and biosorption was proposed to remove toxic metals from the AMD water of a coal mining company of a Brazilian city called Figueira (State of Paraná) (dos Santos et al. 2020). Calcium hydroxide was used for the primary treatment step due to the contaminant's high concentration in the raw wastewater, in which the pH raised from

2.33 to 8.7 causing the precipitation of the metals. The additional treatment step using the microalgae *Scenedesmus* sp., with nutrient supply, reduced significantly the concentration of the toxic metals. The completed treatment achieved high removal efficiencies, greater than 85.8%.

Like algae, agro-industrial wastes have many advantages in wastewater treatment, such as abundant and available sources, low cost, renewable cycle, and environmental friendliness. The diversity of biomass generated by agro-industrial activities also brings another advantage for the biosorption process, because each country or region, with its specific industrial matrix, produces different kind of waste that can be used to solve its peculiar environmental issues. Besides that, their porous structure and active substances such as pectin, cellulose, hemicellulose, and lignin make these materials uptake heavy metals easily, either by physical or chemical sorption (Mo et al. 2018). Olive cake, a residue from the olive oil extraction process, cassava peels, and exhausted yeast, waste from a Brewer industry, and chicken eggshells are examples of agro-industrial wastes applied to the AMD treatment.

Water from an acid mine lake from Turkey was treated using olive oils solid waste as a biosorbent (İlay et al. 2019). The lake water contained expressive amounts of light metals (Mg, Na, Ca, and K), aluminum, and iron, trace elements (Ga, Cr, Cd, Ti, and Pb), among others. Howbeit, just the adsorptive capacity of the olive cake for Fe, Al, Cd, B, and Ti were evaluated. Solid/liquid ratios of 1:5 and 1:10 were studied, and for the high biosorbent dosage, the removal of iron achieved 84.9%; for the other metals, the efficiency was less than 55.3% (Table 6). Thus, the biosorption process using the residual of the olive oil extraction is inexpensive because this biomaterial is abundant in the country; however, the overall adsorption efficiency still needs to be improved, and/or primary and secondary treatments should be performed before the biosorption.

Pondja et al. (2017) investigated the removal of Ca, Mg, Co, Hg, and Mn from a synthetic coal mine water using cassava peels. Solid waste (overburden) of a coal mine from Mozambique was continuously washed by distilled water forming the synthetic wastewater used in the experiments of pH effect and equilibrium. pH had a different effect on the removal efficiency of each metal. Ca, Mg, and Hg were better adsorbed under alkaline conditions; at neutral pH, Mn had the highest removal efficiency (70%) among the others (Table 6). The highest maximum adsorption capacity from the Langmuir fitting to the data was found for Ca ions, 18.9 mg/g, and for the Mg ions, 2.6 mg/g. Thus, the use of low-cost biosorbent such as cassava peels after further studies can be a promising alternative for AMD remediation.

Ramírez-Paredes et al. (2011) studied the biosorption efficiency of exhausted yeast from a brewery industry to treat AMD. The suspended yeast achieved the highest percentages of removal for Zn (76%), followed by Cu (70%), Mn (62%), Ni (48%), and Al (30%), from a synthetic AMD solution (Table 6). For the tests with real AMD, the samples were enriched with each metal in the desired concentration to obtain equilibrium isotherms, in which the highest biosorption efficiencies were achieved for the removal of copper (80%; 0.19 mg/g), nickel (25%; 0.24 mg/g) and manganese (3%; 0.026 mg/g). An ecotoxicity test was performed and indicated that the proposed biosorbent eliminated the ecotoxicity of the acid mine drainage water, achieving the main goal of the remediation of polluted water.

Chicken eggshells were also applied to remove heavy metals from AMD of a Chinese mining area (Zhang et al. 2017), in which the adsorptive performance was evaluated in a continuous fixed-bed system. With a bed height of 20 cm, at 10 mL/min, particle size between 0.425 and 1 mm, the removal percentage of Cd, Pb, Cu, and Fe was 18.9, 77.1, 55.7, 62.4%, respectively (Table 6). The authors justified the low efficiencies for cadmium and copper due to the iron high concentration in AMD where the floccules ($Fe_2(OH)_2CO_3$) obstructed the bed to develop the overall effectiveness. Thus, this waste has a high potential to uptake lead from AMD and could be applied in further studies including a pre-treatment step to remove the iron content before the biosorption.

The natural biosorbents most studied are composed of different biopolymers, such as alginates, cellulose and derivatives, pectin, and the components of fermentation yeasts. Thus, there are also studies applying directly the isolated and purified biopolymer to the remediation of wastewater, instead of the raw biomaterial (Torres 2020). In a dynamic fixed-bed study, Park and Lee (2017) evaluated the biosorption capacity of Ca-alginate for copper and cadmium from AMD. A mass of 381.8 g of Ca-alginate beads filled a Pyrex column with 2.5 cm in diameter and 1 m in height (1 bed volume ~ 133 mL). AMD was pumped in the column in an upward flow. For a flow of 3 mL/min, after 900 bed volumes, the removal efficiencies for Cu and Cd in AMD were higher than 86% and 60%, respectively. In these conditions, the Cu and Cd release is lower than the Korean groundwater quality standard limit for the mine area (Cu: 3 mg/L and Cd: 0.1 mg/L).

2.3 Textile and tannery effluents

Tannery and textile are two important industrial segments that consume great amounts of water and generate a large volume of effluents. These effluents contain many organic and inorganic materials as well as toxic trace elements (Chhonkar et al. 2000). Biosorption has been proposed to remove heavy metals from textile and tannery wastewaters, in which Table 7 exhibits diverse biosorbents that were applied to this purpose.

The indiscriminate discharge of textile effluents can promote environmental impacts due to the release of high concentrations of dyes and toxic metals in water resources like rivers and lakes. Bhardwaj et al. (2014) analyzed effluent samples from different textile industries, and heavy metals such as copper, chromium, cadmium, iron, lead, nickel, zinc, and arsenic were found. Different kinds of biomaterials have been employed to treat textile effluents such as green seaweed leaf, citrus lemon leaf, and chitosan-coated fungal biomass (*Aspergillus niger*) (Table 7).

Latinwo et al. (2015) treated effluent samples from a Nigerian textile company using green seaweed leaf powder. The pH of the samples was 10.4, and the metal concentrations are shown in Table 7. Using a biosorbent dosage of 10 g/L, the equilibrium is achieved in 60 min, removing 87.5, 99.9, 59.7, 57.2, 100, and 86.8% of the Fe, Ca, Mg, K, Ag, and Cr contents, respectively. The biosorption kinetics follows a pseudo-first-order kinetic model, and the biosorption mechanism is controlled by boundary layer surface diffusion. Thus, the fast process and the

Table 7: Textile and tannery effluents treatment through biosorption processes using different biosorbents.

Biosorbent	Metals	Operational Conditions			Adsorption Capacity (mg/g)	Total Removal (%)	References
		pH	Dosage (g/L)	Initial concentration (mg/L)			
Green seaweed leaf powder	Fe	10.8	10	0.0283	0.009	87.5	(Latinwo et al. 2015)
	Ca			0.0815	0.05	99.9	
	Mg			0.8095	0.0013	59.7	
	K			0.0872	0.008	57.2	
	Ag			0.0006	0.0008	100	
	Cr			0.0017	0.0009	86.8	
Citrus lemon leaf powder	Cu	4.0–5.0	40	3.65	-	87.1	(Muniraj et al. 2020)
	Zn			6.65		89.5	
	Ni			2.23		89.6	
	Pb			9.21		88.4	
	Cd			1.02		88.7	
Chitosan coated *Aspergillus niger*	Zn	5.4	-	0.26	-	32.9	(Okoya et al. 2020)
	Pb			0.08		39.0	
	Fe			0.46		84.9	
	Ni			0.08		34.6	
	Mn			0.04		44.2	
	Cu			1.10		42.2	
	Cd			0.11		49.1	
	Cr			0.13		25.1	
	Co			0.07		43.7	
	Mg			0.28		32.0	
Almond shell	Cr	3.7	-	67.5	21.92	65.9	(Yahya et al. 2020)
	Cu			7.0	2.39	70.0	
Modified Chitosan/ bentonite composite	Cr	7.4	50	1055	-	97.8	(Nithya and Sudha 2017)
	Pb			0.43		20.9	
Carbonactivated algae granules of *Chlorella vulgaris*	Cr	6.8	9	9.54	-	97.8	(Mirza et al. 2021)
Carbonactivated algae granules of *Scenedesmus obliquus*	Cr	6.8	9	9.54	-	79.7	(Mirza et al. 2021)

low-cost nature of the biosorbent make the green seaweed biomass a potential biosorbent to be explored more in scale-up studies.

Citrus lemon leaf powder was also applied to reduce the metal toxicity of the textile wastewater (Muniraj et al. 2020). The effluent samples collected in different discharge points in an Indian river have a pH of 11.16. The biosorption of the heavy metals, instead, was more efficient at pH between 4.0 and 5.0 using 40 g/L of biosorbent, in which, for all metals, removal percentages higher than 87% were achieved, and a time of 240 min is required to the sorption equilibrium. The biosorption using citrus lemon leaves is cost-effective and holds excellent potential for the removal of toxic metals, being an available alternative process to the textile effluent treatment.

As well as the abundant biomass from plants, such as the citrus seaweed and citrus leaves cited above, fungal biomass can be also an interesting biosorbent for textile wastewater treatment. Okoya et al. (2020) studied the adsorptive efficiency of an innovative biosorbent composed of chitosan, extracted from snail shells, and the fungal biomass powder (*Aspergillus niger*) to remove the dye and toxic metal content. A column with a diameter of 1.27 cm and a height of 40 cm filled with the biosorbent was used in the study. The concentration of Zn, Pb, Fe, Ni, Mn, Cu, Cd, Cr, Co, and Mg was reduced to 32.9, 39.0, 84.9, 34.6, 44.2, 42.2, 49.1, 25.1, 43.7, and 32.0%, in this order, indicating that the biosorbent was much more effective to remove iron than the other metals. However, chitosan-coated *A. niger* could simultaneously remove dyes, with removal efficiencies higher than 90%, and, in part, toxic metals from textile wastewater.

Similar to textile effluents, wastewaters from tanneries contain a large range of substances derived from hides and skins or are present through the addition of reagents during the processing, including heavy metals such as chromium, the main metallic compound, also cadmium, cobalt, lead, nickel, selenium, and arsenic (Lofrano et al. 2014). To provide a sustainable alternative to the tannery effluents treatment, also, different materials have been proposed (Table 7), such as almond shells, polymeric biocomposites, and algal-based biosorbents.

In a continuous fixed-bed system, almond shell, an available agro-industrial residue, was used to treat an acid tannery effluent (pH = 3.7) composed mainly of chromium (65.7 mg/L) and copper (7.0 mg/L) (Yahya et al. 2020). The tannery effluent was collected from a tanning industry in Nigeria. The column parameters analyzed indicated that the highest column efficiency, 65.9% for Cr and 70.0% for Cu, was attained at a flow rate of 3.0 ml/min, bed height of 7.0 cm, and almond shell powder mass of 16.1 g. The authors pointed out that the biosorption occurs in binding sites in the particle inner surface, and the possible binding groups are carboxyl, hydroxyl, and phenols. The study of a continuous biosorption of chromium and copper on almond shells at the lab scale provided reliable data for further large-scale studies, and consequently for industrial purposes.

The syntheses of polymer/clay biocomposites have attracted much attention in heavy metals remediation from wastewaters because of their low cost, safety, eco-friendly character, and high physical resistance (Begum et al. 2021). Nithya and Sudha (2017) studied the biosorption of the toxic metals chromium and lead on chitosan-g-poly(butyl acrylate)/bentonite nanocomposite. The wastewater samples were collected in a tannery industrial pole in India, one of the countries in which the environment is expressively affected by textile and tannery activities. The maximum removal percentage of Cr (VI) from the tannery effluent using an adsorbent dosage of 5 g in 100 mL, contact time of 240 min, and pH of 3 was 97.81 %; lead, in turn, had a concentration reduction of 20.9%. Due to the efficiency toward chromium, which is the main pollutant in tannery wastewaters, the proposed biosorbent have potential for scale-up studies preceding an industrial application.

Mirza et al. (2021) produced carbon-activated algae granules of *Chlorella vulgaris* and *Scenedesmus obliquus* for the removal of chromium from tannery effluent. The biosorbents preparation occurred by a growing step, in which each alga was cultivated until their adequate development, the biomass drying, and the activation that was performed by photochemical reactions and acid treatment. The maximum removal percentage of chromium, 97.8 and 79.7%, was reached after 120h, using a dosage of 9 g/L of *Chlorella vulgaris* and *Scenedesmus obliquus* biomass, respectively. The authors recommended a scale-up study of the carbon-activated granules of algal strains to enhance the removal of metal contaminants in tannery wastewater.

2.4 Effluent from nuclear power plants

Process of producing electricity in nuclear power plants is carried out through nuclear reactors that are powered by a combustible material, with uranium, polonium, and thorium being the most used elements. This process has the main advantage of being less environmentally destructive, with a minor emission of greenhouse gases. However, the risks to the environment and human health due to the radioactive compounds generated in this process, in addition to the cost and management of waste contaminated by long-lived radionuclides, are worrisome factors (Wai 2011). These wastes are classified according to their radioactivity and can vary from very low radioactivity compounds (Very Low-level waste, LLW) to highly hazardous waste (High-level waste, HLW), containing actinides, lanthanides, and long-lived fission products (Petrangeli 2020).

Wastewater from nuclear power plants has been generated in greater volume along with the growth of this industrial sector, according to the World Nuclear Performance report. In 2019, the production of electricity generated through nuclear reactors increased for the seventh consecutive year (World Nuclear Association 2020). Contaminated effluents can be generated in different stages, along with the entire nuclear fuel cycle, from mining to industrial applications (Valković 2019). Two main approaches are used to control liquid effluents before its discharge: storage, when the effluent contains short-lived radionuclide compounds, until the decomposition of these species; or treat these effluents to remove the radioactive elements present in it (Valković 2019).

Adsorption is one of the most extensively studied alternative method for the uptake of radioactive compounds from nuclear waste (Sengupta and Gupta 2017). Among the most investigated adsorbents, activated carbon and zeolites present a superior performance, mainly due to the thermal and radiolytic stability of these materials (Jiménez-Reyes et al. 2021). Nonetheless, the use of these adsorbents adds a high cost to the adsorption process, which can make them unfeasible for large-scale applications. Therefore, a joint analysis of cost in relation to the performance of the sorbent material must be carried out. In this sense, biosorbents stand out. In addition, the biosorption process generates smaller amounts of toxic by-products. A wide range of studies report remarkable results in the removal of long-lived radionuclides in aqueous medium through different biosorbents (Gupta et al. 2018). Among the studies on the removal of radioactive species, few articles address biosorption using real effluents; this may be mainly due to the biological risk and the difficulty in obtaining such samples. Table 8 displays some studies carried out in recent years aiming the biosorption of radionuclides present in real effluents and in multi-compound systems, simulating the composition of the industrial wastewater.

Among the studies presented in Table 8, only three used samples of real effluents, others performed biosorption in simulated effluent. Regarding the biosorbents, it can be observed that a wide range of biomasses are studied, being mainly derived from algae and agro-industrial residues. From biosorbents derived from marine algae, brown algae stand out as a well-established biomass in several adsorption studies, with high affinity for a wide range of metals. Its affinity with radionuclides is confirmed by several articles reported in the literature in the past two decades (Lee et al. 2014).

Zhou et al. (2016) reported the removal of actinides in synthetic solution using *Giant Kelp* brown algae. The authors outlined that this seaweed has tendency to leach organic compounds when used for the biosorption of metals; thus, the

Table 8: Synthetic and real nuclear wastewater treatment through biosorption processes using different biosorbents.

Biosorbent	Metals	Operating conditions			Adsorption capacity (mg/g)	Total Removal (%)	References
		pH	Dosage (g/L)	Initial Concentration (mg/L)			
Giant Kelp biomass (CaCl₂-modified)	U[b] Th[b]	3.5	0.5	10–50	84.03 59.17	-	(Zhou et al. 2016)
Cystoseira indica (CaCl₂-modified)	U[b] Cu[b] Ni[b]	5	1.0	0–1100	83.69 53.46 14.43	5.3 22.1 29.3	(Keshtkar et al. 2015)
Lemna sp.	U[a]	2.17	20	58.25	2.15	-	(Vieira et al. 2019)
Rice straw carbon (HNO₃ oxidized)	U[b] Th[b]	4.5 ± 0.5	1.0	1–100	24.0 25.0	-	(Yakout and Rizk 2015)
Rice straw carbon (KOH oxidized)	U[b] Th[b]	4.5 ± 0.5	1.0	1–100	35.9 29.5	-	(Yakout and Rizk 2015)
Coconut fiber	U[a] Am[a] Cs[a]	2.5 ± 0.5	20	110 1.8E-03 6.8E-06	0.66 46.3E-06 44.7E-09	-	(Ferreira et al. 2018)
Activated Coconut fiber	U[a] Am[a] Cs[a]	2.5 ± 0.5	20	110 1.8E-03 6.8E-06	1.82 73.4E-06 37.7E-09	-	(Ferreira et al. 2018)
Douglas fir barks	U[a]	4	1.0	0.026 - 0.111	0.08	90	(Jauberty et al. 2011)

[a]Real effluent; [b]Synthetic effluent.

modification of the biosorbent using $CaCl_2$ was carried out. This modification, in addition to increasing the stability of the adsorbent material, has also been shown to significantly improve the removal capacity of the metals of interest. The studies in simulated effluent containing uranium and thorium revealed the competitiveness of these metals for the same active sites on the surface of the biosorbent, with slightly higher selectivity for U(IV). Nonetheless, the adsorptive capacities calculated for the two metals, approximately 84 mg/g for U(IV) and 59 mg/g for Th(IV), are remarkable for a multi-compound system.

Simulating a nuclear wastewater from Iranian regions, Keshtkar and coworkers (2015) evaluated the competitive biosorption of U(IV), Cu(II) and Ni(II) in synthetic solution using the brown algae *Cystoseira indica*. Like Zhou et al. (2016), the researchers also decided to perform chemical modifications using $CaCl_2$ to increase the adsorption efficiency. In a single ion, the modified algae showed greater affinity for Cu(II) than for U(IV), so, as expected, in a ternary system, the presence of Cu and Ni causes inhibitory effects on the removal of U(IV) ions. Factors such as atomic mass and ionic radius may be related to the explanation of the observed effect.

Aquatic species can be a viable option for biosorption systems (Khosravi et al. 2005). Using the macrophyte *Lemna* sp., Vieira et al. (2019) evaluated the removal of U(IV) ions present in real effluent composed mainly of uranium, americium, and cesium. The authors observed that the pH of the actual effluent (2.17) caused considerable disadvantages to the biosorption capacity. Furthermore, the presence of organic compounds and more than one actinide also seem to contribute to this effect. However, because it is a real wastewater, the study presents an innovative result, showing that with the proper operational conditions the treatment of radioactive effluents can be satisfactorily performed by natural adsorbents without modifications.

Agro-industrial wastes are also extensively investigated for the uptake of radioactive metals present in real or synthetic effluents. Chemical and physical modifications are also frequently carried out on these biomaterials to make their removal potential viable. Yakout and Rizk (2015) performed chemical modifications, i.e., oxidation with HNO_3 and KOH, seeking to enhance the adsorptive capacity of activated carbon prepared from rice straw. The biosorption process was carried out in simulated effluent, containing several cations and anions, metals of interest being uranium and thorium. The behavior of single-ion system showed that U(IV) was better removed by rice straw carbon modified with KOH. When this biosorbent was applied to the multi-component system, the authors noted that the effect of the presence of other chemical species was minimal. On the other hand, when Th(IV) was also added to the solution, the removal of U(IV) declined substantially, although the evaluation of equilibrium parameters indicated that there is a formation of stronger and more stable bonds between uranium and the biosorbent; this fact may be associated with the high oxidative state of thorium or with characteristics of uranium behavior. In any case, it can be noted that the removal of U (IV) ions using this biosorbent is favorable even in the presence of other metallic species.

Similarly, coconut fibers were modified via treatment with NaOH by Ferreira et al. (2018) to analyze its application in the biosorption of a real radioactive wastewater. In this study, metals of interest were uranium, americium, and cesium. The authors noted that the activation process improved the removal results obtained.

As observed in the study by Vieira et al. (2019), the adsorptive capacity values of the system are relatively low, which is directly related to the complex matrix of real effluents. This biosorbent demonstrated an inferior performance in the removal of uranium present in real effluent when compared to results evaluated for the macrophyte *Lemna* sp. even though operating conditions in both studies were similar, as well as the effluent composition. In both studies, the adjustment of operating conditions, especially the pH of the aqueous medium, could bring significant improvement to the results obtained.

Douglas fir husks without modifications were investigated by Jauberty et al. (2011) for the removal of uranium present in contaminated wastewater. The biosorbent showed removal of 90% of the U(IV) ions present in the batch system solution. The researchers also evaluated the removal of uranium in a continuous system, an assessment extremely important for the scale-up of this type of process. In the evaluated system, the U(IV) ions were desorbed from the column by a solution of sulfuric acid. Through calculations performed with data collected over 4 weeks of process, the column was able to treat at least 70 m^3 of the tested effluent, being able to uptake more than 8.3 g of U(IV).

The application of biomaterials for the removal of radioactive metals has grown considerably over the past few years. In particular, the application of chemical and physical modifications to biomasses is extensively investigated, seeking to improve the adsorption system by increasing the contact surface or adding new functional groups to its structure. The metal of greatest interest in these studies is often uranium, possibly due to its widespread use and industrial interest. In contrast, metals frequently found in effluents from nuclear power plants such as thorium, americium and cesium are still insufficiently explored in biosorption studies. Similar to other systems, operational conditions directly influence the removal results, especially pH seems to play a determining role in the adsorptive capacity in radionuclide biosorption systems, with pH 4 being the optimized value found in most studies for these metallic species. For the most part, studies presented in the literature perform the evaluation of biosorption process in batch design; few are those where the removal of actinides in continuous mode is carried out, revealing the need for further investigation of such systems, especially with real effluents, aiming the scale-up for industrial applications.

3. Recent advances and future perspectives

Biosorption technology has been considered a promising alternative for the removal of toxic metals in wastewater (Malik 2004). Most toxic metal biosorption studies are still concentrated in batch and continuous modes of operation, on a laboratory scale (Michalak et al. 2013). Laboratory-scale tests are important for determining the biosorbent efficiency, optimal processing conditions, biosorption capacity, and biomass regeneration (Areco et al. 2012, Freitas et al. 2019). Despite this, laboratory-scale studies do not fully address the main implications of wastewater treatment scale-up (Wang and Chen 2009).

The influence of the main parameters that affect the biosorption performance of toxic metals in continuous and batch tests, as well as the mechanisms involved in the capture of toxic metals by different biomasses has been widely discussed in the literature. The next step is the scaling-up of toxic metals biosorption technology. Therefore, research should be directed to applications in real effluents (e.g., mineral extraction, energy and fuel production, surface finishing and electroplating, aerospace, metallurgy, metal molding and coating, electronics and batteries, nuclear industry, milling and mining, pharmaceutical industries, fertilizer industries, tanneries, pulp and paper industries, agriculture, among other industrial activities) in continuous mode on pilot and industrial scale (Wang and Chen 2009, Fu and Wang 2011, Wu et al. 2017, Gupta et al. 2018, Freitas et al. 2019, Kumari and Tripathi 2019).

Industrial wastewater can have different concentrations of other types of pollutants such as suspended material and organic pollutants (Šćiban et al. 2007, Liu et al. 2013). These contaminants will certainly interfere with the biosorption mechanisms, including competitive biosorption between toxic metals, and also between toxic metals and other organic-inorganic pollutants (Kaczala et al. 2009). In addition, the presence of organic matter in real effluents can complex metal ions and make biosorption more difficult (Martín-Lara et al. 2014).

For the treatment of large volumes, in industrial applications, operations in continuous mode on packed fixed-bed columns are most commonly used because they are effective, economical, and more appropriate for biosorption processes (Saeed and Iqbal 2003, Chu 2004, Das et al. 2008). The packed fixed-bed columns also offer process engineering advantages, such as high operational performance and ease of scale expansion (Aksu et al. 1992).

Another important aspect for the industrialization of the biosorption process is the shape and size of biosorbent particles that can influence the head loss in packed fixed-bed column. In general, the shape and size of biosorbent particles are less regular than synthetic materials, which can generate greater head loss in the column. Durability and reuse of biosorbent particles through regeneration cycles in a fixed-bed column is other important aspect, which has also been little investigated in the literature.

Physical-chemical and biological modifications of biosorbent materials to improve their properties such as resistance, selectivity, biosorption capacity, and regeneration are recommended for applications in continuous processes on scale-up (Fomina and Gadd 2014, Freitas et al. 2019). Additionally, biomass immobilization in low-cost support materials can be an alternative to overcome problems related to the regeneration and reuse of the biosorbent (Wang and Chen 2009).

Among the trends for the future of biosorption technology are the development of commercial biosorbents with a high capacity for biosorption and selectivity, in addition to the application of hybrid processes (biological and chemical) (Wang and Chen 2009). The approach of biosorption together with other conventional techniques, such as biological processes, is an alternative to obtain successful hybrid process (Park et al. 2010, Freitas et al. 2019).

Based on the above, through decades of research on the biosorption process, more efforts should now be invested in the application of biosorption in real-world situations. For this, a complete understanding of the behavior of biosorbents in a real water matrix, types of treatment systems, and the process costs involved are also important.

4. Final remarks

Initially, this chapter reported an overview regarding water contamination by toxic metals associated with multiple sources, such as tributary contribution, surface runoff, and anthropogenic activity. Different physical-chemical technologies for removing toxic metals (e.g., chemical precipitation, ion exchange, membrane processes, electrochemical treatment, solvent extraction, coagulation-flocculation, flotation, and adsorption) were addressed. Among these methods, biosorption is considered an economical and environmentally friendly technology for the wastewater treatment contaminated with toxic metals at low concentrations. Several biosorbent materials (e.g., algae, fungi, bacteria, virus, biowaste, plant and animal-based) were addressed, as well as their physical and chemical surface modification to enhance selectivity and improve biosorption capacities. The new classification of biosorption mechanisms was also presented to understand the equilibrium of the process and the interactions between toxic metals and functional groups present on the surface of biosorbents.

This chapter also covered the application of various biosorbents for the removal of toxic metals not only from monocomponent systems, but also from multicomponent systems and wastewaters. The use of several biomaterials for wastewater remediation, focused on the electroplating wastewater, acid mine drainage water, textile and tannery effluents, and residual water from nuclear power plants were addressed. Although several biomaterials have been developed and employed for applications in real effluents, the implementation and commercialization of this technology for toxic metal removal aiming at environmental remediation still requires more detailed studies. Future research on a pilot scale to verify the efficiency of these biomaterials in real effluents are recommended for large-scale applications.

Acknowledgements

The authors are grateful for the financial support provided by Research Supporting Foundation of the State of São Paulo (FAPESP) (Grants # 2017/18236-1 and #2019/11353-8), Brazilian National Research Council (CNPq) (Grant # 406193/2018-5 and 308046/2019-6) and Coordination for the Improvement of Higher Education Personnel (CAPES, financial code – 001).

References

Agrafioti, E., Kalderis, D., and Diamadopoulos, E. 2014. Ca and Fe modified biochars as adsorbents of arsenic and chromium in aqueous solutions. J. Environ. Manage., 146: 444–450.

Akpor, O.B. 2014. Heavy metal pollutants in wastewater effluents: sources, effects and remediation. Adv. Biosci. Bioeng., 2: 37.

Aksu, Z., Sag, Y., and Kutsal, T. 1992. The biosorption of copperod by c. Vulgaris and z. ramigera. Environ. Technol. (United Kingdom), 13: 579–586.

Al-Ghouti, M.A., and Da'ana, D.A. 2020. Guidelines for the use and interpretation of adsorption isotherm models: A review. J. Hazard. Mater., 393: 122383.

Areco, M.M., Hanela, S., Duran, J., and dos Santos Afonso, M. 2012. Biosorption of Cu(II), Zn(II), Cd(II) and Pb(II) by dead biomasses of green alga Ulva lactuca and the development of a sustainable matrix for adsorption implementation. J. Hazard. Mater., 213–214: 123–132. Elsevier B.V.

Avery, S.V., Codd, G.A., and Gadd, G.M. 1993. Biosorption of tributyltin and other organotin compounds by cyanobacteria and microalgae. Appl. Microbiol. Biotechnol., 39: 812–817.

Bădescu, I.S., Bulgariu, D., and Bulgariu, L. 2017. Alternative utilization of algal biomass (*Ulva* sp.) loaded with Zn(II) ions for improving of soil quality. J. Appl. Phycol., 29: 1069–1079.

Bădescu, I.S., Bulgariu, D., Ahmad, I., and Bulgariu, L. 2018. Valorisation possibilities of exhausted biosorbents loaded with metal ions – A review. J. Environ. Manage., 224: 288–297.

Barquilha, C.E.R., Cossich, E.S., Tavares, C.R.G., and da Silva, E.A. 2019. Biosorption of nickel and copper ions from synthetic solution and electroplating effluent using fixed bed column of immobilized brown algae. J. Water Process Eng., 32: 100904. Elsevier.

Begum, S., Yuhana, N.Y., Md Saleh, N., Kamarudin, N.H.N., and Sulong, A.B. 2021. Review of chitosan composite as a heavy metal adsorbent: Material preparation and properties. Carbohydr. Polym., 259: 117613.

Beni, A.A., and Esmaeili, A. 2020. Biosorption, an efficient method for removing heavy metals from industrial effluents: A Review. Environ. Technol. Innov., 17: 100503. Elsevier B.V.

Bhardwaj, V., Kumar, P., and Singhal, G. 2014. Toxicity of heavy metals pollutants in textile mills effluents. Int. J. Sci. Eng. Res., 5: 664–666.

Bridgwater, A.V. 2012. Review of fast pyrolysis of biomass and product upgrading. Biomass and Bioenergy, 38: 68–94.

Bwapwa, J.K., A.T. Jaiyeola, and Chetty, R. 2017. Bioremediation of acid mine drainage using algae strains: A review. South African J. Chem. Eng., 24: 62–70.

Callender, E. 2003. Heavy Metals in the Environment—Historical Trends. pp. 67–105 in Treatise on Geochemistry. Elsevier.

Cardoso, S.L., Costa, C.S.D., Da Silva, M.G.C., and Vieira, M.G.A. 2020. Insight into zinc(II) biosorption on alginate extraction residue: Kinetics, isotherm and thermodynamics. J. Environ. Chem. Eng., 8: 103629. Elsevier.

Carolin, C.F., Kumar, P.S., Saravanan, A., Joshiba, G.J., and Naushad, M. 2017. Efficient techniques for the removal of toxic heavy metals from aquatic environment: A review. J. Environ. Chem. Eng., 5: 2782–2799.

Chatterjee, A., and Abraham, J. 2019. Desorption of heavy metals from metal loaded sorbents and e-wastes: A review. Biotechnol. Lett., 41: 319–333.

Chen, S.H., Ng, S.L., Cheow, Y.L., and Ting, A.S.Y. 2017. A novel study based on adaptive metal tolerance behavior in fungi and SEM-EDX analysis. J. Hazard. Mater., 334: 132–141.

Chhonkar, P.K., Datta, S.P., Joshi, H.C., and Pathak, H. 2000. Impact of industrial effluents on soil health and agriculture-Indian experience: Part II-tannery and textile industrial effluents. J. Sci. Ind. Res. (India), 59: 446–454.

Choi, H.J. 2015. Biosorption of heavy metals from acid mine drainage by modified sericite and microalgae hybrid system. Water. Air. Soil Pollut., 226.

Chojnacka, K., and Mikulewicz, M. 2019. Green analytical methods of metals determination in biosorption studies. TrAC Trends Anal. Chem., 116: 254–265.

Chu, K.H. 2004. Improved fixed bed models for metal biosorption. Chem. Eng. J., 97: 233–239.

Çolak, F., Atar, N., and Olgun, A. 2009. Biosorption of acidic dyes from aqueous solution by *Paenibacillus macerans*: Kinetic, thermodynamic and equilibrium studies. Chem. Eng. J., 150: 122–130.

Cole, A.J., Paul, N.A., de Nys, R., and Roberts, D.A. 2017. Good for sewage treatment and good for agriculture: Algal based compost and biochar. J. Environ. Manage., 200: 105–113.

Cruz, C.C.V., da Costa, A.C.A., Henriques, C.A., and Luna, A.S. 2004. Kinetic modeling and equilibrium studies during cadmium biosorption by dead *Sargassum* sp. biomass. Bioresour. Technol., 91: 249–257.

Da'na, E. 2017. Adsorption of heavy metals on functionalized-mesoporous silica: A review. Microporous Mesoporous Mater., 247: 145–157.

Das, N., Vimala, R., and Karthika, P. 2008. Biosorption of heavy metals—An overview. Indian J. Biotechnol., 7: 159–169.

Dermentzis, K., Christoforidis, A., and Valsamidou, E. 2011. Removal of nickel, copper, zinc and chromium from synthetic and industrial wastewater by electrocoagulation. Int. J. Environ. Sci., 1: 697–710.

dos Santos, K.B., de Almeida, V.O., Weiler, J., and Schneider, I.A.H. 2020. Removal of pollutants from an amd from a coal mine by neutralization/precipitation followed by "*in vivo*" biosorption step with the microalgae *Scenedesmus* sp. Minerals, 10: 1–11.

European Council. 2010. Directive 2010/75/EU Industrial Emissions. Off. J. Eur. Union., L334: 17–119.

European Council. 2020. Directive (EU) 2020/2184, EU (revised) Drinking Water Directive. Off. J. Eur. Communities, 2019: 1–62.

Ferreira, R.V.P., Silva, E.A., Canevesi, R.L.S., Ferreira, E.G.A., Taddei, M.H.T., Palmieri, M.C., Silva, F.R.O., and Marumo, J.T. 2018. Application of the coconut fiber in radioactive liquid waste treatment. Int. J. Environ. Sci. Technol., 15: 1629–1640.

Fomina, M., and Gadd, G.M. 2014. Biosorption: Current perspectives on concept, definition and application. Bioresour. Technol., 160: 3–14. Elsevier Ltd.

Freitas, G.R., Vieira, M.G.A., and Da Silva, M.G.C. 2018. Batch and Fixed Bed Biosorption of Copper by Acidified Algae Waste Biomass. Ind. Eng. Chem. Res., 57: 11767–11777.

Freitas, G.R., Silva, M.G.C., and Vieira, M.G.A. 2019. Biosorption technology for removal of toxic metals: a review of commercial biosorbents and patents. Environ. Sci. Pollut. Res., 26: 19097–19118. Environmental Science and Pollution Research.

Fu, F., and Wang, Q. 2011. Removal of heavy metal ions from wastewaters: A review. J. Environ. Manage., 92: 407–418. Elsevier Ltd.

Gao, Q., Xu, J., and Bu, X.-H. 2019. Recent advances about metal–organic frameworks in the removal of pollutants from wastewater. Coord. Chem. Rev., 378: 17–31.

Golnaraghi Ghomi, A., Asasian-Kolur, N., Sharifian, S., and Golnaraghi, A. 2020. Biosorpion for sustainable recovery of precious metals from wastewater. J. Environ. Chem. Eng., 8: 103996.

Guibal, E., Roulph, C., and Le Cloirec, P. 1992. Uranium biosorption by a filamentous fungus Mucor miehei pH effect on mechanisms and performances of uptake. Water Res., 26: 1139–1145.

Gupta, N.K., Sengupta, A., Gupta, A., Sonawane, J.R., and Sahoo, H. 2018. Biosorption-an alternative method for nuclear waste management: A critical review. J. Environ. Chem. Eng., 6: 2159–2175.

Gupta, V.K., Rastogi, A., Saini, V.K., and Jain, N. 2006. Biosorption of copper(II) from aqueous solutions by *Spirogyra* species. J. Colloid Interface Sci., 296: 59–63.

He, J., and Chen, J.P. 2014. A comprehensive review on biosorption of heavy metals by algal biomass: Materials, performances, chemistry, and modeling simulation tools. Bioresour. Technol., 160: 67–78. Elsevier Ltd.

İlay, R., Baba, A., and Kavdır, Y. 2019. Removal of metals and metalloids from acidic mining lake (AML) using olive oil solid waste (OSW). 4047–4058.

Ilyas, N., Ilyas, S., Sajjad-Ur-Rahman, Yousaf, S., Zia, A., and Sattar, S. 2018. Removal of copper from an electroplating industrial effluent using the native and modified spirogyra. Water Sci. Technol., 78: 147–155.

Jauberty, L., Gloaguen, V., Astier, C., Krausz, P., Delpech, V., Berland, A., Granger, V., Niort, I., Royer, A., and Decossas, J.L. 2011. Bark, a suitable biosorbent for the removal of uranium from wastewater-From laboratory to industry. Radioprotection, 46: 443–456.

Jiménez-Reyes, M., Almazán-Sánchez, P.T., and Solache-Ríos, M. 2021. Radioactive waste treatments by using zeolites. A short review. J. Environ. Radioact., 233: 106610.

Kaczala, F., Marques, M., and Hogland, W. 2009. Lead and vanadium removal from a real industrial wastewater by gravitational settling/sedimentation and sorption onto *Pinus sylvestris* sawdust. Bioresour. Technol., 100: 235–243.

Kan, T., Strezov, V., and Evans, T.J. 2016. Lignocellulosic biomass pyrolysis: A review of product properties and effects of pyrolysis parameters. Renew. Sustain. Energy Rev., 57: 1126–1140.

Kapoor, A. 1995. Fungal biosorption—an alternative treatment option for heavy metal bearing wastewaters: a review. Bioresour. Technol., 53: 195–206.

Keshtkar, A.R., Mohammadi, M., and Moosavian, M.A. 2015. Equilibrium biosorption studies of wastewater U(VI), Cu(II) and Ni(II) by the brown alga *Cystoseira indica* in single, binary and ternary metal systems. J. Radioanal. Nucl. Chem., 303: 363–376.

Khosravi, M., Rakhshaee, R., and Ganji, M. 2005. Pre-treatment processes of to remove Pb(II), Cd(II), Ni(II) and Zn(II) from aqueous solution in the batch and fixed-bed reactors. J. Hazard. Mater., 127: 228–237.

Koedrith, P., Kim, H., Weon, J.-I., and Seo, Y.R. 2013. Toxicogenomic approaches for understanding molecular mechanisms of heavy metal mutagenicity and carcinogenicity. Int. J. Hyg. Environ. Health, 216: 587–598.

Kumari, V., and Tripathi, A.K. 2019. Characterization of pharmaceuticals industrial effluent using GC–MS and FT-IR analyses and defining its toxicity. Appl. Water Sci. 9: 1–8. Springer International Publishing.

Kurniawan, T.A., Chan, G.Y.S., Lo, W.-H., and Babel, S. 2006. Physico–chemical treatment techniques for wastewater laden with heavy metals. Chem. Eng. J., 118: 83–98.

Kwak, I.S., Won, S.W., Choi, S.B., Mao, J., Kim, S., Chung, B.W., and Yun, Y.S. 2011. Sorptive removal and recovery of nickel(II) from an actual effluent of electroplating industry: Comparison between *Escherichia coli* biosorbent and Amberlite ion exchange resin. Korean J. Chem. Eng., 28: 927–932.

Lata, S., Singh, P.K., and Samadder, S.R. 2015. Regeneration of adsorbents and recovery of heavy metals: a review. Int. J. Environ. Sci. Technol., 12: 1461–1478.

Latinwo, G.K., Jimoda, L.A., Agarry, S.E., and Adeniran, J.A. 2015. Biosorption of some heavy metals from textile wastewater by green seaweed biomass. Univers. J. Environ. Res. Technol., 5: 210–219.

Lee, K.Y., Kim, K.W., Baek, Y.J., Chung, D.Y., Lee, E.H., Lee, S.Y., and Moon, J.K. 2014. Biosorption of uranium(VI) from aqueous solution by biomass of brown algae *Laminaria japonica*. Water Sci. Technol., 70: 136–143.

Liu, C., Fiol, N., Poch, J., and Villaescusa, I. 2016. A new technology for the treatment of chromium electroplating wastewater based on biosorption. J. Water Process Eng., 11: 143–151. Elsevier Ltd.

Liu, T., Yang, X., Wang, Z.-L., and Yan, X. 2013. Enhanced chitosan beads-supported Fe0-nanoparticles for removal of heavy metals from electroplating wastewater in permeable reactive barriers. Water Res., 47: 6691–6700.

Lofrano, G., Carotenuto, M., Gautam, R.K., and Chattopadhyaya, M.C. 2014. Heavy Metals in Tannery Wastewater and Sludge: Environmental Concerns and Future Challenges. pp. 249–260. *In*: Sanjay Sharma [ed.]. Heavy metals in water. Royal Society of Chemistry, Cambridge.

Malik, A. 2004. Metal bioremediation through growing cells. Environ. Int., 30: 261–278.

Martín-Lara, M.A., Blázquez, G., Trujillo, M.C., Pérez, A., and Calero, M. 2014. New treatment of real electroplating wastewater containing heavy metal ions by adsorption onto olive stone. J. Clean. Prod., 81: 120–129.

Michalak, I., Chojnacka, K., and Witek-Krowiak, A. 2013. State of the art for the biosorption process—a review. Appl. Biochem. Biotechnol., 170: 1389–1416.

Mirza, S.S., Eida, M., Jabeen, F., Iqtedar, M., Mahmood, A., Akmal, M., and Sabir, M. 2021. Biosorption of chromium from tannery effluent using carbon-activated algae granules of *Chlorella vulgaris* and *Scenedesmus obliquus*. Int. J. Environ. Sci. Technol., doi: 10.1007/s13762-020-03033-z. Springer Berlin Heidelberg.

Mo, J., Yang, Q., Zhang, N., Zhang, W., Zheng, Y., and Zhang, Z. 2018. A review on agro-industrial waste (AIW) derived adsorbents for water and wastewater treatment. J. Environ. Manage., 227: 395–405. Elsevier.

Modak, J.M., and Natarajan, K.A. 1995. Biosorption of metals using nonliving biomass—A review. Mining, Metall. Explor., 12: 189–196.

Moino, B.P., Costa, C.S.D., da Silva, M.G.C., and Vieira, M.G.A. 2017. Removal of nickel ions on residue of alginate extraction from *Sargassum filipendula* seaweed in packed bed. Can. J. Chem. Eng., 95: 2120–2128.

Mosa, A., El-Ghamry, A., and Tolba, M. 2018. Functionalized biochar derived from heavy metal rich feedstock: Phosphate recovery and reusing the exhausted biochar as an enriched soil amendment. Chemosphere, 198: 351–363.

Muniraj, K., Raju, G., Asha, B., and Manikandan, G. 2020. Citrus lemon leaf powder as a biosorbent for the removal of liquid phase toxic metals from textile effluent. Desalin. Water Treat, 196: 422–432.

Naja, G.M., and Volesky, B. 2017. Treatment of metal-bearing effluents: Removal and recovery. Handb. Adv. Ind. Hazard. Wastes Manag., 1067–1112.

Nishikawa, E., da Silva, M.G.C., and Vieira, M.G.A. 2018. Cadmium biosorption by alginate extraction waste and process overview in Life Cycle Assessment context. J. Clean. Prod., 178: 166–175.

Nithya, R., and Sudha, P.N. 2017. Removal of heavy metals from tannery effluent using chitosan-g-poly(butyl acrylate)/bentonite nanocomposite as an adsorbent. Text. Cloth. Sustain, 2: 1–8. Textiles and Clothing Sustainability.

Okoya, A.A., Adenekan, A., Ajadi, F.A., and Ayodele, S.O. 2020. Assessment of chitosan coated *Aspergillusniger* as biosorbent for dye removal and its impact on the heavy metal and physicochemical parameters of textile wastewater. African J. Environ. Sci. Technol., 14: 281–289.

Park, D., Yun, Y.S., and Park, J.M. 2010. The past, present, and future trends of biosorption. Biotechnol. Bioprocess Eng., 15: 86–102.

Park, S., and Lee, M. 2017. Removal of copper and cadmium in acid mine drainage using Ca-alginate beads as biosorbent. 21: 373–383.

Petrangeli, G. 2020. Radioactive waste. pp. 287–290 in Nuclear Safety. Elsevier.

Pondja, E.A.J., Bashitialshaaer, R., Persson, K.M., and Matsinhe, P. 2017. Bioadsorbents of heavy metals from coal mines area in Mozambique. Cogent Environ. Sci., 3: 1355088. Cogent OA.

Pozdniakova, T.A., Mazur, L.P., Boaventura, R.A.R., and Vilar, V.J.P. 2016. Brown macro-algae as natural cation exchangers for the treatment of zinc containing wastewaters generated in the galvanizing process. J. Clean. Prod., 119: 38–49. Elsevier Ltd.

Pyrzynska, K. 2019. Removal of cadmium from wastewaters with low-cost adsorbents. J. Environ. Chem. Eng., 7: 102795.

Qin, H., Hu, T., Zhai, Y., Lu, N., and Aliyeva, J. 2020. The improved methods of heavy metals removal by biosorbents: A review. Environ. Pollut., 258: 113777.

Rahman, M.L., Fui, C.J., Sarjadi, M.S., Arshad, S.E., Musta, B., Abdullah, M.H., Sarkar, S.M., and O'Reilly, E.J. 2020. Poly(amidoxime) ligand derived from waste palm fiber for the removal of heavy metals from electroplating wastewater. Environ. Sci. Pollut. Res., 27: 34541–34556. Environmental Science and Pollution Research.

Ramírez-paredes, F.I., Manzano-muñoz, T., Garcia-prieto, J.C., Zhadan, G.G., Shnyrov, V.L., Kennedy, J.F., Roig, M.G., and Kennedy, J.F. 2011. Biosorption of heavy metals from acid mine drainage onto biopolymers (Chitin and α (1, 3) β -D-glucan) from Industrial Biowaste Exhausted Brewer's Yeasts (*Saccharomyces cerevisiae* L.), 1272: 1262–1272.

Rangabhashiyam, S., Nandagopal, M.S.G., Nakkeeran, E., and Selvaraju, N. 2016. Adsorption of hexavalent chromium from synthetic and electroplating effluent on chemically modified Swietenia mahagoni shell in a packed bed column. Environ. Monit. Assess. 188. Environmental Monitoring and Assessment.

Robalds, A., Naja, G.M., and Klavins, M. 2016. Highlighting inconsistencies regarding metal biosorption. J. Hazard. Mater., 304: 553–556.

Rojas, S., and Horcajada, P. 2020. Metal–organic frameworks for the removal of emerging organic contaminants in water. Chem. Rev., 120: 8378–8415.

Saadi, R., Saadi, Z., Fazaeli, R., and Fard, N.E. 2015. Monolayer and multilayer adsorption isotherm models for sorption from aqueous media. Korean J. Chem. Eng., 32: 787–799.

Saeed, A., and Iqbal, M. 2003. Bioremoval of cadmium from aqueous solution by black gram husk (*Cicer arientinum*). Water Res., 37: 3472–3480.

Salem, N.M., and Awwad, A.M. 2014. Biosorption of Ni(II) from electroplating wastewater by modified (*Eriobotrya japonica*) loquat bark. J. Saudi Chem. Soc., 18: 379–386.

Schiewer, S., and Volesky, B. 1995. Modeling of the proton-metal ion exchange in biosorption. Environ. Sci. Technol., 29: 3049–3058.

Schiewer, S., and Volesky, B. 1997. Ionic strength and electrostatic effects in biosorption of divalent metal ions and protons. Environ. Sci. Technol., 31: 2478–2485.

Schiewer, S., and Balaria, A. 2009. Biosorption of Pb2+ by original and protonated citrus peels: Equilibrium, kinetics, and mechanism. Chem. Eng. J., 146: 211–219.

Šćiban, M., Radetić, B., Kevrešan, Ž., and Klašnja, M. 2007. Adsorption of heavy metals from electroplating wastewater by wood sawdust. Bioresour. Technol., 98: 402–409.

Sengupta, A., and Gupta, N.K. 2017. MWCNTs based sorbents for nuclear waste management: A review. J. Environ. Chem. Eng., 5: 5099–5114.

Sibi, G. 2016. Biosorption of chromium from electroplating and galvanizing industrial effluents under extreme conditions using *Chlorella vulgaris*. Green Energy Environ., 1: 172–177.

Srivastava, S., Agrawal, S.B., and Mondal, M.K. 2015. Biosorption isotherms and kinetics on removal of Cr(VI) using native and chemically modified *Lagerstroemia speciosa* bark. Ecol. Eng., 85: 56–66.

Suganya, E., Saranya, N., Sivaprakasam, S., Varghese, L.A., and Narayanasamy, S. 2020. Experimentation on raw and phosphoric acid activated *Eucalyptuscamadulensis* seeds as novel biosorbents for hexavalent chromium removal from simulated and electroplating effluents. Environ. Technol. Innov., 19: 100977. Elsevier B.V.

Sundararaju, S., Manjula, A., Kumaravel, V., Muneeswaran, T., and Vennila, T. 2020. Biosorption of nickel ions using fungal biomass *Penicillium* sp. MRF1 for the treatment of nickel electroplating industrial effluent. Biomass Convers. Biorefinery, doi: 10.1007/s13399-020-00679-0. Biomass Conversion and Biorefinery.

Torres, E. 2020. Biosorption: A review of the latest advances. Processes, 8: 1–23.

Tsezos, M. 1985. The selective extraction of metals from solution by micro-organisms. A Brief Overview. Can. Metall. Q., 24: 141–144.

United States Environmental Protection Agency. 2018. Drinking Water Standards and Health Advisories.

Valković, V. 2019. The nuclear fuel cycle. pp. 329–396. *In*: Valković, V. [ed.]. Radioactivity in the Environment. Elsevier.

Veglio', F., and Beolchini, F. 1997. Removal of metals by biosorption: a review. Hydrometallurgy, 44: 301–316.

Vieira, L.C., de Araujo, L.G., de Padua Ferreira, R.V., da Silva, E.A., Canevesi, R.L.S., and Marumo, J.T. 2019. Uranium biosorption by *Lemna* sp. and *Pistia stratiotes*. J. Environ. Radioact., 203: 179–186. Elsevier.

Volesky, B. 1987. Biosorbents for metal recovery. Trends Biotechnol., 5: 96–101.

Volesky, B. and Holan, Z.R. 1995. Biosorption of heavy metals. Biotechnol. Prog., 11: 235–250.

Volesky, B. 2001. Detoxification of metal-bearing effluents: biosorption for the next century. Hydrometallurgy, 59: 203–216.

Volesky, B. 2007. Biosorption and me. Water Res., 41: 4017–29.

Wai, C.M. 2011. Emerging separation techniques: Supercritical fluid and ionic liquid extraction techniques for nuclear fuel reprocessing and radioactive waste treatment. Woodhead Publishing Limited.

Wang, J., and Chen, C. 2009. Biosorbents for heavy metals removal and their future. Biotechnol. Adv., 27: 195–226. Elsevier Inc.

World Health Organization. 2008. Guidelines for Drinking-Water Quality.

World Nuclear Association. 2020. World Nuclear Performance Report 2020 Produced by: World Nuclear Association, 1–68.

Wu, P., Jiang, L.Y., He, Z., and Song, Y. 2017. Treatment of metallurgical industry wastewater for organic contaminant removal in China: status, challenges, and perspectives. Environ. Sci. Water Res. Technol., 3: 1015–1031.

Yahya, M.D., Abubakar, H., Obayomi, K.S., Iyaka, Y.A., and Suleiman, B. 2020. Simultaneous and continuous biosorption of Cr and Cu (II) ions from industrial tannery effluent using almond shell in a fixed bed column. Results Eng., 6: 100113. Elsevier Ltd.

Yakout, S.M., and Rizk, M.A. 2015. Adsorption of uranium by low-cost adsorbent derived from agricultural wastes in multi-component system. Desalin. Water Treat., 53: 1917–1922.

Yang, J., and Volesky, B. 1999. Cadmium biosorption rate in protonated sargassum biomass. Environ. Sci. Technol., 33: 751–757.

Zhang, T., Tu, Z., Lu, G., Duan, X., Yi, X., Guo, C., and Dang, Z. 2017. Removal of heavy metals from acid mine drainage using chicken eggshells in column mode. J. Environ. Manage., 188: 1–8. Elsevier Ltd.

Zhou, L., Wang, Y., Zou, H., Liang, X., Zeng, K., Liu, Z., and Adesina, A.A. 2016. Biosorption characteristics of uranium(VI) and thorium(IV) ions from aqueous solution using CaCl2-modified Giant Kelp biomass. J. Radioanal. Nucl. Chem., 307: 635–644. Springer Netherlands.

Zinicovscaia, I., Yushin, N., Abdusamadzoda, D., Grozdov, D., and Shvetsova, M. 2020. Efficient removal of metals from synthetic and real galvanic zinc–containing effluents by brewer's yeast *Saccharomyces cerevisiae*. Materials (Basel), 13: 3624.

10

Ionic Liquids Applied in Removal of Toxic Metals from Water and Wastewater

Carolina Elisa Demaman Oro, Victor de Aguiar Pedott, Rogério Marcos Dallago and *Marcelo Luis Mignoni**

1. Introduction

Water contamination by toxic heavy metals is an alarming environmental problem. Toxic heavy metals are non-biodegradable contaminants that can be present in agricultural soil and water bodies through chemical and industrial waste and effluent from human activities. These metals include mercury (Hg), cadmium (Cd), copper (Cu), chromium (Cr), nickel (Ni), arsenic (As), zinc (Zn), and lead (Pb). There are several techniques proposed and used for heavy metal removal from the environment, such as chemical precipitation, adsorption, ion-exchange, membrane separation, and electrochemical treatment. However, mentioned techniques have process limitations and are often unable to remove heavy metal from wastewaters (Stojanovic and Keppler 2012).

Therefore, Ionic Liquids (ILs) can be used for the electrodeposition of metals that, until now, were impossible to reduce in aqueous solutions and for the dissolution of several metals and alloys. In addition, they have the ability to design redox chemistry and control the nucleation characteristics of the metal. Ionic liquids have large potential windows, high solubility of metal salts, avoid metal/water and water chemistry, and high conductivity compared to non-aqueous solvents that make them ideal for metal processing (Abbott et al. 2008).

Department of Food Engineering, URI Erechim, 1621 Sete de Setembro Av., Fátima, Erechim - RS, 99709-910, Brazil.
Emails: carolinae.oro@hotmail.com; v.a.pedott@gmail.com; dallago@uricer.edu.br
* Corresponding author: mignoni@uricer.edu.br

Ionic liquids are organic salts with a melting point below 100°C and therefore do not show solvent loss through evaporation, and can be considered green substitutes for conventional organic solvents, which are often toxic, flammable and volatile. Specific combinations of cations and anions are performed for the synthesis of the ionic liquid or solid of interest with unique properties, with dissolution and solubility adjusted according to the final objective (Abebe et al. 2020, Markiewicz et al. 2013). ILs allow applications in chemical and biochemical processes, such as chemical synthesis, biocatalytic transformations, electrochemical device designs, biosensors, analytic devices and separation processes (Toledo Hijo et al. 2016, Verma et al. 2019, Zhang et al. 2017), catalysts (Hao et al. 2019, Yulin et al. 2017), structure-directing agents (Duan et al. 2018, Kumar et al. 2019), lubrificants (Cooper et al. 2019, Kawada et al. 2018) and surfactants (Kapitanov et al. 2019, Vaid et al. 2018). The main attraction of ILs is in their structure, which can be controlled and altered to produce the desired chemical properties for the given process. The selectivity can be improved by the introduction of functional groups that confer specific chemical functionalities through combination of cations and anions which improves the ILs' extraction capacities (Gionfriddo et al. 2018), by the appropriate selection of cations and anions. Thus, the ionic liquid is a task specific solvent due to its exceptional selectivity (Płotka-Wasylka et al. 2017).

While heavy metal ions are harmful to the environment and cause disease when they accumulate in the human body, ionic liquids are compatible with environmental standards. In addition to ILs being less toxic compared to organic solvents, they have low vapor pressure and most importantly, they are recyclable (Dièye et al. 2020). Figure 1 shows a summary of the issues to be addressed in this chapter in relation to the use of ionic liquids applied to remove toxic heavy metals from water and wastewater.

Given the above, it is important to develop and apply reliable and economical methods for the removal of toxic heavy metals from water and to protect the

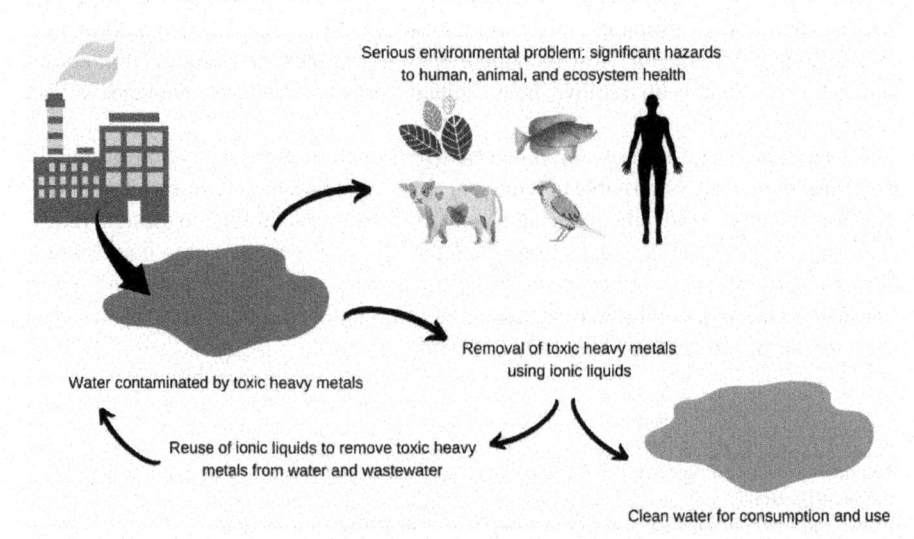

Fig. 1: Summary of topics to be covered in this chapter.

environment. This chapter covers physicochemical and toxicological parameters of toxic metals, emphasizing the removal of these metals using ionic liquids.

2. General properties and characteristics of toxic metals and ionic liquids

The toxicity of heavy metals in water is determined by many factors, such as pH, temperature, redox potential and ionic conditions in the test solution (Eom et al. 2020). Heavy metals are bioaccumulative in aquatic systems and therefore have been the subject of research and monitoring. Furthermore, they are non-biodegradable and exhibit biomagnification.

The literature reports a heavy metal as a metal group or metalloid element with a density greater than 4 g/cm^3 (Vardhan et al. 2019) or 5 g/cm^3 (Sorouraddin et al. 2020), or between 3.5 to 7 g/cm^3 (Akindele et al. 2020) and which is toxic in low concentrations ($\mu g/L$).

Heavy metals are naturally occurring elements in the environment, but their concentrations can also increase due to the inadequate disposal of domestic, industrial and agricultural effluents. In the last decades, human exposure to toxic heavy metals has increased significantly as a result of the use of these metals in different industrial processes (Sorouraddin et al. 2020). Thus, water pollution by toxic metals is basically due to the increase in industrialization and urbanization.

Due to all the problems that heavy metals can cause, their removal from the environment with an appropriate destination is desirable. Therefore, ionic liquids have been used as solvents to improve the extraction conditions of classic processes. Ionic liquids provide selective removal of metals depending on operating conditions, such as time of extraction and balance, metal charge, percentage of water and type of hydrated fusion. Ionic liquids are considered ecological solvents with low vapor pressure, in addition to excellent thermal stability over a wide range of process temperatures (Sethurajan et al. 2019).

3. Toxic metals pollution in water and wastewater

Pollution or contamination of water by heavy metals is characterized by its toxicity and strong bioconcentration. Efficient removal of heavy metals from water is of great environmental and health importance. When these metals are in high concentration in water, they can accumulate in living organisms through food or direct contact (Zhang et al. 2020). However, detection and removal of these heavy metals from the environment is still a major challenge.

Heavy metals generally associated with water pollution are arsenic (As), lead (Pb), chromium (Cr), cadmium (Cd), iron (Fe) and vanadium (V). These metals can cause brain damage, cancer, and system disorders, even at low levels in domestic water (Oyewo et al. 2020).

The most common metals that can accumulate in organisms are cadmium (Cd), copper (Cu), lead (Pb), nickel (Ni), and zinc (Zn). When accumulated, they cause numerous diseases and disorders. In addition, they are considered the most dangerous toxic heavy metals for the environment. These heavy metals can accumulate in the human body indirectly, mainly due to food and water intake, resulting in health disorders (El Saidy et al. 2020). The main characteristics of each of these metals are presented below.

Cadmium is an extremely harmful metal, which can accumulate in the human body and ·cause irreversible damage. In the human body, the accumulation of cadmium leads to certain disorders in the bones, liver, kidneys, and nervous system. The accumulation of this metal can also cause cardiovascular disease (Ahmad et al. 2019, Vardhan et al. 2019).

Copper is an essential element for humans, as it is necessary in the creation of hemoglobin in red platelets and is a micronutrient for humans. However, consumption of food or water contaminated with large amounts of copper can cause severe gastrointestinal problems (Vardhan et al. 2019).

Studies reported in the literature have shown that dermal, inhalation and oral exposure to lead can seriously affect the human nervous system and body development, especially in children and fetuses. For children, lead can affect brain development, decrease red blood cells, and slow reflexes and learning. In adults, it can cause spontaneous abortions, increase pressure, and cause neurological damage (Arshad et al. 2020, Wu et al. 2020).

Nickel is considered one of the dangerous heavy metals used in the industry. It is toxic to living organisms at low concentration levels (values higher than those reported in Table 1) and is classified as a carcinogenic compound. Nickel has been reported to cause chronic problems related to the respiratory tract, lung cancer, and skin dermatitis (El Sheikh et al. 2020).

Zinc is a basic micronutrient and a considerably less dangerous metal. Its main application is in the galvanization of iron and steel items. In drinking water, Zn is usually present in the form of salts or organic complexes (Liu et al. 2019).

Based on these heavy metals and their possible health complications, the maximum values allowed for the presence of metals by the Brazilian Legislation for the discharge of effluents (Conselho Nacional do Meio Ambiente - CONAMA 2011) (430/2011), groundwater to be used for irrigation (Conselho Nacional do Meio Ambiente - CONAMA 2008) (396/2008), and the potability standard for chemicals that pose a health risk (Brazil 2004) (518/2004) are shown in Table 1.

Effluents and untreated industrial wastewater, as well as the use of agricultural fertilizers and pesticides are some of the main contributors to the increase in the concentration of toxic metals in the environment.

4. Physicochemical and toxicological parameters

Heavy metals like copper, zinc, nickel, boron, iron, and molybdenum are essential for plant growth, while lead, mercury, cadmium, and arsenic are not essential. However, even the metals considered essential are toxic and harmful in high concentrations (Vardhan et al. 2019).

The bioavailability of metals to plants result in the accumulation of metals in plant tissues. Ahmad et al. (2019) reported the potential implications and threats of using wastewater for wheat grain production. The authors evaluated the absorption of lead, cadmium, nickel, iron, manganese, copper, chromium, zinc, and cobalt

Table 1: Maximum values allowed for the presence of metals by the Brazilian Legislation for the discharge of effluents (430/2011), groundwater to be used in irrigation (396/2008), and potability standard (518/2004).

Parameter	430/2011 (mg/L)	396/2008 (mg/L)	518/2004 (mg/L)
Arsenic (As)	0.50	-	0.010
Cadmium (Cd)	0.20	0.010	0.005
Chromium (Cr)	0.10 (Cr^{+6}), 1.00 (Cr^{+3})	0.100	0.050
Copper (Cu)	1.00	0.200	2.000
Lead (Pb)	0.50	5.000	0.010
Mercury (Hg)	0.01	0.002	0.001
Nickel (Ni)	2.00	0.200	-
Zinc (Zn)	5.00	2.000	5.000

by wheat grains through irrigation with residual water. The only metal whose concentration exceeded the permitted levels by law and presented a high pollution load index was Cd, which indicated toxicity and contamination of the grains. The authors also concluded that the continued use of treated and untreated wastewater on agricultural land considerably increases the potential toxic metal content of the soil (Ahmad et al. 2019).

Several methods and indices, such as the contamination factor (CF), the geoaccumulation index (Igeo), the enrichment factor (EF), the pollution load index (PLI), and the potential ecological risk (ecological risk index, RI) have been used in order to study soil contamination by heavy metals and to assess the level of pollution. These tools are also used to assess ecological risks and possible sources of heavy metals (Hilali et al. 2020).

5. Removal of toxic metals from water and wastewater

Some industrial processes cause heavy metal to enter water systems, causing environmental problems. The water contamination and its limited availability is increasing nowadays due to population growth. Industries such as textiles, petroleum, metal finishing, automobile, electro plating and leather tanning are the most responsible for metal entries in water systems. Generally, the most encountered heavy metals in water systems are copper, mercury, zinc, lead, cadmium, iron, chromium, cobalt and nickel, where the following ranking represents the toxicity of the heavy metals: Cd > Hg > Pb Cu > Zn > Cr > Co > Fe (Buaisha et al. 2020). The toxicity of metals directly influences the metabolism of microorganisms and the nervous system and cells of human beings and animals (Ma et al. 2016, Ong et al. 2010).

Innumerous treatments for the removal of heavy metals from industrial wastewater are used, such as coagulation/flocculation, ion exchange, flotation, membrane filtration, chemical precipitation, electrochemical treatment, adsorption and biological treatments. Each treatment has advantages and disadvantages; however, the techniques applied depend on the nature of the heavy metals found in the water (Carolin et al. 2017).

In the coagulation/flocculation procedures, heavy metals are precipitated by the formation of carbonates, sulfides, and hydroxides compounds forming colloidal molecules which presents density equal to water (Ghernaout et al. 2015). In order to increase the density of colloidal particles, coagulation treatments are carried out followed by addition of flocculants that agglomerate the coagulated particles forming large particles with higher density which are separated from the solution through filtration (Carolin et al. 2017). The disadvantage of using the coagulation/flocculation technique is the sludge formation with high heavy metals concentration, due to the large utilization of chemicals (Yan et al. 2010).

Ion exchange treatments remove metal ions from water systems through designed ion exchange resins, normally made of cross-linked polymer matrixes with attached functional groups through covalent bonding (Carolin et al. 2017). For a better metal removal efficiency, acidic resins are used, due to the presence of solfonic acid groups, differently from basic resins which presents carboxilic acids as functional groups (Fu and Wang 2011).

Flotation technique consists of dispersions of positively or negatively charged bubbles into the water system where such bubbles remove the metals through migration of the metal in the water system to the bubbles, being suitable for compounds which have different physical and chemical natures (Mahmoud et al. 2015). Such process causes low sludge formation; however, the process has high costs of operation (Carolin et al. 2017).

Membrane filtration is a separation technique based on the particle's size. For metal removal, the complexation of the metal becomes necessary to achieve the required size for membrane separation, which could be microfiltration, ultrafiltration, nanofiltration or reverse osmosis (Fu and Wang 2011). Such process has high efficiency on metal removal, easy operation and compact modules (Fu and Wang 2011). However, membrane filtration has high costs of implementation, fouling and periodic replacement making its application difficult (Carolin et al. 2017).

The chemical precipitation method is cheaper and conventionally used in industries, where the chemicals are added to the solution altering the pH and causing the precipitation of metal species (Fu and Wang 2011). The chemical precipitation technique is mostly used for effluents containing high concentration of Cu (II), Cd (II), Mn (II), and Zn (II) species (Bilal et al. 2013). However, its application causes high amounts of sludge which is difficult to treat and dispose (Kuan et al. 2010).

Electrochemical techniques are normally applied as secondary treatment after precipitation or ion exchange processes (Le et al. 2009). In this technique, electrodes induce the coagulation, deposition or flotation of metals through electron shifting in the solution, being a versatile alternative for water treatment (Le et al. 2009). However, the electrochemical technique demands high maintenance, electrical energy for operation and is limited to certain treatments due to low mass transfer and increases on the effluent temperature (Zhang et al. 2013).

The adsorption process is one of the most indicated for water treatment, due to its low operation cost, low fouling problems and adsorbent regeneration being considerate as an environmentally acceptable method (Carolin et al. 2017). The adsorptions process consists of the use of an absorbent with high surface areas with attached functional groups removing the metals found in the solution through metal-functional group interactions or by physical interests (Ojedokun and Bello 2016). Innumerous types of adsorbents have been developed for metal removal, where the surface area, pore size distribution, functional groups and polarity are the mains factors for choosing the appropriate adsorbent (Ewecharoen et al. 2009).

Biological processes are used as secondary water treatment mainly in activated sludge systems offering an alternative to removal of heavy metals (Maal-bared 2020). The metal removal occurs through biosorption/metabolism-independent processes or bioaccumulation/metabolism-dependent processes (Maal-bared 2020). Biosorption processes involve the use of extracellular polymeric substances (EPS), which has great affinity for heavy metals through sites on the surface of the biopolymer, removing high concentrations of heavy metals from water systems (Binkley and Simpson 2003). Bioaccumulation processes remove heavy metals by involving the metal in microorganism cells through specific or nonspecific uptake systems, where the cell can transform the metal to a less toxic compound (Binkley and Simpson 2003, Maal-bared 2020). However, high metal concentrations in the active sludge could inactivate the microorganism due to the toxicity of the heavy metals, making its application difficult in some cases (Maal-bared 2020).

6. Ionic liquids as extraction agents for toxic metals

Removal of toxic metals using ionic liquids as an alternative solvent is a relatively new technique. Seddon and Hussey reported the first study on the dissolution of transition metal compounds in a non-aqueous polar solution (Hitchcock et al. 1986). In the subsequent decade, innumerous ionic liquids were used as alternative solvents for transition metals extraction (Crofts et al. 1999). In the present times, several ionic liquids with innumerous combinations of cations and anions have been used in metal extraction techniques, showing that most of the metals can be dissolved and extracted in a single ionic liquid or in combinations (Kim et al. 2018, Rajadurai and Lakshmi 2020).

The utilization of ionic liquids as extraction agent for toxic metals aims to be an alternative and greener way for the toxic metal removal process, when comparing to traditional methods. The ionic liquids are known by their greener characteristic as an alternative for conventional solvents due to its characteristic properties, such as low vapor pressure, good chemical and thermal stabilities and variable hydrophobicity, viscosity and solubility, depending on the cation and anion used in the synthesis process (De Los Ríos et al. 2010, Sheldon 2001).

Even with the greener advantages of using ionic liquids as an alternative way for metal extraction, some cations used in the synthesis process may represent severe risks for aquatic environment if ion exchange occurs between the ionic liquid and the contaminated effluent (Cao et al. 2018, Stojanovic and Keppler 2012). The ion exchange causes degradation of the ionic liquids species, diminishing the extraction efficiency and making the application even more expensive (Frade and Afonso 2010, Stojanovic and Keppler 2012). One way to overcome such problems is by increasing hydrophocity of the ionic liquids, by increasing the alkyl chain length of the imidazolium-based ionic liquids or by fluorination of the alkyl chain and designing ILs with suited cations and anions for acting as solvents (Dietz 2006).

The possibility of combining cations and anions for achieving different properties is the key for realizing successful metal extraction processes, improving the process efficiency (De Los Ríos et al. 2010, Gardas and Coutinho 2008). Task Specific Ionic Liquids (TSILs) are a special group of ionic liquids, which presents components with improved solvent properties of classical ionic liquids with solvents already used for metal extraction such as toluene, kerosene and other organic solvents (Stojanovic and Keppler 2012). TSILs act as organic phase and extracting agent avoiding problems such as extractant/solvent miscibility, loss of chelating agent to the aqueous phase and anion exchange mechanism, improving metal extraction and solvent recovery (Lee and Lee 2006, Stojanovic and Keppler 2012).

Abebe et al. (2020) reported a new ionic liquid intentionally designed for biphasic extraction. The ionic liquid used for a liquid/liquid Pb^{2+} extraction from a neutral aqueous phase was N-hexyl-4,4-bipyridinium bis(trifluoromethylsulfonyl) imide([C6byp][Tf2 N]). The ionic liquid demonstrated 98.16% of removal of the metal ion from the aqueous phase and was efficient for at least four cycles with undiminished efficiency (Abebe et al. 2020). The authors demonstrated that the extraction and reuse of the ionic liquid is viable and efficient. Other studies have reported the recovery of the ionic liquid to be able to be reused, as in the case of tri(n-butyl)[2-ethoxy-2-oxoethyl-ammonium](dicyanamide) [BuGBOEt][Dca], which showed a high extraction efficiency (higher than 90%) towards Cu(II), Ni(II), Cd(II), and Pb(II) and was recovered by aqueous EDTA solutions (Zhou et al. 2015). El Sheikh et al. (2020) evaluated trace quantities of nickel Ni(II) ion in water, food and tobacco samples using (1-hexyl-3-methylimidazoliumtris (pentafluoroethyl) trifluoro-phosphate [HMIM][FAP]) as an extraction solvent and quinalizarin (Quinz) as a complexing agent. The author evaluated the influence of innumerous parameters in the extraction procedure such as the amount of ILs, pH, ultrasonification time, sample volume and ions influence, reaching high extraction yields (higher than 95%) at ideal extraction conditions. Table 2 presents the ionic liquids generally applied in the removal of different metals.

7. Main application of ionic liquids as extraction agent

7.1 Liquid-liquid extraction

Ionic liquids are involved in different metal extraction processes, like liquid-liquid extraction, adsorption and membranes processes, where ionic liquids improve the process efficiency and decrease the toxicity caused by the decomposition of classical solvents.

Liquid-Liquid extraction consists in a separation of components by its distribution between two immiscible liquids phases, where the process takes place in mild conditions. Normally, organic solvents were used for extraction processes, which could contaminate the aquatic environment and recovery of the spent organic solvent is difficult (Rajadurai and Lakshmi 2020). To overcome such problems, ionic liquids have been used as alternative solvents for metal extraction (Kim et al. 2018). Generally, the extraction process using ionic liquids is carried out through shaking the metal containing aqueous phase with an immiscible ionic liquid phase; after the extraction process, the ionic liquids with extracted metals are recovered

Table 2: Ionic liquids commonly used in the removal of different heavy metals.

Metal	Ionic Liquid	References
Hg	Imidazolium Based 1-alkyl-3 methylimidazolium, $C_n mim^+$ (n = 4, 6, 8) salts of PF_6	(Guezzen and Amine Didi 2016, Visser et al. 2001)
	Imidazolium Based with BF_4 Tf_2N anions	(Vincenza et al. 2013)
Cd	Imidazolium Based 1-alkyl-3 methylimidazolium, $C_n mim^+$ (n = 4, 6, 8) salts of PF_6	(Visser et al. 2001)
	Thio-glycolic acidR (TiOAC)	(Alguacil et al. 2016)
	Phosphonium Based Ionic Liquids	(Swain et al. 2016)
	Amonium Based Ionic Liquids	(Swain et al. 2016)
Zn	Phosphonium Based Ionic Liquids	(Regel-Rosocka 2009, Regel-Rosocka et al. 2012)
	Imidazolium Based Ionic Liquids	(Wojciechowska et al. 2018)
Fe	Phosphonium Based Ionic Liquids	(Wiśniewski 2013)
	Amonium Based Ionic Liquids	(Swain et al. 2016)
Cr	Imidazolium Based Ionic Liquids	(Goyal et al. 2011, Liu et al. 2010, Zambare and Nemade 2021)
	Butylammonium Based Ionic Liquids	(Eliodório et al. 2021)
	Phosphonium Based ionic Liquids	(Liu et al. 2010)
	Tricaprylmethyl ammonium thiosalicylate	(Rajendran, 2010)
As	Imidazolium Based Ionic Liquids	(Zhang et al. 2019)
As, Cr, Cd, Cu, Zn, Pb and Hg	Phosphonium Based Ionic Liquid	(Thasneema et al. 2021)
Pb, Cd and Zn	Pyridine based ionic liquid	(Wieszczycka et al. 2021)
Cd, Ni, and Zn	Imidazolium Based Ionic Liquids	(Malas et al. 2020)
Zn, Cd, Cu and Fe	Imidazolium and Ammonium-based Ionic Liquids	(De Los Ríos et al. 2012)

using stripping agents (HCl, ammonia based compounds, supercritical CO_2, among others) and reused for other extraction processes (Maria Antonieta Valdés et al. 2014, Rajadurai and Lakshmi 2020).

For metal extraction using ionic liquids as solvents, the coordination bond between the metal ion and ionic liquid has extreme importance in the extraction efficiency, leading to ion pair, electrostatic and hydrogen bond interactions, where the metals are transferred to the ionic liquid phase through ion exchange or solvation mechanisms (Makanyire et al. 2016).

The transport properties in ionic liquids are influenced by hydrophobicity, where the hydrophobic anions are highly responsible for ion exchange while using ionic liquids with hydrophilic cations, leading to an increase in the process efficiency (Rout and Binnemans 2015). The ion exchange isn't the best way for metal removal, due to the fact that some cations found in ionic liquids could be toxic for the aqueous phase, so, for less aggressive metal extraction, designed ionic liquids with non-toxic cations and anions could be used in metal extraction processes (Maria Antonieta Valdés et al. 2014). Functional groups such as thiourea, thioether, phosphate, phosphine, etc. are used in task specific ionic liquids, increasing its affinity toward metal extraction, leading to an increase in the process efficiency (Mehdi et al. 2010, Vincenza et al. 2013).

Solvation ability of ionic liquids allows the migration of metal species toward the ionic liquid phase; however, the weak coordination anions found in ionic liquids cause a decrease of the metal solubility in ionic liquid phase; to overcome such problems, chelating agents are added in the syntheses of ionic liquids to form complexes with the solvated metals (Janssen et al. 2016). The use of chelating agents can cause some problems in the extraction process, like degradation of ionic liquids or ligands and decrease in the number of extraction cycles which makes the process more expensive (Rajadurai and Lakshmi 2020, Stojanovic and Keppler 2012).

The liquid-liquid extraction using some ionic liquids can be improved by heating the solvents' solutions, converting the solution to just one homogeneous phase, which speeds up the metal extraction and the metal containing ionic liquid could be recovered by cooling the solution. As demonstrated by Hoogerstraete (2013), in the extraction of $Cu2+$, $In2+$ and trivalent rare earth ions, using bis-(trifluoromethylsulfonyl)imid [Hbet][Tf2N] ionic liquid, reaching extraction efficiency greater than 95% at best conditions.

Since the discovery of use of ionic liquids as an alternative solvent for metal extraction, many studies have been developed on the subject, where imidazolium, ammonium and phosphonium based ionic liquids are the most used for metal extraction with high efficiency (> 90%) (Depuyt et al. 2017, Khodakarami and Alagha 2020, Kilicarslan et al. 2017, Rzelewska-piekut and Regel-rosocka 2019).

7.2 Adsoption

Adsorption is the main process for metal extraction when using porous materials, where the solute is adsorbed on the surface of adsorbent. The ionic liquids could be used as surfactant, template and crystal growth modifier in the synthesis of different materials as zeolites, mesoporous silicas, activated carbon, metal oxides, clays and activated alumina (Gao et al. 2017, Haouas et al. 2014, Li et al. 2019, Sachse et al. 2015, Zhou and Antonietti 2004). Ionic liquids form extended hydrogen bonds with the materials precursors forming different porous structured materials (Singh and Savoy 2020).

The ionic liquids could act as immobilized matter on inorganic supports as an alternative for their application on metal removal processes, avoiding some drawbacks found on its application in liquid-liquid extraction such as high viscosity, causing problems in mass diffusion, low interface area and large amount of ionic liquid used in liquid-liquid extraction (Lili et al. 2012, Sun et al. 2008).

The advantages of binding interaction between adsorbents and ionic liquids are listed as high interface area, short diffuse distance, lower quantities of ionic liquid used and accelerated transport rate (Lili et al. 2012). The use of ionic liquids leads to a better extraction efficiency of these materials by forming a strong structured interface on the surface of materials, inducing metal adsorption (Ekka et al. 2017, Khulbe and Matsuura 2018). Ionic liquids with functional groups such as imidamide, amine, oxime and hydroxyl increase the metal encapsulating efficiency onto aqueous environments (Aksamitowski et al. 2020).

A wide range of materials with immobilized ionic liquids or with ionic liquids acting as supports could be used for adsorption processes, where silica materials and polymer resins with supported ionic liquids are the most promising materials for metal adsorption processes (Lili et al. 2012, Rajadurai and Lakshmi 2020). Functionalized or chemical modified silica materials show large porous structures, high thermal and mechanical stabilities and allow surface changes; however, its application in basic environments is limited due to degradation of Si-O-Si bonds (Lili et al. 2012, Tian et al. 2010). Polymer resins are used for immobilization of ionic liquids through pendant chloromethyl groups, which facilitate the interaction with ionic liquids in addition to its properties for adsorption processes such as low cost, mechanical robustness, chemical inertness, thermal and chemical stabilities and ready availability proving that the utilization of polymeric resins with supported ionic liquids shows a better process efficiency (Alexandratos and Zhu 2003, Tian et al. 2019).

Innumerous examples of adsorbents modified with ionic liquids are shown in literature, where the most common techniques are impregnation, dry method, column method, wet method and sol-gel method including doping and grafting (Rajadurai and Lakshmi 2020). In grafting technique, the ionic liquids are covalently bonded to the material and in doping process, the molecules interact with positively charged metal ion through electrostatic attraction (Ekka et al. 2017, Rajadurai and Lakshmi 2020).

Studies on the removal of toxic metals using the adsorption process were reported in the literature, for the removal of lead, copper, nickel, chromium, yttrium, mercury, cadmium, among other heavy metals, evidencing that the process could be applied to extraction and adsorption processes (Aksamitowski et al. 2020, Ekka et al. 2017, Navarro et al. 2014, Sun et al. 2016, Zhu et al. 2009).

7.3 Membranes

Over the years, membrane separation processes have been growing in the industrial environment, being one of the most effective methods for selective recovery of solutes from aqueous solutions. Innumerous membranes are used for separation processes such as liquid membranes, polymer inclusion membranes, bulk liquid membranes and emulsion liquid membranes (Yan et al. 2019).

Ionic liquids combined with membranes can use the solvent properties of ionic liquids along with membrane separation performances specifically for supported liquid membrane processes (Makanyire et al. 2016). The ionic liquid acts as carrier agent making the membrane selectively permeable allowing specific solutes to pass through; this process is called facilitated transport (Rajadurai and Lakshmi 2020). This process using classical carriers is limited for application on large scale due to low stability, insufficient lifetime and aging (Jean et al. 2018). To overcome such limitations, ionic liquids could stabilize the carriers by infusing it inside the pores of the membrane bringing several benefits like high viscosity and conductivity and the possibility of using task specific ionic liquids which improve the solubility of the solutes to be removed (Jean et al. 2018).

Liquid membranes with ionic liquids for removal of heavy metals have been extensively studied in the last years, where phosphonium and ammonium based ionic liquids are the most promising for metal extraction (Baczy et al. 2018, Ozevci et al. 2018). The ammonium based ionic liquid Tricaprylmethylammonium chloride (Aliquat 336) is used as carrier agent in literature for removal of Cr^{6+}, Cr^{3+}, Cd^{2+}, Pb^{2+}, Zn^{2+} and Co^{2+} from aqueous solutions (Altin et al. 2011, Baczy et al. 2018, Kagaya et al. 2011, Kebiche-senhadji et al. 2010, Konczyk et al. 2010). Some reports could be find in literature for the removal of Pd2+, Cu2+, Zn2+, Cd2+ from aqueous solutions, using phosphonium based ionic liquids as solvents, such as Trihexyl (tetradecyl) phosphonium chloride, Trihexyl(tetradecyl)Phosphonium and bis(2,4,4-trimethylpentyl)phosphinate (Castillo et al. 2014, Mostazo et al. 2017, Pospiech 2015, Regel-rosocka et al. 2015).

For supported liquid membranes, the pseudo emulsion based hollow fiber strip dispersion technique allows a better process efficiency by increasing the membrane stability (Alguacil and Lopez 2013). The process consists of using a pseudo emulsion solution containing both organic and stripping agents, such solution being responsible for the removal of metal ions in the membrane through counter current flow with the metal containing solution (Alguacil and Lopez 2013). The advantages of using this process are: simultaneous extraction and stripping processes, low consumption of ionic liquids and energy, and large area for mass transfer in hollow fibers (Chaturabul et al. 2015).

Another technique for membrane separation is the pressure-driven, which has great capability to separate solutes based on size. This process consists of using an ionic liquids flow through membrane inner side acting as extractant, removing the metals from wastewater solution that passes onto the other side of the membrane; such technique acquires lower quantities of ionic liquids and residence time, compact installation and low energy consumption (De Los Ríos et al. 2010).

Therefore, for a better separation efficiency, phosphonium based ionic liquids are recommended for membrane applications, due to unique dissolution ability, thermal and chemical stabilities (Rajadurai and Lakshmi 2020).

8. Future outlooks and technological challenges

Ionic liquids have proven to be an efficient alternative for toxic metal removal through innumerous techniques. The improvement of such techniques using ionic liquids have been studied along the years, aiming at reducing costs of ionic liquids and increasing its removal efficiency. Task specific ionic liquids demonstrated to be an effective way for applying processes involving ionic liquids, where ionic liquids with cations and anions, containing low toxicity, are designed for interaction with specific metal ions, avoiding problems as ion exchange of toxic cations encountered in some ionic liquids.

Removal of toxic metals from water systems using ionic liquids in industrial environment becomes necessary to develop technologies for ionic liquids' large-scale production, reducing its costs and developing new task specific ionic liquids, improving its application for different metals. Removal of ionic liquids from industrial effluents still are a drawback for it application, however, new removal techniques could provide a decrease in discard costs and possibility the reutilization of ionic liquids for others applications.

Acknowledgments

The authors thank URI-Erechim, National Council for Scientific and Technological Development (CNPq), Coordination for the Improvement of Higher Education Personnel (CAPES), and Research Support Foundation of the State of Rio Grande do Sul (FAPERGS).

References

Abbott, A.P., Ryder, K.S., and König, U. 2008. Electrofinishing of metals using eutectic based ionic liquids. Trans. Inst. Met. Finish., 86: 196–204. https://doi.org/10.1179/174591908X327590.

Abebe, A., Hilawea, K.T., Amlaku, Y., and Tamrat, B.D. 2020. Synthesis of a new ionic liquid for efficient liquid/liquid extraction of lead ions from neutral aqueous environment without the use of extractants. Cogent Chem., 6. https://doi.org/10.1080/23312009.2020.1771832.

Ahmad, K., Wajid, K., Khan, Z.I., Ugulu, I., Memoona, H., Sana, M., Nawaz, K., Malik, I.S., Bashir, H., and Sher, M. 2019. Evaluation of potential toxic metals accumulation in wheat irrigated with wastewater. Bull. Environ. Contam. Toxicol., 102: 822–828. https://doi.org/10.1007/s00128-019-02605-1.

Akindele, E.O., Omisakin, O.D., Oni, O.A., Aliu, O.O., Omoniyi, G.E., and Akinpelu, O.T. 2020. Heavy metal toxicity in the water column and benthic sediments of a degraded tropical stream. Ecotoxicol. Environ. Saf., 190: 110153. https://doi.org/10.1016/j.ecoenv.2019.110153.

Aksamitowski, P., Wieszczycka, K., and Filipowiak, K. 2020. Silica functionalized by pyridinecarboximidamides as a novel sorbent of heavy metals ions. Sep. Sci. Technol., 55: 2217–2226. https://doi.org/10.1080/01496395.2019.1614062.

Alexandratos, S.D., and Zhu, X. 2003. Amination of Poly(vinylbenzyl chloride) with. Macromolecules 36: 3436–3439. https://doi.org/10.1021/ma0215767.

Alguacil, F.J., and Lopez, F.A. 2013. Journal of industrial and engineering chemistry modeling of facilitated transport of Cr (III) using (RNH 3 + HSO 4 Å) ionic liquid and pseudo-emulsion hollow fiber strip dispersion (PEHFSD) technology. J. Ind. Eng. Chem., 19: 1086–1091. https://doi.org/10.1016/j.jiec.2012.12.003.

Alguacil, F.J., López, F.A., García-Díaz, I., and Rodriguez, O. 2016. Cadmium(II) transfer using (TiOAC) ionic liquid as carrier in a smart liquid membrane technology. Chem. Eng. Process. Process Intensif., 99: 192–196. https://doi.org/10.1016/j.cep.2015.06.007.

Altin, S., Alemdar, S., Altin, A., Yildirim, Y., Alemdar, S., Altin, A., and Yildirim, Y. 2011. Facilitated Transport of Cd (II) through a supported liquid membrane with aliquat 336 as a carrier facilitated transport of Cd (II) through a supported liquid membrane with Aliquat 336 as a carrier. Sep. Sci. Technol., 46: 754–764. https://doi.org/10.1080/01496395.2010.537726.

Arshad, M., Naqvi, N., Gul, I., Yaqoob, K., Bilal, M., and Kallerhoff, J. 2020. Lead phytoextraction by Pelargonium hortorum: Comparative assessment of EDTA and DIPA for Pb mobility and toxicity. Sci. Total Environ., 748: 141496. https://doi.org/10.1016/j.scitotenv.2020.141496.

Baczy, M., Waszak, M., Nowicki, M., Borysiak, S., and Regel-rosocka, M. 2018. Characterization of polymer inclusion membranes (PIMs) containing phosphonium ionic liquids as Zn (II) carriers

Poznan University of Technology, Faculty of Chemical Technology, Institute of Chemical. Ind. Eng. Chem. Res., 1–57. https://doi.org/10.1021/acs.iecr.7b04685.

Bilal, M., Shah, J.A., Ashfaq, T., Gardazi, S.M.H., Tahir, A.A., Pervez, A., Haroon, H., and Mahmood, Q., 2013. Waste biomass adsorbents for copper removal from industrial wastewater-A review. J. Hazard. Mater., 263: 322–333. https://doi.org/10.1016/j.jhazmat.2013.07.071.

Binkley, J., and Simpson, J. 2003. The handbook of water and wastewater microbiology. Elsevier Publishing, London.

Brazil, M. da S. 2004. Portaria N° 518, de 25 de Março de 2004. Brazil.

Buaisha, M., Balku, S., and Özalp, Ş. 2020. Heavy metal removal investigation in conventional activated sludge systems. Civ. Eng. J., 6: 470–477.

Cao, L., Zhu, P., Zhao, Y., and Zhao, J. 2018. Using machine learning and quantum chemistry descriptors to predict the toxicity of ionic liquids. J. Hazard. Mater., 352: 17–26. https://doi.org/10.1016/j.jhazmat.2018.03.025.

Carolin, C.F., Kumar, P.S., Saravanan, A., Joshiba, G.J., and Naushad, M. 2017. Efficient techniques for the removal of toxic heavy metals from aquatic environment : a review. Biochem. Pharmacol., 5: 2782–2799. https://doi.org/10.1016/j.jece.2017.05.029.

Castillo, J., Teresa, M., Fortuny, A., Navarro, P., Sepúlveda, R., and María, A. 2014. Hydrometallurgy Cu (II) extraction using quaternary ammonium and quaternary phosphonium based ionic liquid. Hydrometallurgy, 141: 89–96. https://doi.org/10.1016/j.hydromet.2013.11.001.

Chaturabul, S., Srirachat, W., Wannachod, T., and Ramakul, P. 2015. Separation of mercury (II) from petroleum produced water via hollow fiber supported liquid membrane and mass transfer modeling. Chem. Eng. J., 265: 34–46. https://doi.org/10.1016/j.cej.2014.12.034.

Conselho Nacional do Meio Ambiente - CONAMA, 2011. Resolução N° 430, de 13 de Maio de 2011. Brazil.

Conselho Nacional do Meio Ambiente - CONAMA, 2008. Resolução N° 396, de 3 de Abril de 2008. Brazil. https://doi.org/10.1016/s0021-5198(19)65933-0.

Cooper, P.K., Staddon, J., Zhang, S., Aman, Z.M., Atkin, R., and Li, H. 2019. Nano- and macroscale study of the lubrication of titania using pure and diluted ionic liquids. Front. Chem., 7: 1–9. https://doi.org/10.3389/fchem.2019.00287.

Crofts, D., Dyson, P.J., Sanderson, K.M., Srinivasan, N., and Welton, T. 1999. Chloroaluminate (III) ionic liquid mediated synthesis of transition metal – cyclophane ; complexes : their role as solvent and Lewis acid. J. Organomet. Chem., 573: 292–298.

Dai, C., Zhang, J., Huang, C., and Lei, Z. 2017. Ionic liquids in selective oxidation: catalysts and solvents. Chem. Rev., 117: 6929–6983. https://doi.org/10.1021/acs.chemrev.7b00030.

De Los Ríos, A.P., Hernández-Fernández, F.J., Lozano, L.J., Sánchez, S., Moreno, J.I., and Godínez, C. 2010. Removal of metal ions from aqueous solutions by extraction with ionic liquids. J. Chem. Eng. Data, 55: 605–608. https://doi.org/10.1021/je9005008.

De Los Ríos, A.P., Hernández-Fernández, F.J., Alguacil, F.J., Lozano, L.J., Ginestá, A., García-Díaz, I., Sánchez-Segado, S., López, F.A., and Godínez, C. 2012. On the use of imidazolium and ammonium-based ionic liquids as green solvents for the selective recovery of Zn(II), Cd(II), Cu(II) and Fe(III) from hydrochloride aqueous solutions. Sep. Purif. Technol., 97: 150–157. https://doi.org/10.1016/j.seppur.2012.02.040.

Depuyt, D., Van de Bossche, A., and Binnemans, K. 2017. Metal extraction with a short-chain imidazolium nitrate ionic liquid. Chem. Commun., 53: 5721–5724. https://doi.org/10.1039/c7cc01685a.

Dietz, M.L. 2006. Ionic liquids as extraction solvents: where do we stand ? Sep. Sci. Technol., 41:, 2047–2063. https://doi.org/10.1080/01496390600743144.

Dièye, E.H., Fall, A., Fall, M., Ferreira, C.A., Silveira, M.R.S., and Baldissera, A.F. 2020. Electrosynthesis and electrochemical characterisation of polypyrrole in 1-hexyl-2,3-dimethylimidazolium tetrafluoroborate and 1,2-dimethylimidazolium methylsulfate. Application to the detection of copper in aqueous solutions. Int. J. Environ. Anal. Chem., 1–13. https://doi.org/10.1080/03067319.2020.1791334.

Dong, Z., and Zhao, L. 2018. Covalently bonded ionic liquid onto cellulose for fast adsorption and efficient separation of Cr(VI): Batch, column and mechanism investigation. Carbohydr. Polym., 189: 190–197. https://doi.org/10.1016/j.carbpol.2018.02.038.

Duan, W., Li, A., Chen, Y., Zhuo, K., Liu, J., and Wang, J. 2018. Ionic liquid-assisted synthesis of reduced graphene oxide–supported hollow spherical PtCu alloy and its enhanced electrocatalytic activity toward methanol oxidation. J. Nanoparticle Res., 20. https://doi.org/10.1007/s11051-018-4400-6.

Ekka, B., Dhaka, R.S., Kishore, R., and Dash, P. 2017. Fluoride removal in waters using ionic liquid-functionalized alumina as a novel adsorbent. J. Clean. Prod., 151: 303–318. https://doi.org/10.1016/j.jclepro.2017.03.061.

El Saidy, N.R., El-Habashi, N., Saied, M.M., Abdel-Razek, M.A.S., Mohamed, R.A., Abozeid, A.M., El-Midany, S.A., and Abouelenien, F.A. 2020. Wastewater remediation of heavy metals and pesticides using rice straw and/or zeolite as bioadsorbents and assessment of treated wastewater reuse in the culture of Nile tilapia (Oreochromis niloticus). Environ. Monit. Assess., 192. https://doi.org/10.1007/s10661-020-08760-x.

El Sheikh, R., Hassan, W.S., Youssef, A.M., Hameed, A.M., Subaihi, A., Alharbi, A., and Gouda, A.A. 2022. Eco-friendly ultrasound-assisted ionic liquid-based dispersive liquid-liquid microextraction of nickel in water, food and tobacco samples prior to FAAS determination. Int. J. Environ. Anal. Chem., 102: 899–910. https://doi.org/10.1080/03067319.2020.1727461.

Eliodório, K.P., Pereira, G.J., and de Araújo Morandim-Giannetti, A. 2021. Functionalized chitosan with butylammonium ionic liquids for removal of Cr(VI) from aqueous solution. J. Appl. Polym. Sci., 138: 8–10. https://doi.org/10.1002/app.49912.

Eom, H., Kang, W., Kim, S., Chon, K., Lee, Y.G., and Oh, S.E. 2020. Improved toxicity analysis of heavy metal-contaminated water via a novel fermentative bacteria-based test kit. Chemosphere, 258: 127412. https://doi.org/10.1016/j.chemosphere.2020.127412.

Ewecharoen, A., Thiravetyan, P., Wendel, E., and Bertagnolli, H. 2009. Nickel adsorption by sodium polyacrylate-grafted activated carbon. J. Hazard. Mater., 171: 335–339. https://doi.org/10.1016/j.jhazmat.2009.06.008.

Frade, R.F.M., and Afonso, C.A.M. 2010. Impact of ionic liquids in environment and humans : An overview. Hum. Exp. Toxicol., 29: 1038–1054. https://doi.org/10.1177/0960327110371259.

Fu, F., and Wang, Q. 2011. Removal of heavy metal ions from wastewaters : A review. J. Environ. Manage., 92: 407–418. https://doi.org/10.1016/j.jenvman.2010.11.011.

Gao, M.R., Yuan, J., and Antonietti, M. 2017. Ionic liquids and Poly(ionic liquid)s for morphosynthesis of inorganic materials. Chem. - A Eur. J., 23: 5391–5403. https://doi.org/10.1002/chem.201604191.

Gardas, R.L., and Coutinho, J.A.P. 2008. A group contribution method for viscosity estimation of ionic liquids. Fluid Phase Equilib., 266: 195–201. https://doi.org/10.1016/j.fluid.2008.01.021.

Ghernaout, D., Al-ghonamy, A.I., Boucherit, A., Ghernaout, B., Naceur, M.W., Messaoudene, N.A., Aichouni, M., Mahjoubi, A.A., and Elboughdiri, N.A. 2015. Brownian motion and coagulation process. Am. J. Environ. Prot., 4: 1–15. https://doi.org/10.11648/j.ajeps.s.2015040501.11.

Gionfriddo, E., Souza-Silva, É.A., Ho, T.D., Anderson, J.L., and Pawliszyn, J. 2018. Exploiting the tunable selectivity features of polymeric ionic liquid-based SPME sorbents in food analysis. Talanta, 188: 522–530. https://doi.org/10.1016/j.talanta.2018.06.011.

Goyal, R.K., Jayakumar, N.S., and Hashim, M.A. 2011. Chromium removal by emulsion liquid membrane using [BMIM]+[NTf2]- as stabilizer and TOMAC as extractant. Desalination, 278: 50–56. https://doi.org/10.1016/j.desal.2011.05.001.

Guezzen, B., and Amine Didi, M. 2016. Removal and analysis of mercury (II) from aqueous solution by ionic liquids. J. Anal. Bioanal. Tech., 07. https://doi.org/10.4172/2155-9872.1000317.

Hao, L., Sun, L., Su, T., Hao, D., Liao, W., Deng, C., Ren, W., Zhang, Y., and Lü, H. 2019. Polyoxometalate-based ionic liquid catalyst with unprecedented activity and selectivity for oxidative desulfurization of diesel in [Omim]BF 4. Chem. Eng. J., 358: 419–426. https://doi.org/10.1016/j.cej.2018.10.006.

Haouas, M., Lakiss, L., Martineau, C., El Fallah, J., Valtchev, V., and Taulelle, F. 2014. Silicate ionic liquid synthesis of zeolite merlinoite: Crystal size control from crystalline nanoaggregates to micron-sized single-crystals. Microporous Mesoporous Mater, 198: 35–44. https://doi.org/10.1016/j.micromeso.2014.07.011.

Hilali, A., El Baghdadi, M., Barakat, A., Ennaji, W., and El Hamzaoui, E.H. 2020. Contribution of GIS techniques and pollution indices in the assessment of metal pollution in agricultural soils irrigated with wastewater: case of the Day River, Beni Mellal (Morocco). Euro-Mediterranean J. Environ. Integr., 5: 1–19. https://doi.org/10.1007/s41207-020-00186-8.

Hitchcock, P.B., Mohammed, T.J., Seddon, K.R., Zora, J.A., and Hussey, L. 1986. 1-Methyl-3-ethylimidazolium Hexachlorouranate(IV) and 1-Methyl-3-ethylimidazolium Tetrachlorodioxouranate(V1): Synthesis, Structure, and Electrochem- istry in a Room Temperature Ionic Liquid. Inorganica Chim. Acta, 113: 25–26.

Hoogerstraete, T. Vander. 2013. Homogeneous Liquid−Liquid extraction of metal ions with a functionalized ionic liquid. Phys. Chem. Lett. 4: 1659–1663. https://doi.org/10.1021/jz4005366.

Janssen, C.H.C., Macías-ruvalcaba, N.A., Aguilar-martínez, M., and Kobrak, M.N. 2016. Copper extraction using protic ionic liquids : Evidence of the Hofmeister effect. Sep. Purif. Technol., 168: 275–283. https://doi.org/10.1016/j.seppur.2016.05.031.

Jean, E., Villemin, D., Hlaibi, M., and Lebrun, L. 2018. Separation and purification technology heavy metal ions extraction using new supported liquid membranes containing ionic liquid as carrier. Sep. Purif. Technol., 201: 1–9. https://doi.org/10.1016/j.seppur.2018.02.033.

Kagaya, S.K., Attrall, R.W.C., and Olev, S.D.K. 2011. Solid-phase extraction of cobalt (II) from lithium chloride solutions using a Poly (vinyl chloride)-based polymer inclusion membrane with aliquat 336 as the carrier. Anal. Sci., 27: 653–657.

Kapitanov, I.V., Jordan, A., Karpichev, Y., Spulak, M., Perez, L., Kellett, A., Kümmerer, K., and Gathergood, N. 2019. Synthesis, self-assembly, bacterial and fungal toxicity, and preliminary biodegradation studies of a series of 1-phenylalanine-derived surface-active ionic liquids. Green Chem., 21: 1777–1794. https://doi.org/10.1039/c9gc00030e.

Kawada, S., Watanabe, S., Tadokoro, C., Tsuboi, R., and Sasaki, S. 2018. Lubricating mechanism of cyano-based ionic liquids on nascent steel surface. Tribol. Int., 119: 474–480. https://doi.org/10.1016/j.triboint.2017.11.019.

Kebiche-senhadji, O., Tingry, S., Seta, P., and Benamor, M. 2010. Selective extraction of Cr (VI) over metallic species by polymer inclusion membrane (PIM) using anion (Aliquat 336) as carrier. Desalination, 258: 59–65. https://doi.org/10.1016/j.desal.2010.03.047.

Khodakarami, M., and Alagha, L. 2020. Separation and recovery of rare earth elements using novel ammonium-based task-specific ionic liquids with bidentate and tridentate O-donor functional groups. Sep. Purif. Technol., 232: 115952. https://doi.org/10.1016/j.seppur.2019.115952.

Khulbe, K.C., and Matsuura, T. 2018. Removal of heavy metals and pollutants by membrane adsorption techniques. Appl. Water Sci., 8: 1–30. https://doi.org/10.1007/s13201-018-0661-6.

Kilicarslan, A., Voßenkaul, D., Stoltz, N., Stopic, S., Nezihi, M., and Friedrich, B. 2017. Hydrometallurgy Selectivity potential of ionic liquids for metal extraction from slags containing rare earth elements. Hydrometallurgy, 169: 59–67.

Kim, B.K., Lee, E.J., Kang, Y., and Lee, J.J. 2018. Application of ionic liquids for metal dissolution and extraction. J. Ind. Eng. Chem., 61: 388–397. https://doi.org/10.1016/j.jiec.2017.12.038.

Konczyk, J., Kozlowski, C., and Walkowiak, W. 2010. Removal of chromium (III) from acidic aqueous solution by polymer inclusion membranes with D2EHPA and Aliquat 336. Desalination, 263: 211–216. https://doi.org/10.1016/j.desal.2010.06.061.

Kuan, Y.C., Lee, I.H., and Chern, J.M. 2010. Heavy metal extraction from PCB wastewater treatment sludge by sulfuric acid. J. Hazard. Mater., 177: 881–886. https://doi.org/10.1016/j.jhazmat.2009.12.115.

Kumar, M.A., Krishna, N.V., and Selvam, P. 2019. Novel ionic liquid-templated ordered mesoporous aluminosilicates: Synthesis, characterization and catalytic properties. Microporous Mesoporous Mater., 275: 172–179. https://doi.org/10.1016/j.micromeso.2018.08.033.

Le, X.T., Viel, P., Sorin, A., Jegou, P., and Palacin, S. 2009. Electrochemical behaviour of polyacrylic acid coated gold electrodes: An application to remove heavy metal ions from wastewater. Electrochim. Acta 54: 6089–6093. https://doi.org/10.1016/j.electacta.2009.02.048.

Lee, S., and Lee, S. 2006. Functionalized imidazolium salts for task-specific ionic liquids and their applications. Chem. Commun. 1049–1063. https://doi.org/10.1039/b514140k.

Li, P., Chen, H., Schott, J.A., Li, B., Zheng, Y., Mahurin, S.M., Jiang, D., Cui, G., Hu, X., Wang, Y., Li, L., and Dai, S. 2019. Porous liquid zeolites : hydrogen bonding-stabilized H-porous liquid zeolites: hydrogen bonding-stabilized H-ZSM-5 in branched ionic liquids †. Nanoscale, 11: 1515–1519. https://doi.org/10.1039/x0xx00000x.

Lili, Z., Lin, G., Zhenjiang, Z., Ji, C., and Shaomin, Z. 2012. The preparation of supported ionic liquids (SILs) and their application in rare metals separation. Sci. china Chem., 55: 1479–1487. https://doi.org/10.1007/s11426-012-4632-8.

Liu, Y., Guo, L., Zhu, L., Sun, X., and Chen, J. 2010. Removal of Cr(III, VI) by quaternary ammonium and quaternary phosphonium ionic liquids functionalized silica materials. Chem. Eng. J., 158: 108–114. https://doi.org/10.1016/j.cej.2009.12.012.

Liu, Z., Chen, B., Li, X., Wang, L. ao, Xiao, H., and Liu, D. 2019. Toxicity assessment of artificially added zinc, selenium, and strontium in water. Sci. Total Environ., 670: 433–438. https://doi.org/10.1016/j. scitotenv.2019.03.259.

Ma, Y., Egodawatta, P., Mcgree, J., Liu, A., and Goonetilleke, A. 2016. Science of the total environment human health risk assessment of heavy metals in urban stormwater. Sci. Total Environ., 557-558: 764–772. https://doi.org/10.1016/j.scitotenv.2016.03.067.

Maal-bared, R. 2020. Operational impacts of heavy metals on activated sludge systems: the need for improved monitoring. Environ. Monit. Assess, 192: 1–12. https://doi.org/doi.org/10.1007/s10661-020-08529-2.

Mahmoud, M.R., Lazaridis, N.K., and Matis, K.A. 2015. Study of flotation conditions for cadmium(II) removal from aqueous solutions. Process Saf. Environ. Prot., 94: 203–211. https://doi.org/10.1016/j. psep.2014.06.012.

Makanyire, T., Sanchez-Segado, S., and Jha, A. 2016. Separation and recovery of critical metal ions using ionic liquids. Adv. Manuf., 4: 33–46. https://doi.org/10.1007/s40436-015-0132-3.

Malas, R., Ibrahim, Y., AlNashef, I., Banat, F., and Hasan, S.W. 2020. Impregnation of polyethylene membranes with 1-butyl-3-methylimidazolium dicyanamide ionic liquid for enhanced removal of Cd2+, Ni2+, and Zn2+ from aqueous solutions. J. Mol. Liq., 318: 113981. https://doi.org/10.1016/j. molliq.2020.113981.

Maria Antonieta Valdés, V., Lijanova, I.V., Likhanova, N.V., Vigueras, D.J., Xometl, O.O., Cecati, C., Catarina, C.S., Norte, E.C., and Atepehuacan, S.B. 2014. The removal of heavy metal cations from an aqueous solution using ionic liquids. Can. Soc. Chem. Eng., 1–26. https://doi.org/10.1002/ cjce.22053.

Markiewicz, M., Piszora, M., Caicedo, N., Jungnickel, C., and Stolte, S. 2013. Toxicity of ionic liquid cations and anions towards activated sewage sludge organisms from different sources-Consequences for biodegradation testing and wastewater treatment plant operation. Water Res., 47: 2921–2928. https://doi.org/10.1016/j.watres.2013.02.055.

Mehdi, H., Binnemans, K., Hecke, K. Van, Meervelt, and L. Van. 2010. Hydrophobic ionic liquids with strongly coordinating anions w. Chem. Commun., 46: 234–236. https://doi.org/10.1039/b914977e.

Mostazo, G., Vera, R., Fontàs, C., and Anticó, E. 2017. Polymer inclusion membranes as a new tool for Zn speciation : influence of the membrane composition in diffusional fluxes. Rhodes - Greece.

Navarro, R., Alba, J., Saucedo, I., and Guibal, E. 2014. Hg (II) removal from HCl solutions using a tetraalkylphosphonium ionic liquid impregnated onto Amberlite XAD-7. J. Appl. Polym., 41086: 1–11. https://doi.org/10.1002/app.41086.

Ojedokun, A.T., and Bello, O.S. 2016. Sequestering heavy metals from wastewater using cow dung. Water Resour. Ind., 13: 7–13. https://doi.org/10.1016/j.wri.2016.02.002.

Ong, S., Toorisaka, E., Hirata, M., and Hano, T. 2010. Adsorption and toxicity of heavy metals on activated sludge. Sci. Asia, 36: 204–209. https://doi.org/10.2306/scienceasia1513-1874.2010.36.204.

Oyewo, O.A., Elemike, E.E., Onwudiwe, D.C., and Onyango, M.S. 2020. Metal oxide-cellulose nanocomposites for the removal of toxic metals and dyes from wastewater. Int. J. Biol. Macromol., 164: 2477–2496. https://doi.org/10.1016/j.ijbiomac.2020.08.074.

Ozevci, G., Sert, S., and Eral, M. 2018. Optimization of lanthanum transport through supported liquid membranes based on ionic liquid. Chem. Eng. Res. Desing., 140: 1–11. https://doi.org/10.1016/j. cherd.2018.10.004.

Płotka-Wasylka, J., Rutkowska, M., Owczarek, K., Tobiszewski, M., and Namieśnik, J. 2017. Extraction with environmentally friendly solvents. TrAC - Trends Anal. Chem., 91: 12–25. https://doi. org/10.1016/j.trac.2017.03.006.

Pospiech, B. 2015. Hydrometallurgy Studies on extraction and permeation of cadmium (II) using Cyphos IL 104 as selective extractant and ion carrier. Hydrometallurgy, 154: 88–94. https://doi.org/10.1016/j. hydromet.2015.04.007.

Rajadurai, V., and Lakshmi, B. 2020. Ionic liquids to remove toxic metal pollution, environmental chemistry letters. Springer International Publishing. https://doi.org/10.1007/s10311-020-01115-5.

Rajendran, A. 2010. Applicability of an ionic liquid in the removal of chromium from tannery effluents: A green chemical approach. Pure Appl. Chem., 4: 100–103.

Regel-Rosocka, M. 2009. Extractive removal of zinc(II) from chloride liquors with phosphonium ionic liquids/toluene mixtures as novel extractants. Sep. Purif. Technol., 66: 19–24. https://doi.org/10.1016/j.seppur.2008.12.002.

Regel-Rosocka, M., Nowak, Ł., and Wiśniewski, M. 2012. Removal of zinc(II) and iron ions from chloride solutions with phosphonium ionic liquids. Sep. Purif. Technol., 97: 158–163. https://doi.org/10.1016/j.seppur.2012.01.035.

Regel-rosocka, M., Rzelewska, M., Baczynska, M., Janus, M., and Wisniewski, M. 2015. Removal of palladium (II) from aqueous chloride solutions with cyphos phosphonium ionic liquids as metal ion carriers for liquid-liquid extraction and transport across polymer inclusion membranes. Physicochem. Probl. Miner. Process., 51: 621–631.

Rout, A., and Binnemans, K. 2015. Extraction of trivalent rare-earth ions by mixtures. Dalt. Trans., 44: 1379–1387. https://doi.org/10.1039/c4dt02766c.

Rzelewska-piekut, M., and Regel-rosocka, M. 2019. Separation of Pt (IV), Pd (II), Ru (III) and Rh (III) from model chloride solutions by liquid-liquid extraction with phosphonium ionic liquids. Sep. Purif. Technol., 212: 791–801. https://doi.org/10.1016/j.seppur.2018.11.091.

Sachse, A., Wuttke, C., Díaz, U., and De Souza, M.O. 2015. Mesoporous y zeolite through ionic liquid based surfactant templating. Microporous Mesoporous Mater., 217: 81–86. https://doi.org/10.1016/j.micromeso.2015.05.049.

Sethurajan, M., van Hullebusch, E.D., Fontana, D., Akcil, A., Deveci, H., Batinic, B., Leal, J.P., Gasche, T.A., Ali Kucuker, M., Kuchta, K., Neto, I.F.F., Soares, H.M.V.M., and Chmielarz, A. 2019. Recent advances on hydrometallurgical recovery of critical and precious elements from end of life electronic wastes - a review. Crit. Rev. Environ. Sci. Technol., 49: 212–275. https://doi.org/10.1080/1064338 9.2018.1540760.

Sheldon, R. 2001. Catalytic reactions in ionic liquids. Chem. Commun., 1: 2399–2407. https://doi.org/10.1039/b107270f.

Singh, S.K., and Savoy, A.W. 2020. Ionic liquids synthesis and applications : An overview. J. Mol. Liq., 297: 112038. https://doi.org/10.1016/j.molliq.2019.112038.

Sorouraddin, S.M., Farajzadeh, M.A., and Okhravi, T. 2020. Development of dispersive liquid-liquid microextraction based on deep eutectic solvent using as complexing agent and extraction solvent: application for extraction of heavy metals. Sep. Sci. Technol., 55: 2955–2966. https://doi.org/10.10 80/01496395.2019.1666874.

Stojanovic, A., and Keppler, B.K. 2012. Ionic liquids as extracting agents for heavy metals. Sep. Sci. Technol., 47: 189–203. https://doi.org/10.1080/01496395.2011.620587.

Sun, W., Li, L., Luo, C., and Fan, L. 2016. International journal of biological macromolecules synthesis of magnetic graphene nanocomposites decorated with ionic liquids for fast lead ion removal. Int. J. Biol. Macromol., 85: 246–251. https://doi.org/10.1016/j.ijbiomac.2015.09.061.

Sun, X., Peng, B., Ji, Y., Chen, J., and Li, D. 2008. The solid–liquid extraction of yttrium from rare earths by solvent (ionic liquid) impreganated resin coupled with complexing method. Sep. Purif. Technol., 63: 61–68. https://doi.org/10.1016/j.seppur.2008.03.038.

Swain, S.S., Nayak, B., Devi, N., Das, S., and Swain, N. 2016. Liquid-liquid extraction of cadmium(II) from sulfate medium using phosphonium and ammonium based ionic liquids diluted in kerosene. Hydrometallurgy, 162: 63–70. https://doi.org/10.1016/j.hydromet.2016.02.015.

Thasneema, K.K., Dipin, T., Thayyil, M.S., Sahu, P.K., Messali, M., Rosalin, T., Elyas, K.K., Saharuba, P.M., Anjitha, T., Hadda, and T. Ben 2021. Removal of toxic heavy metals, phenolic compounds and textile dyes from industrial waste water using phosphonium based ionic liquids. J. Mol. Liq., 323: 114645. https://doi.org/10.1016/j.molliq.2020.114645.

Tian, M., Fang, L., Yan, X., Xiao, W., and Row, K.H. 2019. Determination of heavy metal ions and organic pollutants in water samples using ionic liquids and ionic liquid-modified sorbents. J. Anal. Methods Chem., 1–19.

Tian, Y., Yin, P., Qu, R., Wang, C., Zheng, H., and Yu, Z. 2010. Removal of transition metal ions from aqueous solutions by adsorption using a novel hybrid material silica gel chemically modified by triethylenetetraminomethylenephosphonic acid. Chem. Eng. J., 162: 573–579. https://doi. org/10.1016/j.cej.2010.05.065.

Toledo Hijo, A.A.C., Maximo, G.J., Costa, M.C., Batista, E.A.C., and Meirelles, A.J.A. 2016. Applications of ionic liquids in the food and bioproducts industries. ACS Sustain. Chem. Eng., 4: 5347–5369. https://doi.org/10.1021/acssuschemeng.6b00560.

Vaid, Z.S., Rajput, S.M., Shah, A., Kadam, Y., Kumar, A., El Seoud, O.A., Mata, J.P., and Malek, N.I. 2018. Salt-induced microstructural transitions in aqueous dispersions of ionic-liquids-based surfactants. ChemistrySelect, 3: 4851–4858. https://doi.org/10.1002/slct.201800041.

Van Osch, D.J.G.P., Kollau, L.J.B.M., Van Den Bruinhorst, A., Asikainen, S., Rocha, M.A.A., and Kroon, M.C. 2017. Ionic liquids and deep eutectic solvents for lignocellulosic biomass fractionation. Phys. Chem. Chem. Phys., 19: 2636–2665. https://doi.org/10.1039/c6cp07499e.

Vardhan, K.H., Kumar, P.S., and Panda, R.C. 2019. A review on heavy metal pollution, toxicity and remedial measures: Current trends and future perspectives. J. Mol. Liq., 290: 111197. https://doi. org/10.1016/j.molliq.2019.111197.

Verma, C., Mishra, A., Chauhan, S., Verma, P., Srivastava, V., Quraishi, M.A., and Ebenso, E.E. 2019. Dissolution of cellulose in ionic liquids and their mixed cosolvents: A review. Sustain. Chem. Pharm., 13: 100162. https://doi.org/10.1016/j.scp.2019.100162.

Vincenza, M., Spreti, N., Di, P., and Germani, R. 2013. Understanding mercury extraction mechanism in ionic liquids. Sep. Purif. Technol., 116: 294–299. https://doi.org/10.1016/j.seppur.2013.06.006.

Visser, A.E., Swatloski, R.P., Reichert, W.M., Mayton, R., Sheff, S., Wierzbicki, A., Davis, J., and Rogers, R.D. 2001. Task-specific ionic liquids for the extraction of metal ions from aqueous solutions. Chem. Commun., 135–136. https://doi.org/10.1039/b008041l.

Wieszczycka, K., Filipowiak, K., Wojciechowska, I., Buchwald, T., Siwińska-Ciesielczyk, K., Strzemiecka, B., Jesionowski, T., and Voelkel, A. 2021. Novel highly efficient ionic liquid-functionalized silica for toxic metals removal. Sep. Purif. Technol., 265. https://doi.org/10.1016/j.seppur.2021.118483.

Wojciechowska, A., Reis, M.T.A., Wojciechowska, I., Ismael, M.R.C., Gameiro, M.L.F., Wieszczycka, K., and Carvalho, J.M.R. 2018. Application of pseudo-emulsion based hollow fiber strip dispersion with task-specific ionic liquids for recovery of zinc(II) from chloride solutions. J. Mol. Liq., 254: 369–376. https://doi.org/10.1016/j.molliq.2018.01.135.

Wu, X., Cai, Q., Xu, Q., Zhou, Z., and Shi, J. 2020. Wheat (*Triticum aestivum* L.) grains uptake of lead (Pb), transfer factors and prediction models for various types of soils from China. Ecotoxicol. Environ. Saf., 206: 111387. https://doi.org/10.1016/j.ecoenv.2020.111387.

Yan, L., Yin, H., Zhang, S., Leng, F., Nan, W., and Li, H. 2010. Biosorption of inorganic and organic arsenic from aqueous solution by *Acidithiobacillus ferrooxidans* BY-3. J. Hazard. Mater., 178: 209–217. https://doi.org/10.1016/j.jhazmat.2010.01.065.

Yan, X., Anguille, S., Bendahan, M., Moulin, P., Marseille, C., Procédés, E., Epm, M., and Cedex, P. 2019. Ionic liquids combined with membrane separation processes : A review. Sep. Purif. Technol., 222: 230–253. https://doi.org/10.1016/j.seppur.2019.03.103.

Yulin, H., Yao, N., Pan Wu, Y., and Bo Zheng, K. 2017. Recent advances in catalytic condensation reactions applications of supported ionic liquids. Curr. Org. Chem., 21: 462–484. https://doi.org/10. 2174/1385272821666171106144935.

Zambare, R.S., and Nemade, P.R. 2021. Ionic liquid-modified graphene oxide sponge for hexavalent chromium removal from water. Colloids Surfaces A Physicochem. Eng. Asp., 609: 125657. https:// doi.org/10.1016/j.colsurfa.2020.125657.

Zhang, C., Jiang, Y., Li, Y., Hu, Z., Zhou, L., and Zhou, M. 2013. Three-dimensional electrochemical process for wastewater treatment: A general review. Chem. Eng. J., 228: 455–467. https://doi. org/10.1016/j.cej.2013.05.033.

Zhang, J., Liu, C., Xie, Y., Li, N., Ning, Z., Du, N., Huang, X., and Zhong, Y. 2017. Enhancing fructooligosaccharides production by genetic improvement of the industrial fungus *Aspergillus niger* ATCC 20611. J. Biotechnol., 249: 25–33. https://doi.org/10.1016/j.jbiotec.2017.03.021.

Zhang, M., Ma, X., Li, J., Huang, R., Guo, L., Zhang, X., Fan, Y., Xie, X., and Zeng, G. 2019. Enhanced removal of As(III) and As(V) from aqueous solution using ionic liquid-modified magnetic graphene oxide. Chemosphere, 234: 196–203. https://doi.org/10.1016/j.chemosphere.2019.06.057.

Zhang, W., Duo, H., Li, S., An, Y., Chen, Z., Liu, Z., Ren, Y., Wang, S., Zhang, X., and Wang, X. 2020. An overview of the recent advances in functionalization biomass adsorbents for toxic metals removal. Colloids Interface Sci. Commun., 38: 100308. https://doi.org/10.1016/j.colcom.2020.100308.

Zhou, Y., and Antonietti, M. 2004. A series of highly ordered, super-microporous, lamellar silicas prepared by nanocasting with ionic liquids. Chem. Mater. 16: 544–550. https://doi.org/10.1021/cm034442w.

Zhou, Y., Boudesocque, S., Mohamadou, A., and Dupont, L. 2015. Extraction of metal ions with task specific ionic liquids: influence of a coordinating anion. Sep. Sci. Technol. 50, 38–44. https://doi.org/10.1080/01496395.2014.952747.

Zhu, L., Liu, Y., and Chen, J. 2009. Synthesis of N-Methylimidazolium functionalized strongly basic anion exchange resins for adsorption of Cr (VI). Ind. Eng. Chem. Res. 48: 3261–3267. https://doi.org/10.1021/ie801278f.

Index

A

acid mine drainage 183, 186, 188, 197
Adsorbents 147–150, 154, 156–158, 160, 161, 163
adsorption 98, 99, 103–105, 110
Advantages 4, 5
Adverse effects 1, 7
alkalinity 12, 15–18, 21–25
Alleviation mechanism 51
anaerobic digestion 100, 102, 110
antioxidant 34, 35, 37, 41, 42
Arsenic alleviation 56, 59, 62, 63

B

bath recycling 96
Biofertilizer 48–54
biogas 95, 100, 102, 103, 110
Biosorption 172–176, 178, 181, 183–197

C

Cadmium 76, 88–90, 203, 205–207, 213
calcium carbonate 73, 74, 85, 86
carcinogenic 11, 14, 15, 26–28
chromium (III) 98, 99, 105–108, 110
compaction pressure 84, 85
complexation 150, 152, 153, 156, 162, 163
contaminant 174, 180
Contamination 203, 205, 207
Copper 203, 205–207, 213
coupled process 139

D

dechroming 105–107, 111
deep tube wells 11, 15, 16, 18, 19, 21–25, 27, 28
Disadvantages 4, 5
Drosophila 33–43

E

Eco-friendly approaches 4
effluent 147, 151, 152, 156, 159, 173, 174, 182–186, 189–197
electrodialysis 118, 120, 130–132, 138
electroplating 173, 183–186, 196, 197
Environment 203–210, 213, 215
environmental 42
Exposure routes 1, 2, 7
Extraction 203–205, 209–214

F

fertilizer 102, 107, 108, 111
fluorescent lamps 72–74, 76, 88, 90
Foam glass 72–90
foaming agent 74, 78, 80, 83, 84, 90

G

genotoxicity 33, 34, 38, 39, 42, 43
glass waste 72, 76
green revolution 47, 49
Groundwater 10–12, 14, 15, 18–21, 24–28
Growth 50–52, 55, 59–62, 66

H

hazardous wastes 73
health 34, 42
Health risk assessment 10, 11, 14, 26, 27
heating rate 74–76, 86, 87, 90
heterophasic materials 73
hybrid membranes 123, 130, 139

I

in vivo 34, 35, 41, 42
Inorganic pollutants 2, 4
ion exchange 147, 153, 156, 162

Ionic liquids 203–205, 209–215
isotherm 148–150, 152, 154, 157, 158, 162

K

kinetic 148, 150, 154, 157

L

lamps 72–74, 76, 88, 90
landfill 100, 101, 110
lead 73, 74, 76, 79, 85, 86, 88–90, 203, 205–207, 213

M

membrane distillation 132–134, 138
mercury 73, 74, 76, 88–90
Metal 33, 34, 42
microfiltration 120–122, 124, 127, 137
Minimal Risk Levels 3
mutagenicity 33, 34, 36, 38, 39, 41–43

N

Nanofertilizer 48, 50, 51, 53–56, 58–60, 62, 65–67
Nanofertilizer application 50, 59, 60, 65
Nanomaterials 33–36, 41
Nanotechnology 48, 60, 65
Nickel 203, 205–207, 210, 213
nuclear power 172, 183, 192, 195, 197

O

osmosis 118, 120, 121, 126, 127, 129, 130, 135, 137
oxidative stress 34–38, 42

P

particle size 74, 83, 86, 87, 90
pervaporation 127, 134–136, 138
Physico-chemical parameters 11, 12, 22, 25, 28
precipitation 97–99, 110

R

Removal 203–205, 207–215

S

Selectivity 204
shallow tube-wells 11, 12, 19, 21–23, 27
shavings 95, 100, 103–108, 110
sludge 95, 98, 100–105, 107
Solid Waste 3, 4, 7
sorption 149, 150, 162
Sustainable agriculture 48, 49, 51, 52, 67

T

tannery 183, 189–192, 197
textile 172, 183, 189–191, 197
thermal conductivity 76, 87, 88, 90
toxic metals 122, 172, 173, 183, 185–189, 191, 195–197, 203, 205–207, 209, 213, 215
toxicity 34, 35

U

ultrafiltration 120–124, 127, 132, 137, 139

V

volumetric expansion 74, 75, 77, 84–87, 90

W

waste disposal 101
waste recycling 72
Wastewater 119–126, 128–139, 172, 173, 183, 185, 186, 189, 191, 197
Water Quality Index 10, 11, 13, 25, 28
Water treatment techniques 7

Z

Zinc 59, 53, 54, 66, 203, 205–207